CISM COURSES AND LECTURES

The series presents lecture notes, monographs, edited works and proceedings in the field of Mechanics, Engineering, Computer Science and Applied Mathematics.
Purpose of the series is to make known in the international scientific and technical community results obtained in some of the activities organized by CISM, the International Centre for Mechanical Sciences.

INTERNATIONAL CENTRE FOR MECHANICAL SCIENCES

COURSES AND LECTURES - No. 429

SMART STRUCTURES

APPLICATIONS AND RELATED TECHNOLOGIES

EDITED BY

AFZAL SULEMAN
INSTITUTO SUPERIOR TÉCNICO

Springer-Verlag Wien GmbH

This volume contains 230 illustrations

Originally published by Springer-Verlag Wien New York in 2001
SPIN 10855601

In order to make this volume available as economically and as rapidly as possible the authors' typescripts have been reproduced in their original forms. This method unfortunately has its typographical limitations but it is hoped that they in no way distract the reader.

ISBN 978-3-211-83681-1 ISBN 978-3-7091-2686-8 (eBook)
DOI 10.1007/978-3-7091-2686-8

PREFACE

In the past decade, technological developments in material and computer sciences have evolved to the point where their synergistic combination have culminated in a new field of multi-disciplinary research in smart structures, systems and related technologies. The advances in material sciences have provided a comprehensive and theoretical framework for implementing multifunctionality into materials, and the development of high speed digital computers has permitted the transformation of that framework into methodologies for practical design and production. The concept is elementary: a highly integrated sensor system provides data on the structure and its environment to a processing and control system which in turn signals integrated actuators to modify the system properties in an appropriate fashion. The multifunctional electro-magneto-thermo -mechano-rheological materials have presented an exceptional promise in the fields of active vibration suppression, shape control, noise attenuation, structural health monitoring, smart machines and micro-electro-mechanical systems with application in aircraft, aerospace, automobile, civil structures and consumer industry.

This book is a result of the Advanced School on Smart Structures and Materials: Theory and Applications, which took place in the Centre International des Sciences Mécaniques (CISM), Udine, Italy, during June 19-23, 2000. The course, lectured by the authors of the different parts of this book, brought together a large number of participants ranging from doctoral and postgraduate students to researchers, developers and young faculty, concerned with advanced theoretical and design issues in smart structures, materials and related technologies. The lecture notes used to support the course were carefully revised by the authors, taking into account the discussions generated and the interests of the target readers. The book covers the fundamentals in smart structures, materials, systems and related technologies and it provides a comprehensive overview of the current state of the art in the field. The mathematical foundations, engineering design tools and experimental techniques illustrated with practical applications are presented.

I am indebted to the lecturers of the Advanced School not only for putting together excellent presentations that greatly motivated the active participation of those that attended the course but also for their contribution to the lecture notes and to this book. I am grateful to all participants in the Advanced School for their excellent contributions to the discussions that took place during and after the course. A word of acknowledgement is also due to the CISM Scientific Council for supporting the Advanced School and recognizing the importance of topics related with smart structures, systems and related technologies in the framework of the Mechanical Sciences. Finally, a special thank you is due to Prof. Arantes e Oliveira and Prof. Carlos Mota Soares not only for the initial discussions that led to the proposal and organization of the Advanced School in Smart Structures, Materials and related technologies but also for their continuing support.

Afzal Suleman

CONTENTS

CHAPTER I

ACTIVE SHAPE CONTROL

1. Smart structures – an overview
2. Modelling of piezoelectric actuators in shells, plates and beams
3. A finite element adaptive plate model
4. An adaptive aeroelastic wing
5. An adaptive truss structure

A. Suleman
Associate Professor
IDMEC-Instituto Superior Tecnico
Department of Mechanical Engineering
Av Rovisco Pais
1049-001 Lisboa - PORTUGAL
Tel: 351-21-841 7324
Fax: 351-21-847 4045
e-mail: suleman@ist.utl.pt

1. Smart Structures – an Overview

A. Suleman[1], E. Prasad[2], R. Blackow[2] and D. Waechter[2]

1. IDMEC-Instituto Superior Técnico, Lisbon, Portugal
2. Sensor Technology Ltd., Ontario, Canada

Abstract Smart structures represent a new engineering approach or design philosophy that integrates the actions of sensors, actuators and control circuit elements into a single system that can respond adaptively to environmental changes in a useful manner. These integrated systems possess an inherent adaptation - a functionality that adds significant value to materials, technologies or end products - that in-turn enables system performance enhancements that are not possible with traditional approaches. This section documents the state-of-the-art evaluation of the embryonic field of multifunctional materials and adaptive structures.

1.1 Introduction

Materials have profoundly influenced history such that time periods have been classified based on the materials being used i.e.stone age, the bronze age and the iron age. The current composite materials are providing a platform to the dawn of a new era, the multifunctional materials age, which will capitalize on novel material characteristics to exploit emerging technologies for the synthesis of smart structures, systems and related technologies. It is anticipated that these innovative materials and related technologies will be utilized in nanotechnology, biomimetics, micro-electro-mechanical systems, and smart machines and structures, to name a few. Self-repair, self-diagnosis, self-multiplication and self degradation are also some of the characteristics anticipated to be a feature of smart systems and technologies

An adaptive structure is a system that consists of intrinsic sensors, actuators and control mechanisms to sense external stimuli, respond to these stimuli in a predetermined manner and revert back to its original state with the removal of the stimuli. Active structures utilize advanced functional materials whose properties can be used for sensing an external stimulus and/or responding to the stimulus. Such materials are commonly used in numerous functional devices for various applications. A review of the state-of-the-art in adaptive structure technologies is presented, and evaluation criteria for sensors, actuators and control systems are presented for use in adaptive structural systems.

1.2 Multifunctional Materials and Adaptive Structures

Multifunctional materials are defined as those materials that have intrinsic or extrinsic capabilities to respond to an external stimulus in a useful manner. The external stimulus being sensed could involve a change in a material's environmental condition, such as light, temperature, pressure, humidity, electric field, magnetic field, etc. The response to an environmental change would involve a change in one or more of the material's physical properties; such as, size, shape, colour, structure, conductivity, magnetization or polarization, etc.

There are numerous examples of adaptive structures that have been developed. They are broadly classified as those related to precise shape control, active vibration damping, adaptive noise cancellation, and real-time structural health monitoring applications.

1.3 Enabling Technologies for Adaptive Structures

An adaptive structure has three basic components: sensors, actuators, and control systems. The performance of these components dictates the overall performance of the adaptive structure. Considerable research and development activities have been directed towards realizing appropriate sensors, actuators, and control systems for customized adaptive structure applications. The technologies related to sensors and actuators are primarily based on physical phenomena, device concepts and designs, materials properties, and materials processing technologies. The technologies related to control systems are based on control theories, electronics and computers.

1.4 Sensors

The application of an energy field on a material generates a change in its characteristics. In sensor materials, a slight change in energy field can produce a significant change in one or more characteristic, and this change can be detected. The response of a sensor material can be either mechanical, electrical, optical, magnetic, thermal or chemical. For adaptive structural applications, the most useful response is electrical, which can be monitored and analysed by a control system. Most of sensor technologies suitable for adaptive structure applications are well established, and sensors with different specifications can be procured from many suppliers. Optical strain sensors have been developed within the past few years for aerospace and space applications.

Strain Sensors

Strain sensors measure the change in the length of an object by detecting the change in its mechanical, optical, acoustical, and/or electrical properties . The strain sensors commonly used in adaptive structures applications include Strain Gauges, Optical Fibre Strain Sensors, and Piezoelectric Strain Sensors.

Strain Gauges are often used to measure surface strains. They operate under the principle of piezoresistance, whereby a material undergoes a change in resistance when it is strained. The fractional change in resistance over the fractional change in length is defined as the gauge factor. Other important characteristics of strain gauges are gauge length, type and amount of strain being measured (bending, axial, shear or torsional), and the operating temperature range. The most common technique for measuring resistance change with strain gauges is in a Wheatstone bridge configuration, with four strain gauges used as resistors in each of the arms of the bridge. The most common types of strain gauges are wire, foil and semiconductor strain gauges.

Fibre Optic Strain Sensors provide excellent strain measurements since they are immune to electromagnetic interference. They can be bonded to the surface of a structure or embedded directly into a composite structure. Fibre Optic Strain Sensors can be used in high temperature applications (800°C for silica fibres and 1700°C for sapphire fibres). They use the change in intrinsic properties of the fibre material, such as frequency, wavelength, phase, mode index, polarization state, index of refraction, attenuation coefficient, etc. to measure strain. However, signal processing is complicated and expensive, which prevents the use of fibre optic strain gauges in most applications. Fibre optic strain measurements use two interferometry techniques, Fabry-Perot and Mach-Zehnder interferometry.

Table 1.1 - Comparison of Polymer and Ceramic Piezoelectric Materials

Property	PVDF	PZT
Density (10^3 kg/m^3)	1.78	7.5
Relative Permittivity	12	1200
d_{31} 10^{-12} C/N	23	110
g_{31} 10^{-3} Vm/N	216	10
k_{31} (% @ 1 kHz)	12	30

Piezoelectric Strain Sensors have the ability to convert mechanical energy into electrical energy. When a piezoelectric material is strained, the charge generated is proportional to the thickness of the material and the amount of force being applied. As a result, the thicker the ceramic the more sensitive the piezoelectric sensor. However, the sensor is usually limited to smaller dimensions due to the fact it has to interact with the structure without affecting its properties. The most common piezoelectric charge coefficients used to characterize a piezoelectric material are d_{31} and d_{33}. The d_{31} coefficient describes how a piezoelectric disc will expand or contract in the x-y plane when electric field is applied in the z-direction, while the d_{33} coefficient describes its expansion and contraction in the z-direction. Another useful factor is the electromechanical coupling coefficient (k), which is related to the stored mechanical input energy to converted stored electrical energy, their ratio being k^2. Piezoelectric sensors have been utilized in a number of adaptive structure applications. Table 1.1 compares properties of PVDF and PZT materials. PVDF is a flexible piezoelectric polymer with a voltage coefficient that is 20-times higher than PZT ceramic. PVDF sensors tend to operate best in dynamic situations, because the induced charge imbalance created by straining the material dissipates with time. For applications requiring smaller frequency ranges, ceramic piezoelectric materials are still a better option. The advantage that the film sensors have over the ceramic sensor materials is their flat response over a large frequency range, typically up to the GHz range.

Displacement Sensors

Displacement sensors are used to measure magnitude of displacements experienced by a structure at any location caused by static loading or vibration excitations. A number of displacement sensors can be used. These include Potentiometer Displacement Transducers, Linear Voltage Differential Transducers, Eddy Current Transducers, and Variable Capacitance Transducers.

Potentiometer Displacement Transducers consist of a uniform coil of wire or a film of high resistivity material, whose resistance is directly proportional to the length of the potentiometer. The potentiometer output is proportional to the voltage between one end of the potentiometer and a moving contact coupled to the object being monitored and hence is proportional to the displacement of the object. These sensors have a number of drawbacks associated with wear, thermal stability, and nonlinearity produced by improper impedance matching of the power supply, potentiometer, and measurement circuitry.

Force and Acceleration Sensors

Force on an object can be measured with a Piezoelectric Force Sensor or a Load Cell, while acceleration can be measured with a Piezoelectric Accelerometer. Piezoelectric Force Sensors utilize piezoelectric transducers to produce an electrical output which is proportional to the applied force. The force transducer is mounted in series with the force transmission path in order to expose the piezoelectric element directly to the forces that are to be measured. Since the piezoelectric is preloaded, the force sensor can measure both tensile and compressive forces. A high stiffness ensures a high resonant frequency and that the sensor will have a minimum effect on the structural integrity. A Load Cell consists of a strain gauge coupled to a stronger metallic plate that will bend as a result of a load. The strain gauge measures the amount of plate strain, which can be directly related to the applied force. These cells can be tuned for different forces by varying the properties of the metal plate used in the cell. Piezoelectric Accelerometers rely on the piezoelectric effect to generate an electrical output that is proportional to the applied acceleration. The amount of charge generated across the piezoelectric is proportional to the applied force, and the applied force is proportional to acceleration. Piezoelectric accelerometers come in a number of different mechanical configurations. These configurations are defined by the nature in which the inertial force of an accelerated mass acts upon the piezoelectric material. There are three primary configurations commercially available: Shear, Compression, and Flexural Beam.

Temperature Sensors

Thermocouple is the most common temperature sensor. It consists of two dissimilar metal wires joined together at one end by a spot weld. When a thermocouple joint is heated, a voltage change proportional to the change in temperature can be measured across the two free ends of the thermocouple. Although the degree of the effect varies for different metals, a number of bi-metal combinations have been standardized for different temperature ranges. Thermocouples are an effective, relatively inexpensive way to measure temperature in a number of applications.

Thermistors operate under the principle that the resistivity of a material varies with temperature. Resistive temperature detectors (RTDs) can be produced in the form of a metal coil or film. Negative temperature coefficient (NTC) thermistors are p-type ceramic semiconductors that show a decrease in resistivity as the temperature is increased. Positive temperature coefficient (PTC) thermistors are n-type ceramic semiconductors that increase their resistance to the flow of electricity as the temperature increases.

Regardless of the nature of the measurement, the performance of any sensor can be evaluated in terms of a number of key sensor performance characteristics: sensitivity, gauge length, bandwidth, response time, operational temperature, accuracy, power, weight and cost.

1.5 Actuators

Actuation can be defined as the controlled release of energy. The actuation process converts a reservoir or 'driving' energy into the 'actuating' energy. The driving energy may be in the form of an electrical field, magnetic field, thermal energy, mechanical strain, or stored as a hydrocarbon fuel. The 'actuating' energy may be in the form of light, heat, radiation, or mechanical strain. For adaptive structure applications, micro-positioning actuators based on mechanical actuation technologies are considered suitable. Various actuator materials and devices are being developed for these applications. These actuator technologies can be classified on the basis of the actuation phenomenon or the types of materials utilized.

Shape-Memory Alloy

Shape-memory alloys are 'passively' adaptive materials. They undergo a thermoelastic austenite-to-martensite transformation when cooled below a critical temperature. In the martensite state, the alloy is plastic and can be easily deformed to a desired shape. When heated above the critical temperature, the martensite phase transforms into an austenite phase, and the material recovers its original pre-deformed shape. This ability of the material to remember its original shape has given it the name shape-memory alloy. If the shape-memory alloy is once again cooled below the critical temperature and no macroscopic shape change occurs, it is said to display the 'one-way effect'. This 'free recovery' process can be repeated numerous times; however, the material would need to be deformed during each low-temperature cycle. An 'extrinsic two-way effect' can be achieved if an appropriate load is applied throughout the temperature cycle. During this process, the load would deform the martensite phase at low temperatures, the shape-memory alloy would regain its original shape at high temperature due to austenite phase transition, and martensite phase transformation during the next cooling cycle would allow the load to deform the alloy once again. An 'intrinsic two-way effect' does exist in some shape-memory alloys, in which the material remembers a low-temperature shape and a high-temperature shape. In these materials, no load is needed to force the material to remember the second shape. Although this effect can occur in both the loaded and unloaded states, the effect is small and not well understood.

Shape-memory alloy actuators have been used in many automotive applications. These thermal actuation applications include shock absorber valve control, thermal compensation in automatic transmission gearshift, and rattle reduction in gear boxes. The use of shape-memory alloy actuators for thermal actuation offers several benefits, including high forces, large movements, small adaptable size, high work-per-unit-volume, and different modes of actuation, including linear, bending, and torsion modes. Shape-memory actuators can utilize a single piece of metal in washer, tube, strip, wire, or helical coil forms. For applications involving cyclic loads of compressive and tensional forces, helical springs stressed in torsion are often utilized. These actuators are constrained within a structural assembly that prevents the alloy from recovering its original shape during heating. This produces a large stress in the shape-memory alloy that is then imparted to the structure, inducing a strain. Since the phase transition in shape-memory alloys occurs over a temperature range, thermal cycling produces thermal hysteresis. These alloys also show fatigue in cyclic applications; the larger the magnitude of shape recovery during a thermal cycle, the shorter is the life. Therefore, the magnitude of strain is kept low in these actuators to extend their working life.

Vibration damping and control using shape-memory alloy actuators can be achieved by utilizing their high damping characteristics, the hysteretic damping, the variable stiffness of the shape memory alloy embedded structures, and the shape-memory effect itself. These materials have high strength and large mechanical loss factors and can be formed into springs for passive vibration damping applications. The natural damping ability of shape-memory alloys could be enhanced with the pseudo elastic effects observed in some shape-memory alloys.

Embedding of shape-memory wires into an elastic structure can modulate the elastic structure's effective stiffness. This can alter the natural resonance frequency of the component and avoids system resonance. The shape-memory effect can also be used directly for vibration damping. When shape memory alloy wires are attached to the base of a cantilever beam, they can damp vibrations in the beam by producing counter strains to those vibrations. Since the driving energy is lost in the form of heat, this method can only be used for low-frequency applications (2-6 Hz) to avoid exceeding operating temperature of shape-memory alloy wires.

Piezoelectric Actuators

Piezoelectric actuators are amongst the most widely used actuators in adaptive structure applications. Piezoelectric materials become electrically charged when subjected to a mechanical stress or produce mechanical strain when an electric field is applied across them. In case of piezoelectric materials, the strains produced are about 0.1% at a field of about 10 kV/cm. The strain (or displacement) vs. electric field characteristic is nearly linear in the operating region, but the strain saturates at very high fields. Piezoelectric materials are actively adaptive materials. The displacement can be positive (tension) or negative (compression) depending on the direction of the field. Under alternating electric fields, these materials will go through tension and compression in a cyclic fashion and produce vibrations. The converse is also true. Thus, these materials can detect strain, movement, force, pressure, or vibration by developing appropriate electrical responses. They can also generate strain, movement, force, pressure, or vibration through the application of a suitable electric field. Since these devices can both detect and transmit sound, they are also called transducers and are used extensively in acoustic communication systems.

The measure of the piezoelectric response to an applied electric field is the piezoelectric charge constant (d). This is a measure of the amount of strain produced by an applied electric field. The most common piezoelectric charge constants used to compare different piezoelectric materials are d_{33}, d_{31} and d_h. The d_{33} coefficient describes the strain relationship in the direction of poling of the piezoelectric material, or along the axis of alignment of piezoelectric molecules (i.e. direction of electric field). The d_{31} coefficient describes the strain relationship perpendicular to the direction of poling or electric field.

The electromechanical coupling coefficient k is used in electromechanical conversions. This factor relates the stored mechanical input energy to the converted stored electrical energy. The planar coupling coefficient k_p denotes the coupling between the electric field in the d_{33} direction and the simultaneous mechanical action in the d_{31} direction. The thickness coupling coefficient k_t denotes the coupling between the electric field in the d_{33} direction and the mechanical action in the same direction. The square of the electromechanical coupling coefficient (k^2) is the ratio of the stored mechanical input energy to the converted stored electrical energy, which represents the efficiency of the conversion (or transduction) process.

The most basic piezoelectric element would consist of a piezoelectric disc, washer, or block. The amount of strain produced in the material is dependent on the thickness of the element and the magnitude of the voltage applied across the thickness. These elements are used in a number of applications: fish finders, sonar transducers, ultrasonic transducers, lighters, etc. Piezoelectric tubes offer an alternative method of producing strains. In the piezoelectric tube the strain is a function of the length of the tube and the thickness of the tube wall. These tubes can operate radially or extensionally. Such piezoelectric tube ceramics are used as actuator elements in semiconductor lithography, scanning tunnelling microscopes, tunable optical fibres, etc.

A number of actuator applications require different functional characteristics such as larger displacements, load carrying capacity or bandwidth. These demands have led to the development of new actuator designs. In order to increase the amount of strain in a piezoelectric actuator, a number of thin piezoelectric elements can be stacked in the actuator. In this stack configuration, the total displacement of the stack would equal the number of elements in the stack multiplied by the displacement of one element. Use of thin elements allows high electrical fields to be applied across the thickness of the elements, which increases strain. Stacks actuators with displacements of 100 μm can generate blocked forces equal to several tonnes.

The displacement capabilities of piezoelectric materials in many sonar applications are enhanced by mechanically coupling these elements with flexible metallic frames. Flextensional actuators, which mechanically couple a piezoelectric element with the flexural properties of a second material, mechanically amplify the volume change of the piezoelectric. Other mechanically coupled composites have also been developed to take advantage of the directional connectivities in the composite structures. These include bimorphs, polymer-piezoelectric composites, moonies, and the new rainbow and thunder technologies.

The bimorph, which consists of a piezoelectric element coupled with a second piezoelectric element, operates as an actuator through the generation of stresses in the coupled bi-layer structure. There are two possible ways of connecting these elements together. In series, the bimorph is constructed in such a way that the polarizations of the two wafers are opposite to the each other. When a potential is applied across the two wafers, regardless of its polarity, one wafer will try to elongate, while the other will try to contract. These opposite actions create the bimorph's characteristic bending action. In the parallel bimorph, the polarizations of the two wafers are aligned in the same direction but the electrode between the wafers is common to both wafers, and the outside electrodes are driven by the same potential. Again the resulting bending action is created by one wafer trying to expand while the other tries to contract. Although these elements do not have large load carrying capabilities, bimorphs can generate displacements of approximately 1 mm. These bimorph actuators have been incorporated into high impact, high speed dot matrix printers.

The moonie consist of two crescent-shaped metallic discs bonded on either side of a piezoelectric disc. The moonie functions as an actuator by coupling the radial expansion and contractions of a piezoelectric disc with the flexible properties of the two metallic end caps. These devices can produce displacements as high as 30 μm and have higher force generation capacities than bimorphs.

The rainbow and thunder devices are structurally similar to a bimorph in that they both have bilayer structures. The major difference between the two devices is the fact that the bilayer system in the rainbow and thunder are pre-stressed. These low-frequency, low-force devices produce displacements in the range of 20 to 50 μm for frequencies under 40 Hz. Although these devices do not match the displacements offered by bimorphs, they have larger force capacities than bimorphs and large enough displacements to offer an alternative to the bimorph.

Another piezoelectric-based device that has been developed recently to increase the displacement is the piezoelectric motor. The propagating-wave type (surface-wave or 'surfing' type) piezoelectric motor combines two standing waves with a 90-degree phase difference both in time and in space and is controllable in both rotational directions by means of the travelling elastic wave induced by the thin piezoelectric ring. A ring-type slider in contact with the 'rippled' surface of the elastic body bonded onto the piezoelectric is driven in both directions by exchanging the sine and cosine voltage inputs. These motors have the benefits of having virtually no backlash and moderate torques but are not meant for long-term use, due to wear and heat generation.

Piezoelectric actuators are being extensively investigated for adaptive structure applications. An example of a commercially available adaptive structure using piezoelectric actuators is Toyota's piezo TEMS (Toyota Electronic Modulated Suspension), which was developed to improve the handling and stability of the automobile and for the passenger's comfort.

Piezoelectric actuators are also being investigated for surface-bonded and embedded applications in adaptive structures. A number of researchers have been experimenting with embedding piezoelectric into carbon fibre composites for shape control experiments.

Electrostrictive Actuators

Electrostriction is a phenomenon observed in all dielectric materials. When an electric field is applied across the dielectric material, the dipoles align themselves with the field. This process induces an internal strain and the material changes dimensions. When the electric field is removed, the dipoles re-orient and the material regains its original dimensions. If the direction of the electric field is reversed, the dipoles again align themselves with the field and induce strain. Since the strain (S) is proportional to the square of the electric field (E), i.e. $S = ME^2$, it is always positive (i.e. the material is under tension) and irrespective of the polarity of the applied field. In order to obtain bidirectional operation (tension and compression) from electrostrictive materials, typically an electric field bias is used.

These materials are evaluated in terms of their electro-mechanical properties, which include the strain produced for a given applied voltage (which is similar to piezoelectric strain constant and is given by $d = dS/dE$), the force generation capabilities, the degree of hysteresis, the dielectric constant (K), the dissipation factor (D), and the thermal operational range. The dielectric constant is a measure of the charge storage capabilities of the electrostrictive element. The larger this value, the higher the electric potential that can be applied across the material before breakdown or current flow through the material. The dissipation or dielectric loss factor is a measure of the conductive losses through the material.

Although electrostrictive effect is present in all dielectric materials, the magnitude of strain is usually very small and not useful. Electrostrictive effect was known to occur in a number of piezoelectric materials such as Rochelle salt, barium titanate and modified PZT, but the effect was always small compared to the piezoelectric effect. However, recently lead magnesium niobate (PMN) has created a lot of interest due to its high electrostrictive strain capabilities and very low hysteresis properties. This material produces strains that are comparable to strains produced by piezoelectric PZT and the hysteresis is negligible.

Electrostrictive materials offer an important advantage in high-speed actuation. These materials do not exhibit hysteresis and, as a result, are promising materials for high-frequency, high-power, and precision applications, such as high-power sonar transducers, deformable mirrors, laser-based machining, and optical systems. Like piezoelectric ceramics, electrostrictive materials can be used to actuate in a number of different geometries. In fact, with a few exceptions, the geometries that are utilized for piezoelectric will also apply to electrostrictive materials.

Magnetostrictive Actuators

Magnetostriction is similar to electrostriction except that the driving field is magnetic. Magnetostrictive materials possess magnetic anisotropy in their atomic structure. They exhibit a change in dimension when placed in a magnetic field. The change in dimension (strain) is a result of a re-orientation of the atomic magnetic moments, or small magnetic domains. As the magnitude of the magnetic field is increased, more domains become aligned until magnetic saturation occurs when all magnetic domains get aligned with the applied magnetic field. If the magnetic field is reduced, the strain decreases and, upon removal of the applied field, the material regains its original shape. The strain produced is proportional to the square of the magnetic field. If the field is reversed, the same strain behaviour is observed. Thus magnetostrictive strain is also always positive.

The most common form of a magnetostrictive material for actuator application is a rod or a cylinder. A typical actuator configuration consists of an active magnetostrictive element tightly wrapped in an exciting coil. Magnetostrictive materials have been used in the development of a direct-drive rotary motor. Magnetostrictive actuators have also been examined for their use on trailing-edge flaps to actively control helicopter vibration by modifying the shape of the flap. An active vibration isolation platform using a magnetostrictive actuator has also been developed to reduce shock levels on a platform that is supported by the base, which is inducing the shocks. This system succeeded in reducing most of the shock disturbances, but some residual vibrations were still present in the platform. In another vibration isolation mount design, magnetostrictive actuators were used in low-frequency vibration isolation applications involving small deflections, but the digital processing speeds limited the system's response times. In the prototype actuator design, when a 7-A current was put through a 50-turn coil surrounding the Terfenol-D rod, 5.22 μm extensions were observed, corresponding to a 0.021 % strain. It is to be noted that the strains produced in a magnetostrictive material are smaller than those produced in piezoelectric and electrostrictive materials, and the hysteresis is higher than the electrostrictive material.

Electrorheological Actuators

Rheology is the science of the flow and deformation of matter. An electrorheological (ER) fluid is a fluid whose viscosity can be modified through the application of an electric field. These materials exist in a wide variety of colloidal suspensions of dielectric solids in non-conducting liquids. These highly viscous colloidal suspensions are composed of fine particles (0.1-1.0 μm) that are uniformly distributed throughout the fluid.

ER fluids handle shear stresses very well. When evaluating the shear stress vs. strain characteristics of these fluids, it is possible to categorize their behaviour into pre-yield and post-yield regimes. These regimes depend on the electric field strength and the material's rheological property can be tuned by varying the magnitude of the field. Thus, ER materials also belong to the 'very adaptive' material category. Yield strains for these materials range from below one to several percent.

A number of potential applications of ER fluids have been quoted in the literature. They fall into four categories: static mode (release mechanisms), variable flow controls (valves), vibration damping devices (shock absorbers), and shear modes (clutch devices). ER fluid-flow between two chambers or cavities can be controlled by adjusting the viscosity of the fluid. In vibration damping applications, the ER fluid usually operates in either the shear or extensional configuration. The shear configuration is used primarily when the fluid undergoes shear strains, while the fluid operates under extension or compression stresses in the extensional configuration.

The first adaptive structure using ER fluids was proposed in a 1990 patent. By incorporating an ER fluid into an otherwise passive hollow structure, the response of the entire structure can be 'tuned'. Using a sandwiched beam or panel with ER fluid would give the assembly the ability to adjust its stiffness and damping parameters. Adaptive structures incorporating such hollow beams or panels filled with ER fluids will have the ability to modify their dynamic characteristics under varying conditions. By applying real-time control schemes with such composites, intelligent noise and vibration control systems can be achieved. Finally, during shear mode operations, the ER fluid mechanically couples two surfaces by increasing or decreasing its viscosity with the application or removal of an electric field, which can be used in clutching devices.

Although each type of actuator material described in the previous section has an independent set of coefficients for comparing actuator quality, a general set of evaluating criteria is needed for the comparison of actuators of different types. Typically, the performance of an actuator is evaluated in terms of a number of key actuator performance characteristics: displacement, force generation, hysteresis, response time, bandwidth, temperature range, accuracy, power and weight.

1.6 Applications of Adaptive Structures

Although the field of adaptive structures is relatively new, it has gained wide attention due to the impact it can have on the performance, reliability, and stability of many systems used in space, aerospace, communication, transportation, and industrial, civil, and consumer applications. Over the past few years, numerous promising applications of adaptive structures have been conceived and demonstrated.

Adaptive structure applications can be classified into four areas of utilization of adaptive structure technologies. These applications are reviewed in this section. It is to be noted that the review is primarily focussed on piezoelectric actuator based adaptive structures since they are the most widely used actuators in adaptive structures. The four categories of adaptive structure applications are:

- Active Noise Cancellation
- Active Vibration Damping
- Adaptive Shape Control
- Structural Health Monitoring

Active Noise Cancellation

Noise is sound that causes discomfort to the hearer. It is caused by irregular vibrations. The dynamic range of sound intensity the human ear can perceive is 10^7:1, this range is made manageable through the use of a logarithmic scale, which has the dB as its basic unit. The human threshold of hearing on this scale is 0 dB.

Industrial noise sources tend to have several distinct characteristics. One is turbulence or grinding, this is evenly distributed across the frequency bands and is referred to as "broadband noise". Another is "narrow band noise," which concentrates most of its acoustic energy at specific frequencies.

When the noise is from equipment with rotary motions, a set of tones or "tonal noise" is produced. The sources for noise are numerous. For example, noise can radiated from engines, exhaust systems, fans, and blowers.

Active noise cancellation (ANC) was developed as a way to reduce if not eliminate some of these different types of noise. Active noise control works on the basic principle of destructive interference, where the undesirable sound wave is countered with a sound wave of equal amplitude and 180 degrees out-of-phase with the undesirable sound. The result is that the sound waves cancel, and the undesirable sound is eliminated. This principle is implemented in adaptive structures, including noise cancellation headsets, transformer quieting systems, interior noise reduction in automobiles and aircraft, and numerous other applications.

Active noise control technology consists of sensors, signal conditioning, a control system, power amplifiers, and actuators. In a typical ANC system operating in a steady harmonic noise field, the noise would be sensed with an array of microphones, the signals from which would be analyzed by a combination of a central processor unit and a digital signal processor.

A number of different sensors and actuators are used in these applications. The sensors utilized in these adaptive ANVC systems include eddy current proximity sensors, tachometers (velocity sensors), accelerometers, microphones, and piezoelectric sensors. For the most part, proximity sensors and tachometers are used for determining the rotational characteristics of the noise source. Microphones are the most common sensors used in the monitoring of cavity noise. Finally, piezoelectric actuators are used to monitor the noise associated with panel vibrations such as those present in transformer structures.

The most common actuator used in the active control of sound is a relatively simple loudspeaker with a variable baffle. The loudspeakers are primarily involved in the cancellation of noise or sound waves by applying a "mirror image" sound wave to counter to the undesirable noise. By the superposition theory, the amplitudes of these waves cancel and a quiet zone is created. A second approach to active noise control involves cancellation of the noise at the source. In these applications, piezoelectric ceramics and electromagnetic shakers are popular, especially when used in the control of panel vibrations. These applications are considered to be examples of active noise control since the goal is to reduce the amount of auditory acoustic noise.

Active Vibration Damping

Vibrations can cause damage due to material fatigue or compromise performance of precision instruments. They pose health risks to humans from industrial noise and potential failure of heavy equipment and structures. Stray vibration may cause misreading and reduced sensitivity in navigational gyroscopes, communication antennas, etc. They can also effect tolerances and surface finish in precision machining due to chatter vibrations in machine tools. Thus, elimination of undesired vibrations is very important from various considerations. Traditionally, vibrations are absorbed through passive methods, which employ isolating springs and dampers. These passive methods are usually unable to provide sufficient bandwidth or attenuation.

Use of adaptive structure technologies is promising for active vibration damping applications. Adaptive structures either create nullifying disturbances (similar to active noise control) or use piezoelectric ceramics and resonant shunts to dissipate vibration energy. Active vibration control

(AVC) differs from active noise control (ANC) since the vibrations which are being damped by AVC are within the structural material, while the vibrations being eliminated are in ANC are in the air (sound). However, the basic principles of ANC can be applied to AVC and the control hardware and software architectures are generally similar in both these adaptive structure technologies.

Chatter in cutting tools adversely affects the surface finish of the machined piece and limits the speed of machining operations. Excellent surface finish is critical in many precision fittings and in reflecting mirror surfaces in precision optics or space-based antennas for improved pointing accuracy. Active reduction of chatter in the cutting tools during machining operation has been demonstrated to considerably improve the surface finish.

Active vibration damping is also being used in some automobile suspensions to improve the handling and stability of the vehicle as well as passenger comfort. Toyota's Electronic Modulated Suspension (TEMS) is based on a road stability sensor and a shock adjuster, which utilizes piezoelectric sensors and actuators and an electronic control system.

Many active strut members have been developed for single-point active vibration isolation in various load-bearing strut applications. Boeing Corporation has developed an anti-vibration engine mount for aircraft application. These mounts reduce the dynamic response of the airframe to engine vibrations. A similar concept has been patented by GEC-Marconi for the reduction of vibrations in marine vessels. Bridgestone Corporation has presented a similar concept for the vibration isolation of large machinery such as milling machines. General Motors has also developed an isolation mount for automobile engine. Anti-vibration automobile suspensions have also been developed by Nissan, Hitachi, and others. Robert Bosch GmbH has described the application of such mounts in agricultural machinery. A constant length active strut member is described in a patent by Kaman Aerospace Corporation, which is designed not for vibration isolation but mainly to compensate for thermal deformations in large space structures. Singer Company has described the use of such a strut in ring-laser gyroscopes.

Active vibration damping is also being investigated for reducing vibrations in helicopter rotors, gear boxes, and airframes. By reducing the vibration levels in helicopters, pilots can fly longer, and the manoeuvrability of the helicopter and its fuel efficiency can be increased. A number of different methods of reducing these vibration are being investigated theoretically, which utilize placement of piezoelectric sensors and actuators along the airframe to suppress vibrations at selected locations and modelling of the rotor/airframe by finite element analysis using the airframe mass, stiffness, the type of flexural deformations in the airframe, rotor/airframe coupling, etc. The simulated results show a significant decrease in vibration. Recently, some of this theoretical research has been put into practice at Sikorsky Aircraft, which is working on reducing vibration-induced problems in gear meshing and helicopter rotors.

Another area of utilization of active vibration damping is the reduction of panel flutter in aircraft. Aircraft panel flutter increases aerodynamic drag and reduces fuel efficiency and manoeuvrability of the aircraft. Many different theoretical solutions for the suppression of airfoil flutter have been investigated using different models for panel structures and controlling schemes. These investigations predict substantial increases in airfoil performance with piezoelectric-based active vibration control.

Uncontrolled vibrations, oscillations, twisting, bending, etc are a major problem in the assembly, deployment, or operation of these large space structures, space robotics and other systems, space based antennas, and space telescopes. Space structures and systems are generally lightweight, flexible and deployable. Lack of rigidity, low gravity, and absence of aerodynamic resistance make it difficult to damp any vibrations, oscillations, twisting, bending, etc. once they are initiated. These vibrations, oscillations, and other modes of elastic deformations are generated by the operation of equipment on the space structure, periodic orientation/position/altitude correction initiated by the space guidance system and rocket thrusters, impact by micro-asteroids, micro-vibrations due to structural thermal cycling, etc.

Adaptive Shape Control

Adaptive shape control of flexible structures is very attractive for correcting image distortions in mirrors and telescopes, improving the performance of aerodynamic and hydrodynamic lifting surfaces (e.g., aircraft and submarines), and improving pointing accuracy in optical/laser systems, antennas, and large reflectors. The majority of these applications incorporate piezoelectric (or electrostrictive) actuators into the structure whose shape is to be controlled . Shape control of a relatively thin structure can be achieved by applying both tensional and compressive forces at different locations in the structure with single-element or stack actuators, which produce bending or by using distributed bimorph bending actuators. The degree to which the material will actually bend depends on a number of factors such as dimensions of the structure, its physical properties (e.g. Young's modulus, stiffness, etc.), distribution of actuators, and the amount of tension, compression, or bending forces being applied. The actuators can either be surface bonded or embedded in multilayer composite structures. Strain or displacement sensor are employed to detect surface deformations, and their output is used by the control system to initiate desired actuation from each actuator to achieve the shape control.

Deformable mirrors have been used in many applications, such as variable pointing of lasers, frequency excitation and optical switching, correction of mirror curvature distortion due to self-weight and thermal distortion in large and precision optical systems, adaptive optics to correct thermal and atmospheric distortions in high power space telescopes and self-dimming deformable rear view mirrors for automobiles.

There is considerable interest in the aerospace and defence sectors to use adaptive structure technologies for adaptive shape control of aircraft wings, helicopter rotor blades, propellers, and missile fins. Adaptive shape control of future aircraft wings is expected to reduce aerodynamic drag and improve speed and fuel efficiency of aircraft. Adaptive shape control has also been investigated for use in missile fins to actively navigate them towards their target. Composite structures are extensively being investigated for use in future aircraft and space applications. Use of composite structures for adaptive shape control offers the advantage of either surface bonding or embedding the sensors and actuators. Thus, many investigators are studying the fabrication of test articles (plates, beams, etc.) incorporating sensors and actuators, modelling the test articles for shape control applications, and control systems for these applications. Adaptive aircraft conformal skins are also being extensively investigated for future applications to incorporate antennas on the aircraft surface. Once such adaptive skins are incorporated, their adaptive shape control/correction will also become important for the performance of antenna systems.

It is to be noted that the need for shape control in large space structures comes from the lack of rigidity in the structure due to the lightweight, flexible, and deployable design used in these structures. Moreover, the space environment is nearly gravity-free, offers little aerodynamic resistance, and exposes the structures to daily thermal cycles with large temperature gradients. Thus, realizing the desired shape in the structure and precisely maintaining it throughout the operation is difficult. Therefore, adaptive shape control techniques become very attractive for meeting the performance characteristics of precision space systems.

Large space structures basically consist of truss structure assemblies and structural panels. Elastic deformations in these structures may be minimized by providing desired stiffness to the structural components through the use of distributed network of sensors and actuators. Active strut members are proposed for use at joints and load bearing points to provide precision dimensional correction to the truss structure. Shape control of panels or segmented reflector surfaces may be achieved by deformation or correction as required by using strategically placed sensors and actuators and an appropriate control system.

Structural Health Monitoring

Structural health monitoring is becoming very important for aerospace industry. With the tight monetary constraints placed on the military and civilian aircraft operators, there is a continual need to reduce the cost of ownership of aircraft. However, this has to be achieved without detriment to the airworthiness of these aircraft or to those involved in their operation. To meet these requirements, maintenance managers and aircraft schedulers must be in a position to make timely and accurate decisions about the serviceability of their equipment, structure and systems; decisions that will maximize availability for a given life cycle cost.

Health and Usage Monitoring Systems (HUMS) are used in aerospace industry to monitor the condition of engine, power train, avionics control systems, and aircraft structure so that the best possible decisions can be made based on a detailed and accurate assessment of health of individual components, complete systems, and entire fleets. HUMS are advanced computerized information support systems that monitor critical information of selected aircraft components, analyse the data, and provide a basis for effective resource management and maintenance scheduling.

Health (Condition) Monitoring is any technique that can convey information about the mechanical condition of the aircraft structure, engine, drive train, or other critical components. It includes gas-path aerodynamics performance analysis, vibration analysis, acoustic signature monitoring, oil and wear debris analysis, exhaust gas analysis, component damage analysis, expert systems, corrosion monitoring, and such emerging techniques as intake-debris and engine-distress monitoring.

Bibliography

1. T. Tagaki, "A Concept of Intelligent Materials and the Current Activities of Intelligent Materials in Japan", Proc. of the 1st European Conf. on Smart Structures and Materials, Glasgow (1992).

2. A. Ahmad, T.A. Wheat, S.E. Prasad, S. Varma, H.D. Wiederick, B. Mukherjee, S. Sherritt and D.F. Jones, "Smart Ceramic Materials and Systems", Proceedings of the International Symposium on Developments and Applications of Ceramics and New Metal Alloys, Aug. 29-Sep. 2, Quebec City, Quebec, Canada, 577-588 (1993).

3. J.V. Pagnaelli, "PTC Ceramics Heat Automotive Controls", Automotive Engineering **92** (6), 31-35 (1984).

2. Modelling of Piezoelectric Actuators in Shells, Plates and Beams

A. Suleman

IDMEC-Instituto Superior Técnico, Lisbon, Portugal

Abstract This lecture describes models to represent the interactions between piezoceramic actuators and elastic structures in the form of cylindrical shells, plates and beams. The piezo loads are determined in terms of material properties of the actuator and structure, the geometry of the actuator placement. and the applied actuator voltages. Also. a review on the current research in finite element models of the electromechanical coupling is presented.

2.1 Introduction

In order to simulate the behaviour of the piezo actuators in control applications, it is necessary to develop accurate models of the mechanics of the induced strain actuation. Here, the formulation of general analytical models to describe the interactions between a pair of piezoceramic actuators and elastic substructures consisting of a beam, plate or thin shell are presented. The objective of the presentation is to attain a physical appreciation of the interactions between the piezo actuators and the substructure using analytical mathematical derivations and arguments.

2.2 Analytical Models

Piezoelectric Constitutive Relations

Piezoelectric materials are assumed to be linear and the actuation strain is analogous to the thermal strain. Piezoelectric ceramics can be idealized as an orthotropic material like a composite unidirectional laminate. The constitutive relations are based on the assumption that the total strain in the actuator is the sum of the echanical strain induced by the stress, the thermal strain due to the temperature and the controllable actuation strain due to the applied electric voltage. Thus, the coupled electromechanical constitutive relations are

$$\bar{\sigma} = \mathbf{e}\,\mathbf{E} + \mathbf{D}\varepsilon + \overline{kT}$$
$$\bar{\varepsilon} = \mathbf{d}\bar{E} + \mathbf{S}\bar{\sigma} + \overline{\alpha T}$$

where $\bar{\sigma}$ is the stress vector, $\bar{\varepsilon}$ is the strain vector. $\bar{E}$ is the vector of applied electric field, $\mathbf{e}$ is the matrix of piezoelectric coefficients [N/V-m]; $\mathbf{d}$ is the matrix of piezoelectric strain coeficients [m/V]. $\mathbf{D}$ is the matrix of stiffness coefficients, $\mathbf{S}$ is the matrix of compliance coefficients, $\bar{k}$ is a vector of thermoelastic coefficients. $\overline{\alpha}$ is the vector of thermal coefficients of expansion and $\overline{T}$ is the vector of temperature variation.

Shell Equations

The material used in this lecture is based on the published paper by Banks and Smith [1]. For completeness, important equations are reproduced in this text. For a detailed derivation and explanation of the equations, the reader should consult reference [1]. Consider a thin circular cylindrical shell of radius R, thickness h and having the axial coordinate x as shown in Figure 1. The variable z measures the distance of a point on the shell from the corresponding point on the middle surface ($z=0$) along the normal to the middle surface.

Strain-Displacement Relations

Using the shell theory assumptions in conjunction with the equations of three-dimensional elasticity theory gives:

$$e_x = \varepsilon_x + z\kappa_x$$

$$e_\theta = \frac{1}{1+\frac{z}{R}}(\varepsilon_\theta + z\kappa_\theta)$$

$$\gamma_{x\theta} = \frac{1}{1+\frac{z}{R}}\left[\varepsilon_{x\theta} + z\left(1+\frac{z}{2R}\right)\tau\right] \tag{2.1}$$

where e_x and e_θ are normal strains at an arbitrary point within the cylindrical shell and $\gamma_{x\theta}$ is the shear strain. Here, ε_x, ε_θ and $\varepsilon_{x\theta}$ are the normal and shear strains in the middle surface and κ_x, κ_θ and τ are the mid-surface changes in curvature and twist.

Eqs. (2.1) represent the exact form of the kinematic equations. For shallow shells, the term $\frac{z}{R}$ can be neglected resulting in

$$e_\theta = (\varepsilon_\theta + z\kappa_\theta)$$

$$\gamma_{x\theta} = \left[\varepsilon_{x\theta} + z\left(1+\frac{z}{2R}\right)\tau\right] \tag{2.2}$$

In terms of the axial, tangential and radial displacements u, v and w, respectively, the expressions for the mid-surface strains and changes in the curvature for the cylindrical shell are

$$\varepsilon_x = \frac{\partial u}{\partial x}; \qquad \kappa_x = -\frac{\partial^2 w}{\partial x^2};$$

$$\varepsilon_\theta = \frac{\partial v}{\partial \theta} + \frac{w}{R}; \qquad \kappa_\theta = -\frac{1}{R^2}\frac{\partial^2 w}{\partial \theta^2} + \underline{\frac{1}{R^2}\frac{\partial v}{\partial \theta}}; \tag{2.3}$$

$$\varepsilon_{x\theta} = \frac{\partial v}{\partial x} + \frac{1}{R}\frac{\partial u}{\partial \theta}; \qquad \tau = -\frac{2}{R}\frac{\partial^2 w}{\partial x \partial \theta} + \underline{\frac{2}{R}\frac{\partial v}{\partial x}}$$

The underlined terms are neglected in the shallow shell theory.

Stress-Strain Relations

It is assumed that the shell material is elastic and isotropic. Hooke's law in conjunction with the assumption that the transverse shear stresses as well as the normal strain components are small compared to other stresses and strains yields

$$\sigma_x = \frac{E}{1-\nu^2}(e_x + \nu e_\theta)$$

$$\sigma_\theta = \frac{E}{1-\nu^2}(e_\theta + \nu e_x) \tag{2.4}$$

$$\sigma_{x\theta} = \sigma_{\theta x} = \frac{E}{2(1+\nu)}\gamma_{x\theta}$$

where σ_x, σ_θ are normal stresses, and $\sigma_{x\theta}$, $\sigma_{\theta x}$ are tangential shear stresses. The constants E and ν are the Young's modulus and Poisson's ratio for the shell structure, respectively.

Force and Moment Resultants

Integrating the stresses over the face of the element, the force resultants can be expressed as

$$\begin{bmatrix} N_x \\ N_{x\theta} \\ Q_x \end{bmatrix} = \int_{-h/2}^{h/2} \begin{bmatrix} \sigma_x \\ \sigma_{x\theta} \\ \sigma_{xz} \end{bmatrix} dz \tag{2.5}$$

and

$$\begin{bmatrix} N_\theta \\ N_{\theta x} \\ Q_\theta \end{bmatrix} = \int_{-h/2}^{h/2} \begin{bmatrix} \sigma_\theta \\ \sigma_{\theta x} \\ \sigma_{\theta z} \end{bmatrix} dz \tag{2.6}$$

Similarly, the moment resultants are

$$\begin{bmatrix} M_x \\ M_{x\theta} \end{bmatrix} = \int_{-h/2}^{h/2} \begin{bmatrix} \sigma_x \\ \sigma_{x\theta} \end{bmatrix} z\, dz \tag{2.7}$$

and

$$\begin{bmatrix} M_\theta \\ M_{\theta x} \end{bmatrix} = \int_{-h/2}^{h/2} \begin{bmatrix} \sigma_\theta \\ \sigma_{\theta x} \end{bmatrix} z\, dz \tag{2.8}$$

The orientation of the various forces and moments is shown in Figure 1. The integrals are determined accordingly to yield

$$N_x = \frac{Eh}{1-\nu^2}\left[\frac{\partial u}{\partial x} + \nu\left(\frac{1}{R}\frac{\partial v}{\partial \theta} + \frac{w}{R}\right)\right]; \qquad M_x = -\frac{Eh^3}{12(1-\nu^2)}\left[\frac{\partial^2 w}{\partial x^2} + \frac{\nu}{R^2}\frac{\partial^2 w}{\partial \theta^2}\right]$$

$$N_\theta = \frac{Eh}{1-\nu^2}\left[\frac{1}{R}\frac{\partial v}{\partial \theta} + \frac{w}{R} + \nu\frac{\partial u}{\partial x}\right]; \qquad M_\theta = -\frac{Eh^3}{12(1-\nu^2)}\left[\nu\frac{\partial^2 w}{\partial x^2} + \frac{1}{R^2}\frac{\partial^2 w}{\partial \theta^2}\right] \tag{2.9}$$

$$N_{x\theta} = \frac{Eh}{2(1+\nu)}\left[\frac{1}{R}\frac{\partial v}{\partial \theta} + \frac{\partial v}{\partial x}\right]; \qquad M_{x\theta} = -\frac{Eh^3}{12R(1+\nu)}\frac{\partial^2 w}{\partial x \partial \theta}$$

Shell Equations

The equations of the dynamic equilibrium of the element are obtained by balancing the internal force and moment resultants as shown in Figure 1 with any externally applied forces and moments. Let

$$\vec{q} = \hat{q}_x \hat{i}_x + \hat{q}_\theta \hat{i}_\theta + \hat{q}_n \hat{i}_n \quad \text{and} \quad \vec{m} = \hat{m}_x \hat{i}_x + \hat{m}_\theta \hat{i}_\theta$$

represent the surface forces and moments due to an external field which is acting on the middle surface. Hence, $\vec{q}$ and $\vec{m}$ have units of force and moment per unit area, respectively.

Considering equilibrium of forces in the x, θ and z direction, yields

$$\begin{aligned} &R\frac{\partial N_x}{\partial x} + \frac{\partial N_{\theta x}}{\partial \theta} + R\hat{q}_x = 0; \\ &\frac{\partial N_\theta}{\partial \theta} + R\frac{\partial N_{x\theta}}{\partial x} + Q_\theta + R\hat{q}_\theta = 0; \\ &R\frac{\partial Q_x}{\partial x} + \frac{\partial Q_\theta}{\partial \theta} - N_\theta + R\hat{q}_n = 0. \end{aligned} \tag{2.10}$$

respectively. The transverse shearing force Q_θ is considered to be negligible in the second of Eq. (10) and is subsequently neglected when determining the final equilibrium equations. Similarly, with 0 as the reference origin, the balancing of moments with respect to θ, x and z yields

$$
\begin{aligned}
&R\frac{\partial M_x}{\partial x}+\frac{\partial M_{\theta x}}{\partial \theta}-RQ_x+R\hat{m}_\theta=0;\\
&\frac{\partial M_\theta}{\partial \theta}+R\frac{\partial M_{x\theta}}{\partial x}-RQ_\theta+R\hat{m}_x=0;\\
&N_{x\theta}-N_{\theta x}-\frac{M_{\theta x}}{R}=0.
\end{aligned}
\tag{2.11}
$$

respectively. By referring to the integral definitions of $N_{x\theta}$, $N_{\theta x}$ and $M_{\theta x}$, it can be seen that the third expression in (11) is identically satisfied due to the symmetry of the stress tensor.

Time enters the equilibrium equations through the inertial terms; hence for time dependent problems, the force $\hat{q}_x$ is replaced by $-\rho h\frac{\partial^2 u}{\partial t^2}+\hat{q}_x$, where ρ is the mass density of the shell. Similar substitutions are made for $\hat{q}_\theta$ and $\hat{q}_n$. By combining Eqs. (10) and (11), one arrives at the time-dependent equilibrium equations for a thin cylindrical shell with radius of curvature R and thickness h

$$
\begin{aligned}
&R\rho h\frac{\partial^2 u}{\partial t^2}-R\frac{\partial N_x}{\partial x}-\frac{\partial N_{\theta x}}{\partial \theta}=R\hat{q}_x;\\
&R\rho h\frac{\partial^2 v}{\partial t^2}-R\frac{\partial N_\theta}{\partial \theta}-R\frac{\partial N_{x\theta}}{\partial x}=R\hat{q}_\theta;\\
&R\rho h\frac{\partial^2 w}{\partial t^2}-R\frac{\partial^2 M_x}{\partial x^2}-\frac{1}{R}\frac{\partial^2 M_\theta}{\partial \theta^2}-2\frac{\partial^2 M_{x\theta}}{\partial x\,\partial \theta}+N_\theta=R\hat{q}_n+R\frac{\partial \hat{m}_\theta}{\partial \theta}
\end{aligned}
\tag{2.12}
$$

Plate Equations

Let us consider a thin rectangular plate whose edges lie along the coordinate lines $x=0,a$ and $y=0,b$ and also assume that the plate is subjected to both longitudinal and transverse loading via the surface forces and moments $\hat{q}_x,\hat{q}_\theta,\hat{q}_n$ and $\hat{m}_x$, $\hat{m}_\theta$. With u,v and w denoting the displacements in the x,y and normal directions, respectively, the Kirchhoff plate equations are given by

$$
\begin{aligned}
&\rho h\frac{\partial^2 u}{\partial t^2}-\frac{\partial N_x}{\partial x}-\frac{\partial N_{xy}}{\partial y}=\hat{q}_x;\\
&\rho h\frac{\partial^2 v}{\partial t^2}-\frac{\partial N_y}{\partial y}-\frac{\partial N_{xy}}{\partial x}=\hat{q}_y;\\
&\rho h\frac{\partial^2 w}{\partial t^2}-\frac{\partial^2 M_x}{\partial x^2}-\frac{\partial^2 M_y}{\partial y^2}-\frac{\partial^2 M_{xy}}{\partial x\,\partial y}-\frac{\partial^2 M_{yx}}{\partial y\,\partial x}=\hat{q}_n+\frac{\partial \hat{m}_x}{\partial y}+\frac{\partial \hat{m}_y}{\partial x};
\end{aligned}
\tag{2.13}
$$

where

$$
\begin{aligned}
&N_x=\frac{Eh}{1-\nu^2}\left[\frac{\partial u}{\partial x}+\nu\frac{\partial v}{\partial y}\right]; \qquad M_x=-\frac{Eh^3}{12(1-\nu^2)}\left[\frac{\partial^2 w}{\partial x^2}+\nu\frac{\partial^2 w}{\partial y^2}\right]\\
&N_y=\frac{Eh}{1-\nu^2}\left[\frac{\partial v}{\partial y}+\nu\frac{\partial u}{\partial x}\right]; \qquad M_y=-\frac{Eh^3}{12(1-\nu^2)}\left[\nu\frac{\partial^2 w}{\partial x^2}+\frac{\partial^2 w}{\partial y^2}\right]\\
&N_{x\theta}=\frac{Eh}{2(1+\nu)}\left[\frac{\partial v}{\partial x}+\frac{\partial u}{\partial y}\right]; \qquad M_{xy}=-\frac{Eh^3}{12R(1+\nu)}\frac{\partial^2 w}{\partial x\partial y}
\end{aligned}
\tag{2.14}
$$

The first two equations describe the longitudinal movement of the plate while the third equation describes the transverse motion of the plate.

Beam Equations

The motion of an undamped thin beam of length l and width h can be determined from the dynamics of thin plate theory by considering only the vibrations in the x-direction along with the usual transverse vibrations (in the z direction). This yields the Euler-Bernoulli beam equations

$$
\begin{aligned}
&\rho h\frac{\partial^2 u}{\partial t^2}-\frac{\partial N_x}{\partial x}=\hat{q}_x;\\
&\rho h\frac{\partial^2 w}{\partial t^2}-\frac{\partial^2 M_x}{\partial x^2}-=\hat{q}_n+\frac{\partial \hat{m}_y}{\partial x};
\end{aligned}
\tag{2.15}
$$

where

$$
N_x=Eh\frac{\partial u}{\partial x}; \qquad M_x=-\frac{Eh^3}{12}\frac{\partial^2 w}{\partial x^2}=-EI\frac{\partial^2 w}{\partial x^2}
\tag{2.16}
$$

2.3 Piezoelectric patch actuation contribution

For a thin cylindrical shell, the loads can be written in terms of the line forces and moments $\hat{N}_x, \hat{N}_\theta, \hat{M}_x$ and M_θ and the normal surface load $\hat{q}_n$. In the problem under consideration, these quantities results from the actuation of piezoceramic patches of thickness t which are assumed to be perfectly bonded to a cylindrical shell of thickness h with midsurface radius R. The patches are assumed to have their edges parallel to lines of constant x and θ. Because the patches generate no shear strains, the external load $\hat{q}_n$ is taken to be $\hat{q}_n = 0$. The external line moments and forces are simply

$$\begin{aligned} \hat{N}_x &= N_{x_{piezo}}; & \hat{M}_x &= M_{x_{piezo}} \\ \hat{N}_\theta &= N_{\theta_{piezo}}; & \hat{M}_\theta &= M_{\theta_{piezo}} \end{aligned} \qquad (2.17)$$

where $N_{x_{piezo}}, M_{x_{piezo}}, N_{\theta_{piezo}}$ and $M_{\theta_{piezo}}$ are the respective in-plane forces and moments generated by the piezo actuators. The surface moments and forces take the form

$$\hat{m}_x = -\frac{1}{R}\frac{\partial M_{\theta_p}}{\partial \theta}; \qquad \hat{m}_\theta = -\frac{\partial M_{x_p}}{\partial x}$$

$$\hat{q}_x = -S_{1,2}(x)\,\hat{S}_{1,2(\theta)}\frac{\partial N_{x_p}}{\partial x}; \qquad \hat{q}_\theta = -S_{1,2}(x)\,\hat{S}_{1,2(\theta)}\frac{\partial N_{\theta_p}}{\partial \theta}$$

For an actuator patch with uniform thickness and limiting values x_1, x_2, θ_1 and θ_2, the presence of the indicator function

$$S_{1,2}(x) = \begin{cases} 1 & x < (x_1 + x_2)/2 \\ 0 & x = (x_1 + x_2)/2 \\ -1 & x > (x_1 + x_2)/2 \end{cases} \qquad (2.18)$$

can be interpreted from the fact that the forces generated by the patch in the x-direction are anti-symmetric (equal in magnitude but opposite in sign) about the line $x = (x_1 + x_2)/2$. The same holds true for the forces in the θ direction with $\hat{S}_{1,2}(\theta)$ being defined in analogous manner to $S_{1,2}(x)$.

In order to determine $N_{x_{piezo}}, M_{x_{piezo}}, N_{\theta_{piezo}}$ and $M_{\theta_{piezo}}$ and hence the loads on the shell, it is useful to write the normal strains $\varepsilon_x, \varepsilon_\theta$ and midsurface changs in curvature κ_x, κ_θ of the middle surface ($z=0$) of the cylindrical shell. Note that due to the presence and actuating nature of the piezos when a voltage is applied, the normal strains and changes in the curvature are no longer given by the expression in Eq. (3), which were derived for a homogeneous thin cylindrical shell. At this point, the strains and curvatures are considered to be unknown and are to be determined by formulating stress-strain relations in the patch and shell followed by the balancing of moments and forces in the combined structure. In this manner, expressions for these midsurface characteristics (and hence the resulting patch moments and forces) can be found in terms of the material properties of the shell and patch, the radius of curvature of the shell, and the voltages being applied to the patches.

Now, the strain relationships for a shallow cylindrical shell with mid-surface radius R is given by

$$
\begin{aligned}
e_{x_{piezo}} &= \varepsilon_x + z\,\kappa_x; \\
e_{\theta_{piezo}} &= \varepsilon_\theta + z\kappa_\theta;
\end{aligned} \tag{2.19}
$$

where $e_{x_{piezo}}$ and $e_{\theta_{piezo}}$ are normal strains at an arbitrary point on the patch. When a voltage is applied and the patch is actuated, in accordance with the basic shell theory, equal strains are induced in the x and θ directions and the radius of curvature is not changed in either direction. Patches satisfying this assumption could be made, for example, by taking a portion of a thin walled tubular piezoceramic element. For the outer patch, the magnitude of the induced free strains is given by:

$$
e_{p_{outer}} = e_{x_{p_{outer}}} = e_{\theta_{p_{outer}}} = \frac{d_{31}}{h} V_{outer} \tag{2.20}
$$

where d_{31} is a piezoceramic strain constant and V_{outer} is the applied voltage. Note that when the voltage is applied to a patch with edge coordinates x_1, x_2, θ_1 and θ_2, the point $(x,\theta) = [(x_1 + x_2)/2,\ R(\theta_1 + \theta_2)/2]$ will not move whereas the axially symmetric points on either side will move an equal amount in opposite directions. This requires the use of the sign functions.

Assuming that the two patches have the same Young's modulus, E_p and Poisson's ratio ν_p, the stress distribution in the outer patch is given by

$$
\sigma_{x_{p\ outer}} = \frac{E_p}{1-\nu_p^2}\left[e_x + \nu_p e_\theta - (1+\nu_p)e_{p-outer}\right]
$$

$$
\sigma_{\theta_{p\ outer}} = \frac{E_p}{1-\nu_p^2}\left[e_\theta + \nu_p e_x - (1+\nu_p)e_{p-outer}\right] \tag{2.21}
$$

with the negative signs resulting from conservation of forces. Similar expressions are used for the induced free strain and stress distribution in the inner patch. By integrating the stresses over the face of a fundamental element, it follows that the moment and force resultants for the patches can be expressed as

$$
M_{x_{p-outer}} = \int_{h/2}^{h/2+t} \sigma_{x_{p\ outer}} \left(1+\frac{z}{R}\right) z\,dz; \qquad M_{x_{p\ inner}} = \int_{-h/2-t}^{-h/2} \sigma_{x_{p\ inner}} \left(1+\frac{z}{R}\right) z\,dz
$$

$$
M_{\theta_{p-outer}} = \int_{h/2}^{h/2+t} \sigma_{\theta_{p-outer}}\, z\,dz; \qquad M_{\theta_{p-inner}} = \int_{-h/2-t}^{-h/2} \sigma_{x_{p\ inner}}\, z\,dz \tag{2.22}
$$

$$N_{x_{p-outer}} = \int_{h/2}^{h/2+t} \sigma_{x_{p-outer}} \left(1+\frac{z}{R}\right) dz; \qquad N_{x_{p-inner}} = \int_{-h/2-t}^{-h/2} \sigma_{x_{p-inner}} \left(1+\frac{z}{R}\right) dz$$

$$N_{\theta_{p-outer}} = \int_{h/2}^{h/2+t} \sigma_{\theta_{p-outer}} \, dz; \qquad N_{\theta_{p-inner}} = \int_{-h/2-t}^{-h/2} \sigma_{x_{p-inner}} \, dz$$

with units of moment per unit length and force per unit length, respectively. Note that the midsurface strains and curvature changes are unknown and can be determined by balancing moments and forces in the combined structure.

Throughout this development, the expressions in Eq. (2.22) apply to patches covering the full circumference of the shell and having infinite axial length. The equations can be modified for finite actuator patches by expressing the total line moments and forces as

$$M_{x_p} = \left(M_{x_{p-inner}} + M_{x_{p-outer}}\right)\left[H_1(x) - H_2(x)\right]\left[H_1(\theta) - H_2(\theta)\right]$$

$$M_{\theta_p} = \left(M_{\theta_{p-inner}} + M_{\theta_{p-outer}}\right)\left[H_1(x) - H_2(x)\right]\left[H_1(\theta) - H_2(\theta)\right]$$

$$N_{x_p} = \left(N_{x_{p-inner}} + N_{x_{p-outer}}\right)\left[H_1(x) - H_2(x)\right]\left[H_1(\theta) - H_2(\theta)\right] S_{1,2}(x)\, \hat{S}_{1,2}(\theta) \tag{2.23}$$

$$N_{\theta_p} = \left(N_{\theta_{p-inner}} + N_{\theta_{p-outer}}\right)\left[H_1(x) - H_2(x)\right]\left[H_1(\theta) - H_2(\theta)\right] S_{1,2}(x)\, \hat{S}_{1,2}(\theta)$$

where H is the Heaviside function and $H_i(x) = \left[H(x - x)_i\right]$; $i = 1,2,...$ with a similar definition in θ. The indicator function derives from the property that for homogeneous patches having uniform thickness, opposite but equal strains are generated about the point $(x,\theta) = \left[(x_1 + x_2)/2,\ \ R(\theta_1 + \theta_2)/2\right]$ in the two coordinate directions. Balancing the force and moment gives:

$$M_{x_{p-inner}} + M_{x_{p-outer}} = -\frac{Eh^3}{12(1-\nu^2)}\left[\kappa_x + \nu\kappa_\theta + \frac{\varepsilon_x}{R}\right]$$

$$M_{\theta_{p-inner}} + M_{\theta_{p-outer}} \approx -\frac{Eh^3}{12(1-\nu^2)}\left[\kappa_\theta + \nu\kappa_x - \frac{\varepsilon_\theta}{R}\right]$$

$$N_{x_{p-inner}} + N_{x_{p-outer}} = -\frac{Eh}{1-\nu^2}\left[\varepsilon_x + \nu\varepsilon_\theta + \frac{h^2}{12}\frac{\kappa_x}{R}\right] \tag{2.24}$$

$$N_{\theta_{p-inner}} + N_{\theta_{p-outer}} \approx -\frac{Eh}{1-\nu^2}\left[\varepsilon_\theta + \nu\varepsilon_x + \frac{h^2}{12}\frac{\kappa_\theta}{R}\right]$$

A similar analysis can be used to determine the forces and moments that are due to the actuation of piezoceramic patches which have been bonded to a flat plate or beam.

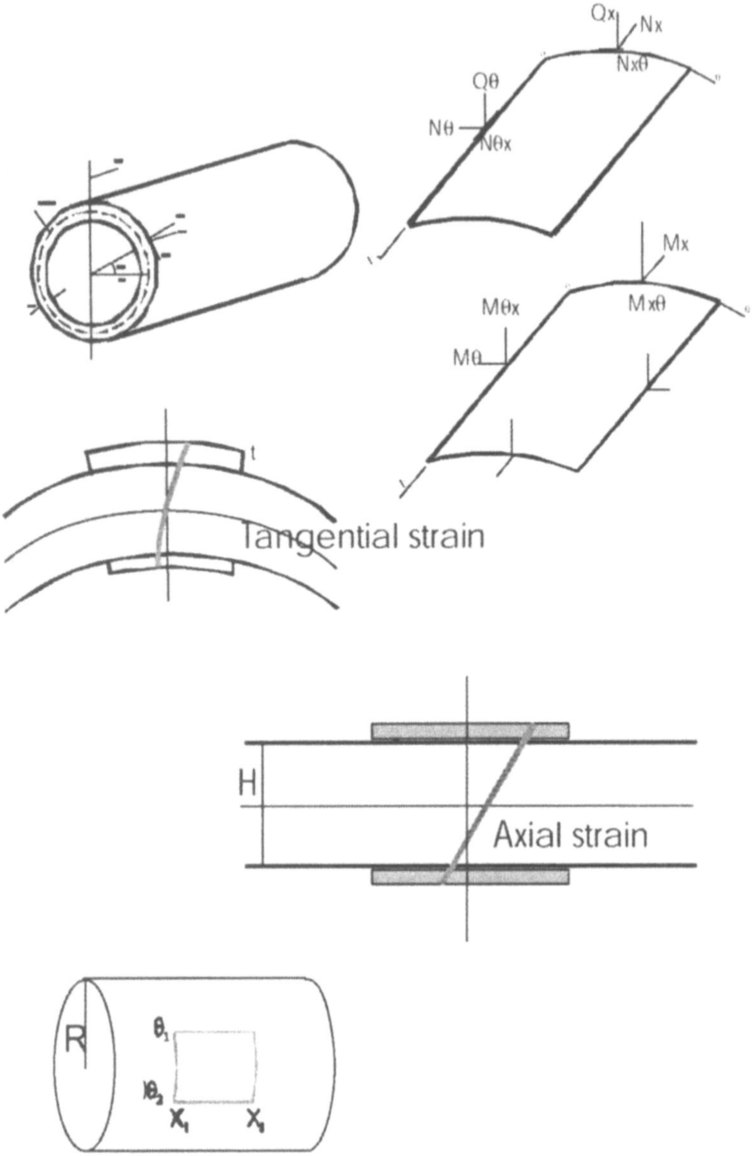

Figure 1 – Cylindrical Shell

2.4 Finite Element Models

The finite element modeling of adaptive structures has attracted numerous researchers and has become a major area of research. Early investigations were devoted to 3D elements with nodal electric potential dof. They take account of the surface characteristics and full electro-mechanical coupling, inherent to piezoelectric patches. However, it was found that these elements were too thick to represent thin structures. Hence, attention was directed to 2D elements, despite the difficulty of the conventional mid-plane formulations to take into account potentials on upper/lower surfaces. This motivated the recent development of one and two-dimensional elements with electric degrees of freedom, as pointed out by Benjeddou [2]

Solid Elements

Three-dimensional piezoelectric solid elements contain additional electric dof. They are either tetrahedral or brick elements. The electric potential is assumed linear except for the 20-node hexahedron for which it is quadratic. Thermal piezoelectric effects have been considered for 8-node and 20-node hexahedral elements. Internal dof have been added to enhance the behavior of the 8-node brick element when used to model very thin structures.

Early elements did not deal with layered structures. These were handled using the equivalent single layer model Detailed description of some piezoelectric solid elements is given in Table 1.

Shell Elements

Only a few piezoelectric shell elements have been reported in the literature (Table 2). A 4-node shell element extending the shallow shell shear deformation theory has been proposed, using an equivalent single layer model for a three-layer shell. Upper and lower nodal electric potential dof were selected to represent surface characteristics of the piezoelectric layer. Reduced integration was used to avoid shear locking. An 8-node quadrilateral shell element free of electric dof was also formulated using the 3D-degenerated shell theory where the piezoelectric effect was treated as an initial strain problem.

An axisymmetric 3-node triangular shell element has been developed to study mooney transducers. An 8-node quadrilateral shell element has been also proposed. These elements do not deal with multi-layers. A layer-wise constant shear angle 12-node degenerated 3D shell element has been formulated. Displacements and electric potential were supposed in plane quadratic and through-thickness linear.

Plate Elements

The representation of the surface electric characteristics of the piezoelectric layers is somewhat difficult for conventional mid-plane two-dimensional formulations. Only quadrilateral elements have been proposed for finite element modeling of electromechanical plates (Table 3). In-plane variable dof 4-node element has been formulated using discrete layer theories for laminated piezoelectric plates. The potential and in-plane displacements are piecewise linear, whereas the deflection can be either constant or linear. This formulation has the advantage to represent the induced potential due to the numerical through-thickness finite element subdivisions.

A 4-node element has been proposed for active control of Kirchhoff-Love plate bending vibrations. Another one has been reported using a combination of a 3D piezoelectric element, defined by pseudo-nodes for top/bottom surfaces, and an ACLD element. Also, an 8-node quadrilateral element has been formulated using a higher-order shear deformable displacement theory, but assuming through-thickness linear variation of the electric potential. Previous elements have nodal electric potential dof. A simple Mindlin plate element with one potential dof per piezo-electric layer, and using uniform reduced numerical integration and hourglass stabilization has also been proposed.

A quadratic 9-node element, formulated on the basis of a higher-order shear deformation theory has been proposed. It fulfills zig-zag effect for in-plane displacements, interlaminar equilibrium, and top/bottom transverse shear stress conditions. Beside, the electric potential was assumed quadratic in the plate thickness and is described at the layer level by top, central and bottom electric potential dof. Stress and electric dof were condensed at the layers system level.

Beam Elements

There are few piezoelectric beam elements with electric dof (Table 4). It is may be due to the fact that the condensation of electric dof can be easily done on the continuum formulation level for one-dimensional space equations.

Hu-Washizu variational principle has benn extended to include electric loads in order to formulate a Timoshenko beam element with offset nodes. Beside the mechanical dof, upper/lower electric potential dof per piezoelectric element were considered.

Sandwich theory has been applied to formulate a 2-node element of an original 4-layer ACLD system. The shear effect is neglected for the piezo. and basic beam. The axial displacements of the latter layers and their deflection are retained as mechanical dof. However, the shear effect is considered for piezoelectric cores of sandwich piezoelectric beams. The mean and relative axial displacements of the core or the faces and their deflection are retained as mechanical dof.

Timoshenko and Bernoulli-Euler theories-based beam finite elements have been proposed. They were extended to include Saint-Venant and warping torsions in PZT/Epoxy composites for bending. The deflection was assumed quadratic and rotational and twisting angles are also dof. A quadratic variation of axial displacement and shear angle, in conjunction with a cubic Hermite deflection were also assumed for a 3-node beam element. It was based on thermal analogy and a 3layer theory with shear-free piezo/beam layers. An elastic part of the shear angle was used as additional DOF to represent a time-domain viscoelastic model for ACLD treatments.

2.5 Concluding Remarks

In this lecture, general analytical and finite element models describing piezoelectric modeling and the interaction between piezoelectric patches and an underlying structure have been presented. The lecture notes are based on the published papers by Banks and Smith [1] and Benjeddou [2].

References

1. Banks, H.T. and Smith, R.C., "The modeling of piezoceramic patch interactions with shells, plates and beams", NASA Report 189740, ICASE report No. 92-66, 1992.

2. Benjeddou, A., "Advances in Finite Element Modelling of Smart Structural Elements", NATO Advanced Study Institute, Vol IV, Troia, Portugal, 1998.

Table 1. Piezoelectric solid finite elements [2]

Shape	Authors	Interpolation	Nodal dof	Total dof	Comments
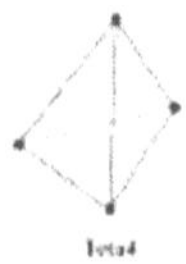	Allik, Hughes (1970)	u,v,w: linear φ: linear	u,v,w φ	16	Static condensation of the electric dof
	Ghandhi, Hagood (1997)	u,v,w: linear φ: linear	u,v,w φ,	16+ internal dof	Nonlinear constitutive equations.
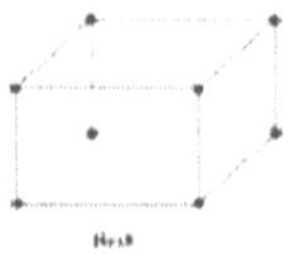	Tzou, Tseng (1990) (1991)	u,v,w: linear + quadratic modes,φ: linear	u,v,w+ Intern. Dof: φ	32	Static condensation of the incompatible modes at the element level
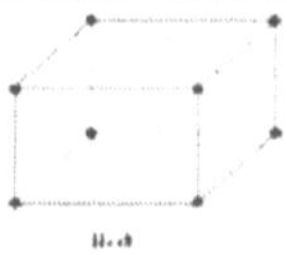	Ha, Keilers, Chang (1992)	u,v,w: linear φ:linear+quad incom.modes.	u,v,w+ Intern. Dofs	32	Equivalent single layer model
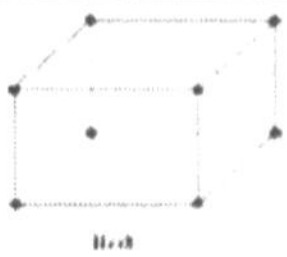	Chin, Varadan (1994)	u,v,w: linear φ: linear	u,v,w φ	32	Lagrange methods for matrix formulation.
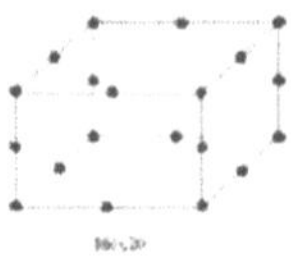	Allik, Webman (1974)	u,v,w: quadratic φ: quadratic	u,v,w φ	80	Application to response of sonar transducers.
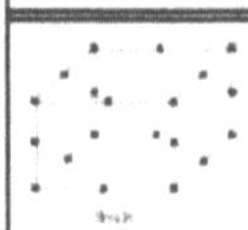	Koko, *et al.* (1997)	u,v,w: quadratic φ: quadratic θ: quadratic	u,v,w φ θ	100	Thermo-piezoelectric constitutive eq

Table 2. Piezoelectric shell finite elements.[2]

Shape	Authors (year)	Approx.	Nodal Dofs	Total Dofs	Comments
 Shell 4	Lammering (1991) (1992)	u,v,w: linear β_x, β_y :linear φ:linear	u,v,w β_x, β_y	28	Shallow shell shear theory
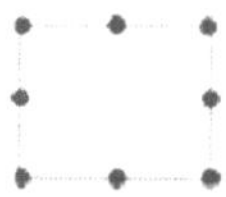 Shell 8	Thirupathi, Seshu, Naganathan (1997)	u,v,w: quadratic β_x,β_y: quadratic	u,v,w β_x, β_y	40	3D-degenerated shell theory; Piezo effect as initial strain problem
 Shell 8-axi8	Guo, Cawley Hitchings (1992)	u,v,w: quadratic φ: quadratic	u,v,w φ	32	Generalized stress, strain, displacement, force variables (mechanical + electric)
	Varadan, Chin (1993)	u, w: linear φ:linear	u, w φ	9	Potential energy including both elastic and electric energies Lagrange formulation of motion equations.
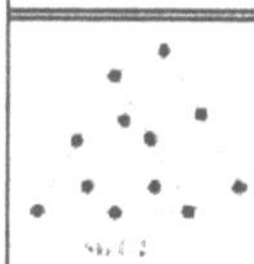	Tzou, Ye (1994)	u,v,w,φ in-plane: quadratic thickness: linear	u,v,w φ	48	Layer-wise constant shear angle theory.

Table 3. Piezoelectric plate finite elements [2]

Shape	Authors (year)	Approx.	Nodal Dofs	Total Dofs	Comments
 Quad 4	Suleman, Venkayya (1995)	u,v,w: bilinear $\theta_x,\theta_y,\theta_z$: bilinear φ: linear	u,v,w $\theta_x,\theta_y,\theta_z$ $\varphi_1,\ldots,\varphi_{np}$	24 + np	Mindline plate element (C^0) + hour-glass stabilization.: 1 potential Dof per piezoelectric layer
 Quad 4	Veley, Rao (1995-96)	u,v,w: linear φ: linear	u,v,w φ	16	Combination of a 3D piezoelectric element and a ACLD element (Quad/Hex/Quad)
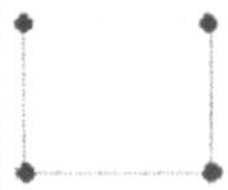 Quad 4	Chen, Wang, Liu (1997)	w: cubic φ:linear	w, $w_{,x}$, $w_{,y}$, φ	16	Kirchhoff plate element (bending only).: Guyan reduction for the electrical dof
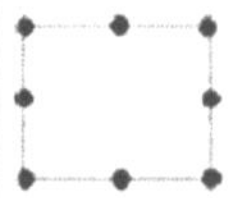 Quad 8	Ray, Bhattacharyya Samanta (1994)	u,v,w, I_x, I_y, I_z, m_x, m_y, m_z, n_x, n_y, φ_1, φ_n quadratic	all	104	11 variables higher-order shear deformable theory; Linear potential in thickness
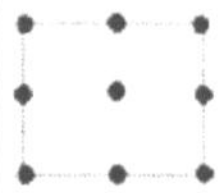	Carrera (1997)	u,v,w: quadratic	u,v,w β_α D_α	varia-ble	Multilayered plates including piezo-layers.; Higher order shear deformation theory.
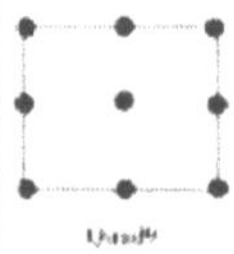	Yin, Shen (1997)	u,v,w: β_x, β_y ψ=-E_3 all quadratic	u,v,w β_x, β_y ψ=-E_3	54	Mindlin plate theory (C^0) Linear voltage but transverse field dof

Table 4 . Piezoelectric beam finite elements [2]

Shape	Authors (year)	Approx.	Nodal Dofs	Total Dofs	Comments
 Beam2 with offset nodes	Shen (1994) (1995)	u: linear w: cubic Hermite β: linear	u w,w_x β	8+ V_{to}, V_b	Timoshenko beam theory using an extended Hu-Washizu principle (Mixed) Upper/lower potential dofs per element
 Beam2	Carpenter (1997)	u: linear v: cubic Hermite φ: linear	U v, v_x φ	8	Euler-Bernoulli strain model Includes electric field work

Table 5. Piezoelectric finite elements found in the covered literature [2]

Elements	Shape and approximations	Availability
Solid	4-nodes linear tetrahedron	available
	8-nodes linear hexahedron	available
	20-nodes quadratichexahedron	available
Shell	3-nodes linear axisymmetric flat triangle	available
	8-nodes quadratic axisymmetric quadrangle	available
	4-nodes linear flat quadrangle	available
	8-nodes 3D -degenerated quadratic quad	not available
	12-nodes 3D-degenerated quadratic prism	available
Plate	3-nodes linear triangle	not available
	4-nodes linear quadrangle	available
	8-nodes quadrangle	available
	9-nodes quadrangle	available
Beam	2-nodes linear element	available
	3-nodes quadratic element	not available

3. A Finite Element Adaptive Plate Model

A. Suleman

IDMEC-Instituto Superior Técnico, Lisbon, Portugal

Abstract A simple quadrilateral shell finite element has been formulated for the analysis of adaptive laminated composite plates. The finite element is quadrilateral in shape and has 4 nodal points with 24 degrees of freedom.

3.1 Equations of Motion

Figure 3.1 shows the adaptive composite configuration and the direction of polarization of the piezoelectric laminates.

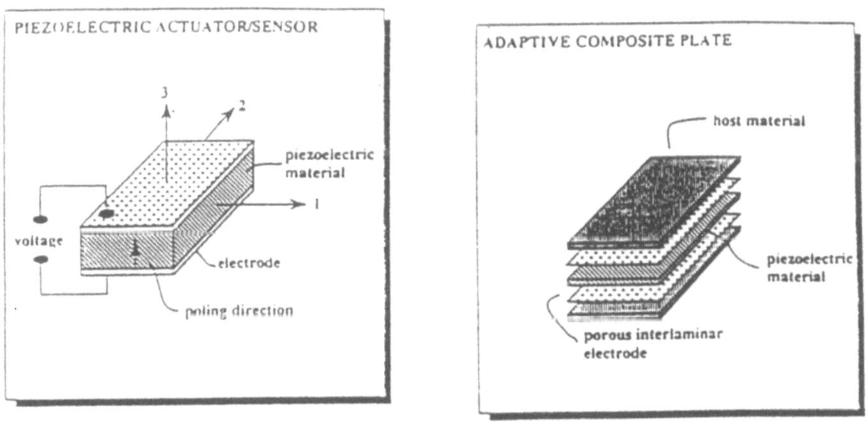

Figure 3. 1 – Adaptive composite configuration and the direction of polarization of the piezoelectric laminates

To derive the equations of motion for the laminated composite plate with piezoelectrically coupled electromechanical properties, we use the generalized form of Hamilton's principle

$$\delta \int_{t_1}^{t_2} [T - \Pi + W_e]\, dt = 0 \tag{3.1}$$

where T is the kinetic energy, Π is the potential energy, and W_e is the work done by the electrical field. The kinetic and potential energies can be written in the form

$$T = \int_V \frac{1}{2} \rho\, \dot{\bar{u}}^T \dot{\bar{u}}\, dV; \qquad \Pi = \int_V \frac{1}{2} \bar{S}^{e^T}\, \bar{T}^e\, dV$$

where $\overline{S}^{c}$ and $\overline{T}^{c}$ are the generalized elastic strain and stress vectors. The work done by the electrical forces can be written as

$$W_e = \int_{V_p} \frac{1}{2} \overline{S}^{e^T} \, \overline{T}^{e} \, dV_p$$

where $\overline{S}^{e}$ is a vector of electrical field (volts/meter) in the piezoelectric material, and $\overline{T}^{e}$ is a vector of electrical displacements (charge/area). In the context of the present analysis, it is assumed that the work dome by the electromagnetic forces is negligible.

3.2 Electromechanical Constitutive Relations

For piezoelectrics the properties are defined relative to the local poling direction. Available piezoelectric materials have the direction of poling associated with the transverse direction and the material is approximately isotropic in the other two directions. In matrix form the equations governing these material properties can be written as

$$\overline{T}^{e} = \mathbf{e}^{T} \, \overline{S}^{c} + \varepsilon \, \overline{S}^{e}$$
$$\overline{T}^{c} = \mathbf{c} \, \overline{S}^{c} - \mathbf{e} \, \overline{S}^{e}$$

where $\overline{T}^{e}$ is the electric displacement vector; $\mathbf{e}$ is the dielectric permittivity matrix; $\overline{S}^{c}$ is the elastic strain vector; ε is the dielectric matrix at constant mechanical strain; $\overline{S}^{e}$ is the electric field vector; $\overline{T}^{c}$ is the elastic stress vector and $\mathbf{c}$ is the matrix of elastic coefficients at constant electric field strength.

3.3 Stress-Strain Relations

The composite laminate shell is presumed to consist of perfectly bonded laminae. Moreover, the bonds are presumed to be infinitesimally thin. Thus, following the classical lamination theory, the state of stress in the element is given by

$$\overline{S} = \left\{ \overline{T}^{m} \quad \overline{T}^{b} \quad \overline{T}^{ts} \quad \overline{T}^{e} \right\}$$
$$= \left\{ T_x^{m} \quad T_y^{m} \quad T_{xy}^{m} \quad T_x^{b} \quad T_y^{b} \quad T_{xy}^{b} \quad T_{xz}^{ts} \quad T_{yz}^{ts} \quad D_1 \quad \ldots \quad D_{n_p} \right\}$$

Eight generalized strains and one electrical field parameter per lamina describe the state of deformation for a Mindlin plate with electromechanical properties. Thus, the augmented generalized strain vector takes the form

$$\overline{S} = \left\{ \overline{S}^{m} \quad \overline{S}^{b} \quad \overline{S}^{ts} \quad \overline{S}^{e} \right\}$$
$$= \left\{ S_x^{m} \quad S_y^{m} \quad S_{xy}^{m} \quad S_x^{b} \quad S_y^{b} \quad S_{xy}^{b} \quad S_{xz}^{ts} \quad S_{yz}^{ts} \quad -E_1 \quad \ldots \quad -E_{n_p} \right\}$$

The stress-strain relationship takes the form

$$\overline{T} = \begin{Bmatrix} \overline{T}^c \\ \overline{T}^e \end{Bmatrix} = \begin{bmatrix} \mathbf{c} & \mathbf{c} & \mathbf{0} & \mathbf{e} \\ \mathbf{c} & \mathbf{c} & \mathbf{0} & \mathbf{e} \\ \mathbf{0} & \mathbf{0} & \mathbf{g} & \mathbf{0} \\ \mathbf{e}^T & \mathbf{e}^T & \mathbf{0} & \varepsilon \end{bmatrix} \begin{Bmatrix} \overline{S}^m \\ \overline{S}^b \\ \overline{S}^{ts} \\ \overline{S}^e \end{Bmatrix} - \begin{Bmatrix} \overline{\alpha}^m \\ 0 \\ 0 \\ 0 \end{Bmatrix} \Delta T$$

where $\mathbf{c}$ is the transformed moduli matrix for each lamina including the piezoelectric layers. The transverse shear stiffness matrix $\mathbf{g}$ is defined in terms of the transverse strain energy through the thickness. ΔT is the temperature gradient across the laminate and $\overline{\alpha}^m$ are the coefficients of thermal expansion for each lamina.

3.4 Strain-Displacement Relations

The large deformation strain displacement relation for a general plate element undergoing both extension and bending at any point through the thickness is the sum of membrane and change of curvature strain components:

$$\overline{S}^b = \begin{Bmatrix} S_x^b \\ S_y^b \\ S_{xy}^b \end{Bmatrix} = \begin{Bmatrix} u_{,x} \\ v_{,y} \\ u_{,y} + v_{,x} \end{Bmatrix} + \frac{1}{2} \begin{Bmatrix} w_{,x}^2 \\ w_{,y}^2 \\ 2w_{,x}w_{,y} \end{Bmatrix} + z \begin{Bmatrix} -w_{,xx} \\ -w_{,yy} \\ -2w_{,xy} \end{Bmatrix}$$

The shape functions used for the 4-noded shallow shell element are:

$$\mathbf{N} = \frac{1}{4}\left(\overline{C} + \overline{\xi}\xi + \overline{\eta}\eta + \overline{H}\xi\eta\right).$$

There are five degrees of displacement degrees of freedom at each node for the elastic behaviour, and there is one potential degree of freedom per layer for the piezoelectric effect. Thus

$$\overline{q}_i^s = \{u \quad v \quad w \quad \theta_x \quad \theta_y \quad \}_i;$$
$$\overline{q}^e = \{\phi_1 \quad \dots \quad \phi_{n_p}\}$$

for $i = 1, 2, \dots, 8$.. The strain-displacement relations are based on the von Karman deformation theory and the electric field-potential relations $\overline{S}^e = -\nabla\phi$. The potential degrees of freedom are constant throughout the plane of the piezoelectric layer and they are assumed to vary linearly through the thickness. Thus the matrix relating the generalized strains to the nodal displacements and electric potentials can be written as follows:

$$\overline{S} = \begin{Bmatrix} \overline{S}^s \\ \overline{S}^e \end{Bmatrix} = \begin{bmatrix} \mathbf{b}^s & \mathbf{0} \\ \mathbf{0} & \mathbf{b}^e \end{bmatrix} \begin{Bmatrix} \overline{q}^s \\ \overline{q}^e \end{Bmatrix}$$

where for the shell element

$$\mathbf{b}_i^s = \begin{bmatrix} \frac{\partial N_i^s}{\partial x} & 0 & 0 & 0 & 0 \\ 0 & \frac{\partial N_i^s}{\partial y} & 0 & 0 & 0 \\ \frac{\partial N_i^s}{\partial y} & \frac{\partial N_i^s}{\partial x} & 0 & 0 & 0 \\ 0 & 0 & 0 & 0 & z\frac{\partial N_i^s}{\partial x} \\ 0 & 0 & 0 & -z\frac{\partial N_i^s}{\partial y} & 0 \\ 0 & 0 & 0 & -z\frac{\partial N_i^s}{\partial x} & z\frac{\partial N_i^s}{\partial x} \\ 0 & 0 & \frac{\partial N_i^s}{\partial x} & 0 & N_i^s \\ 0 & 0 & \frac{\partial N_i^s}{\partial x} & -N_i^s & 0 \end{bmatrix}; \; i = 1,\cdots,n_{el}$$

The element stiffness and mass matrices are first evaluated by expressing the integrals in the local natural coordinates ξ and η of the element and then performing numerical integration using the Gaussian quadrature. The element matrices are then assembled to obtain the global $\mathbf{K}_{ss}$ and $\mathbf{M}_{ss}$ matrices. Substituting for the generalized stress and strain expressions into Equation (1), we obtain the mass, elastic stiffness and piezoelectric stiffness matrices:

$$\mathbf{M}_{cc}^j = \int_{V_j} \rho \mathbf{N}^T \mathbf{N} \, dV_j,$$

$$\mathbf{K}_{cc}^j = \int_{V_j} \mathbf{b}^{c^T} \mathbf{c} \mathbf{b}^c \, dV_j,$$

$$\mathbf{K}_{ce}^j = \int_{V_j} \mathbf{b}^{c^T} \mathbf{e} \mathbf{b}^e \, dV_j,$$

$$\mathbf{K}_{ee}^j = \int_{V_j} \mathbf{b}^{e^T} \mathbf{e} \mathbf{b}^e \, dV_j, \qquad \textit{for} \qquad j = 1,\cdots,n_{el}$$

3.5 Geometric Stiffness

In the passive control methodology adopted in this study, the in plane forces generated by the piezoelectric actuators are accounted for through the nonlinear geometric stiffness matrix. The terms in the geometric stiffness matrix for an element are linear functions of the components of stress in the element. For plate and shell elements it is usual to consider only the membrane stresses. The elements of the geometric stiffness matrix $\mathbf{K}_g$ can be derived from a potential energy function Π_g given by the expression

$$\mathbf{K}_g = N_x \int_A \mathbf{N}_{,x}^T \mathbf{N}_{,x} dA + N_y \int_A \mathbf{N}_{,y}^T \mathbf{N}_{,y} dA + N_{xy} \int_A \mathbf{N}_{,x}^T \mathbf{N}_{,y} dA + N_{xy} \int_A \mathbf{N}_{,y}^T \mathbf{N}_{,x} dA$$

The membrane stress components are initially determined by prescribing a voltage to the piezoelectric patches and subsequently the calculated stresses are used to setup the geometric stiffness matrix.

For the entire structure, using the standard assembly technique for the finite element method and applying the appropriate boundary conditions, we obtain the complete equations of motion for a thermo -piezoelectrically coupled electromechanical composite panel in a flow field

$$\overbrace{\begin{bmatrix} \mathbf{M}_{cc} & \mathbf{0} \\ \mathbf{0} & \mathbf{0} \end{bmatrix}}^{\text{inertia}} \begin{Bmatrix} \ddot{\overline{U}}^c \\ \ddot{\overline{U}}^e \end{Bmatrix} + \overbrace{\begin{bmatrix} \mathbf{K}_{cc} & \mathbf{0} \\ \mathbf{0} & \mathbf{0} \end{bmatrix}}^{\text{linear stiffness}} \begin{Bmatrix} \overline{U}^c \\ \overline{U}^e \end{Bmatrix} + \overbrace{\begin{bmatrix} \mathbf{0} & \mathbf{K}_{ce} \\ \mathbf{K}_{ec} & \mathbf{K}_{ee} \end{bmatrix}}^{\text{piezo stiffness}} \begin{Bmatrix} \overline{U}^c \\ \overline{U}^e \end{Bmatrix} +$$

$$\overbrace{\begin{bmatrix} \mathbf{K}_{\Delta T} & \mathbf{0} \\ \mathbf{0} & \mathbf{0} \end{bmatrix}}^{\text{thermal stiffness}} \begin{Bmatrix} \overline{U}^c \\ \overline{U}^e \end{Bmatrix} + \overbrace{\begin{bmatrix} \mathbf{K}_{g} & \mathbf{0} \\ \mathbf{0} & \mathbf{0} \end{bmatrix}}^{\text{nonlinear stiffness}} \begin{Bmatrix} \overline{U}^c \\ \overline{U}^e \end{Bmatrix} = \begin{Bmatrix} P_{\Delta T} \\ \mathbf{0} \end{Bmatrix}$$

where

$\mathbf{M}_{cc}$ is the mass matrix;

$\mathbf{K}_{cc}$ is the linear elastic stiffness matrix

$\mathbf{K}_{ce}$ is the electromechanical coupling matrix;

$\mathbf{K}_{ee}$ is the piezo dielectric matrix;

$\mathbf{K}_{\Delta T}$ is the initial thermal stress stiffness matrix

$\mathbf{K}_{g}$ is the nonlinear stiffness matrix

The equations of motion for a laminated adaptive composite panel with piezoelectric actuators and sensors subjected to aerodynamic and thermal loads have been presented.

3.6 Static Actuation and Sensing Results

To illustrate the actuation and sensing mechanisms, two examples are presented in this section. The numerical results were compared to experiments and finite element simulations performed by Tzou and Tseng(1990), and Crawley and Lazarus(1991)

The first validation test case was based on an experiment conducted by Tzou and Tseng (1990). The experimental apparatus consists of a cantilevered piezoelectric bimorph beam with two PVDF layers bonded together and polarized in opposite directions (Figure 2a). The finite element model was divided into five equal elements, each with two piezoelectric layers bonded together. This produced a model with 53 total degrees of freedom (Figure 2b).

First, the actuation mechanism, governed by $\overline{U}_c = \mathbf{K}_{cc}^{-1} \mathbf{K}_{ce} \overline{U}_e$, is investigated. The top and bottom surface of the beam were subjected to an electric potential across the thickness of the beam and the resulting displacements were determined. A unit voltage (*1 V*) produces a tip deflection of 3.45×10^{-7} *m* as shown by the results tabulated in Figure 2(c). It is observed that there is no difference between the results evaluated by the composite finite element model and the theoretical results. The slightly lower tip deflection observed in the experiment could be caused by non-perfect bonding, voltage leakage, energy dissipation, etc. The total number of degrees of freedom used in this analysis (63 - 53 structural and 10 electrical) is considerably lower than the model studied by Tzou and Tseng (144 - 108 structural and 36 electrical), resulting is lower computational memory and time requirements.

The bimorph beam is also studied for sensing voltage distribution for a given static deflection. This sensing mechanism is governed by $\overline{U}_e = \mathbf{K}_{ee}^{-1} \mathbf{K}_{ec} \overline{U}_c$. When external tip loads are applied to produce a given deflection pattern, the electrical degrees of freedom output a sensing voltage. The results in Figure 2(d) show that a voltage of *290 V* is produced for an imposed tip deflection of *1 cm*. The results are in good agreement with the solid finite element solution. However, since the finite plate element used in this study guarantees the continuity of strains due to bending, i.e. rotation at nodes, the accuracy of sensing may be higher than the brick element that guarantees only displacement continuity at nodes.

Furthermore, it can be observed that while the results for Tzou and Tseng are given in terms of nodal voltages, the present theory produces elemental voltages, constant over each piezoelectric layer.

The second validation case was based on the experiments conducted by Crawley and Lazarus (1991). The experimental apparatus consists of a cantilevered laminated composite graphite/epoxy plate with distributed G1195 piezoceramic (PZT) actuators bonded to the top and bottom surfaces (Figure 3a). The finite element model consists of 160 elements with a total of 880 degrees of freedom (Figure 3b)

During actuation, a constant voltage with an opposite sign was applied to the actuators on each side of the plate. The deflections of the center line and both edges were measured by proximity sensors. Figure 3(c) shows the comparison of the deflection due to longitudinal bending for a $[0/\pm 45]_s$ layup between the present plate formulation, the solid brick element finite element model and the experimental results. All the solutions are in close agreement, with lower deflection observed for the solid element formulation due to shear locking effects associated with such finite element models. The discrepancy observed in the experimental results may be attributed to shear losses in the bonding layers.

Next the sensing mechanism was tested (Figure 3d) . During this simulation, the center row of piezoceramics was considered as sensors while the outer two rows were used as actuators. A constant voltage of *100 V* was applied to one row of actuators with a positive sign on the top surface and a negative sign on the bottom surface. The same voltage was applied to the other row but in this instance the polarity was reversed, thus inducing a twisting motion to the plate. Furthermore, a constant mechanical load of *0.2 N* was applied at the tip of the plate. Thus, the output sensor voltages were numerically determined for the combination of electrical and mechanical loads.

References

1. Suleman and V.B. Venkayya, "A Simple Finite Element Formulation for a Laminated Composite Plate with Piezoelectric Layers", *Journal of Intelligent Materials Systems and Structures*, Vol. 6, No. 6, pp. 776-782, Nov. 95.

2. H.S. Tzou, and C.I. Tseng, "Distributed Piezoelectric Sensor/Actuator Design for Dynamic Measurement/Control of Distributed Parameter Systems: a Piezoelectric Finite Element Approach", *Journal of Sound and Vibration*, Vol. 138, No. 1, 1990, pp. 17-34.

3. E.F. Crawley, and K.B. Lazarus, "Induced Strain Actuation of Isotropic and Anisotropic Plates", *AIAA Journal*, Vol. 29, No. 6, 1991, pp. 944-951.

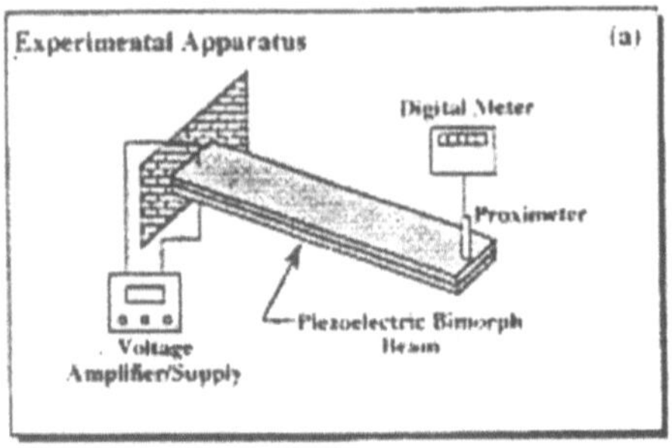

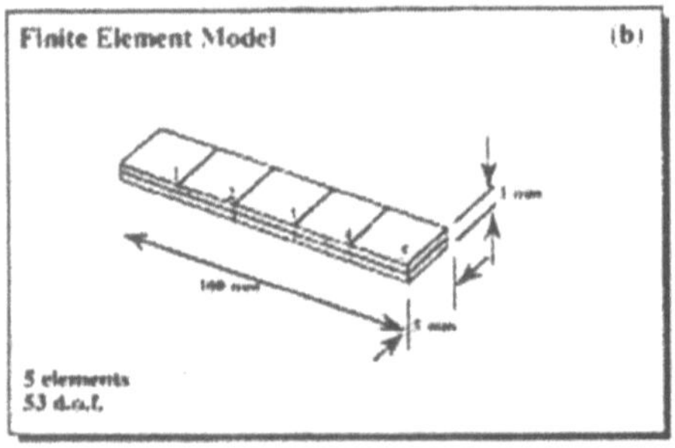

Actuation Mechanism (c)

$\times 10^{-7}$ m

Position	1	2	3	4	5
Theory	0.14	0.55	1.24	2.21	3.45
Beam FE[b]	0.12	0.51	1.16	2.10	3.30
Present	0.14	0.55	1.24	2.21	3.45
EXP[a]	-	-	-	-	3.15

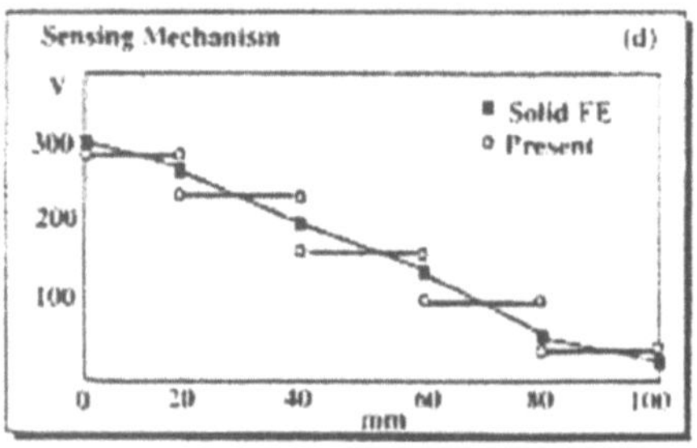

Figure 3.2 - Test study comparison between the present formulation and Tzou and Tseng (1990) showing: (a) the experimental apparatus for the piezoelectric bimorph beam; (b) The finite element model; (c) The static actuation mechanism; and (d) the sensing mechanism

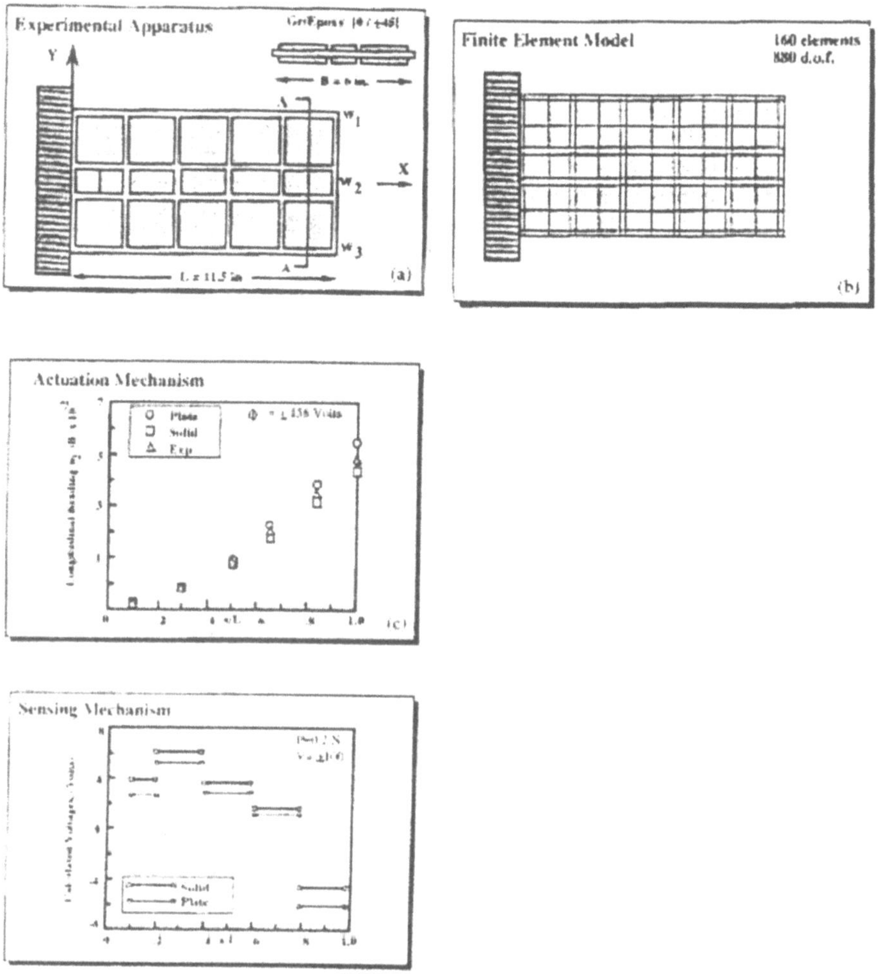

Figure 3.3 - The experimental apparatus showing the Gr/Epoxy cantilevered plate studied by Crawley and Lazarus (1991): (a) experimental apparatus; (b) The finite element model; (c) The static actuation mechanism; and (d) the static sensing mechanism.

4. An Adaptive Aeroelastic Wing

A. Suleman[1], A.P. Costa[2], P. Moniz[1] and C. Crawford[2]

1. IDMEC-Instituto Superior Técnico, Lisbon, Portugal

2. Academia da Força Aérea, Laboratório de Aeronáutica, Sintra, Portugal

Abstract Experimental buffeting and gust alleviation response results are presented for a sweptback flat wing model using piezoelectric actuators to control bending and wing tip camber shape. Model response was sensed by strain gages, accelerometers and piezoelectric sensors. The open-loop and closed-loop results of this experimental study show that piezoelectric actuation is effective in attenuating vibration and the wake-induced buffet response over the range of parameters investigated. The airfoil camber shape control increases remarkably the efficiency of the system, by using a controlled change of lift in order to produce the required bending moments.

4.1 Introduction

The primary passive solutions to dynamic aeroelastic problems consist of increasing the stiffness and balancing the mass and these have been used as early as 1922. They still comprise the basic passive means of improving the response and stability of an aircraft. Increased stiffness can also be achieved by using advanced composite tailoring which can greatly alter the stability characteristics of a given wing.

Active control concepts to improve the aeroelastic performance of wings have since emerged. The active aeroelastic control solutions consist of using aerodynamic control surfaces as they are readily available on conventional aircraft. One of the first active aeroelastic control experiments began testing in 1972 in NASA Langley Research Center's Transonic Dynamics Tunnel. The model was a clipped delta wing with a leading edge and a trailing edge actuator. Since that time, control experiments using flap actuators have studied a variety of control design techniques and objectives. The Active Flexible Wing project used the flexibility of the wing in conjunction with active controls to provide greater maneuverability.

In the last couple of decades, active aeroelastic control has evolved and a new actuation concept has emerged for structural control. This is the direct strain actuation using adaptive materials and structures. These electro-magneto-thermo-mechanical materials have presented an exceptional promise when compared to conventional ones. For deformation of thin structural elements, the most widely used multifunctional materials are piezoelectric actuators. Piezoelectrics have several advantages over hydraulic actuators because they have a higher frequency bandwidth of operation and because they act directly in the structure by straining it. Piezoelectrics have higher bandwidths than are possible in shape memory alloys, they are more compact than magnetostrictive devices and they are bidirectional by nature unlike electrostrictive materials.

There has been considerable analytical and computational work performed to determine the feasibility of applying piezoelectric actuators to control wing flutter and buffeting. However, the experimental verification effort has been rather limited to a few studies reported in the literature. The DARPA/USAF Wright Laboratory program was to design, build and test wind tunnel models to quantify performance improvements that could be achieved by incorporating smart materials such has PZT's for actuation and sensing systems in aircraft wings. The Smart Wing program performed wind tunnel tests at the NASA Langley's Transonic Dynamic Tunnel to demonstrate the use of smart actuator systems in a realistic modeled aircraft operational environment. The wind tunnel test quantified aerodynamic improvements of two concepts: the use of embedded SMA wires in the trailing edge to provide a smooth variable contoured control surface, and SMA torque tubes built into the wing structure which enabled the wing to be twisted or torqued.

Several active vibrations suppression concepts have also been investigated by a program shared between Daimler-Benz Aerospace Military Aircraft (DASA), Daimler-Benz Forschung (DBF) and Deutsche Forchungsantalt fur Luftund Raumfahrt (DLR). Here, a thin surface of piezo actuators is set out to flatten the dynamic portion of the combined static and dynamic maximum bending moment loading case directly in the shell structure.

Crawley and de Luis [1] conducted one of the first studies on the use of piezoelectrics in vibration control of flexible structures. In the area of active buffet load alleviation, the first experiment demonstrating the feasibility of using piezoelectric actuators in aeroelastic control was conducted by Heeg et al [2].

Shape control of a flexible wing structure has a great potential to improve the aerodynamic lifting-surface performance. Significant reductions in the shock-induced drag can be achieved by small adaptive modifications to the wing cross-sectional profile. The results presented here are the experimental wind tunnel results obtained to actively suppress buffeting by controlling the shape of the wing camber using piezoelectric actuators. Two methodologies are presented and compared. One uses the piezoelectric actuators in the traditional way, placing them at the root to generate bending moments to compensate the mechanical vibration along the span of the wing. The other method consists of using the piezoelectrics to achieve active airfoil camber shape control, in order to control lift, so that the change of lift can be used to generate the same type of moments and hopefully result in the use of fewer piezo actuators and less expended energy.

4.2 Two-Dimensional Adaptive Wing

The laboratory, a photograph and a sketch of the test model are shown in Figures 4.1and 4.2, respectively. The dimensions were determined based on the wind tunnel size, blowing air velocity and the limitations of the piezoceramic actuators. Structurally, it was aimed at having a fairly flexible wing with low bending and torsion mode frequencies [3].

Six piezoceramic sensor patches $38\times25\times0.2\ mm$ are adhered to the top and bottom of the plate near the root. The actuators (ACX-QPN40N) are divided in two groups, the ones that produce bending moments along the span of the wing, two pairs near the leading edge and one near the trailing edge, all near the root in order to maximise the bending moment. The other group, which is responsible for the camber shape control, is located near the wing tip in order to produce changes in the bending moment produced by the changes in lift. The output signal of the PZT sensors was made proportional to the vertical displacement of the model.

A finite element modal analysis was performed to generate natural frequencies and mode shapes needed to determine the appropriate placement of the piezoelectric actuating plates for maximum control authority and to design the controller. The vibration mode shapes and frequencies are shown in Figure 4.3.

Figure 4.1 – Overview of the laboratory setup.

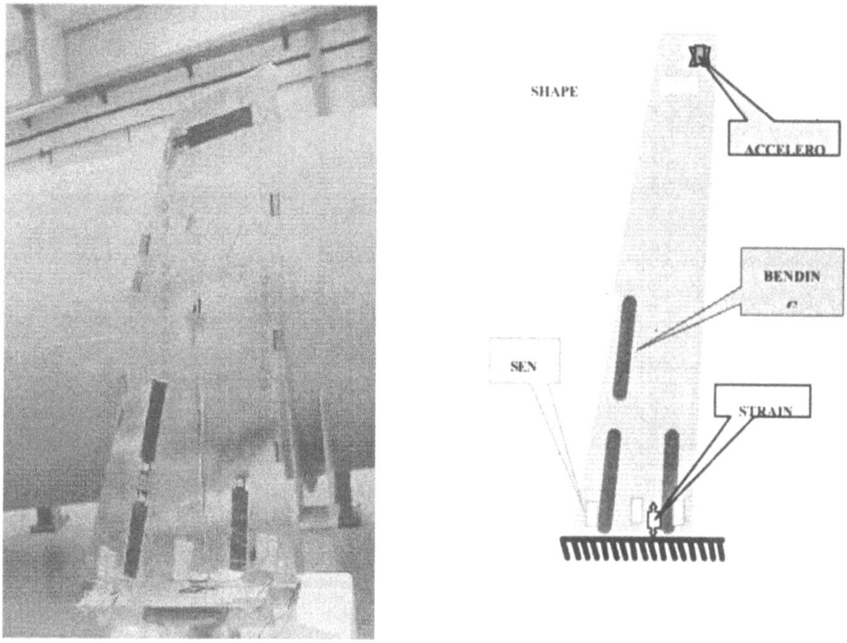

Figure 4.2 (a) Flat plate wing test article; (b) Schematic of the test article

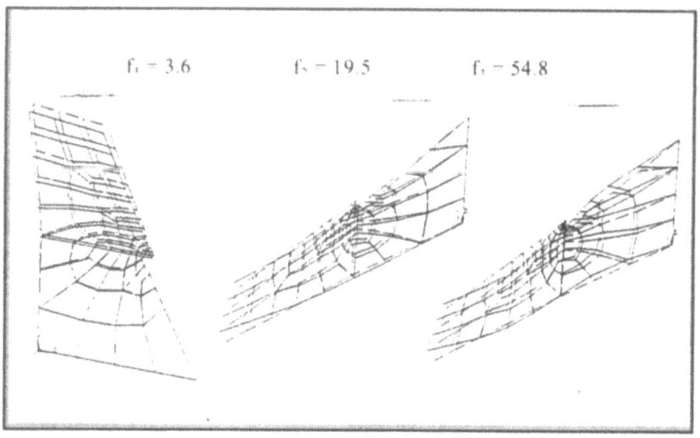

Figure 4.3 - Wing Finite element model modal frequencies and mode shapes used in the controller design.

Table 4.1 – Flat plate wing results

	Piezo damping		Piezo damping & shape control	
	Amplitude	**Frequency**	**Amplitude**	**Frequency**
Control law off	585 mm	1Hz	620 mm	1Hz
Control law on	395 mm	0.9 Hz	325 mm	0.66 Hz
Improvement	32 %	10%	47.5%	34%

For the present study the wind tunnel was modified so that a rigid plate was placed at the upstream end of the test section. The wake from this obstacle was used to generate buffeting flow. The position of this object could be easily adjusted so that the resulting wake would impinge on the model mounted downstream. The configuration finally selected was the one that produced the largest buffet response of the model. Therefore, all of the buffet response data presented herein were obtained for the obstacle in the same location and orientation. Tests were carried out with only the aluminium plate inside the wind tunnel, in order to determine the speeds at which the buffet would occur (a speed of 5.5 m/s was considered optimal).

With the control law gain set to the desired value the tunnel speed was increased to and then held constant at a pre selected value. Buffet response measurements were made at a velocity of 5.5 m/s. The output signal from the PZT sensor was channelled to a transfer function analyser that was used to calculate the auto-correlation function of the response signal. When the control laws are applied, the average amplitude due to buffeting decreased has it can be seen in Table 4.1.

An appreciable buffeting reduction has been obtained, especially when using airfoil shape control, which combined with the root actuators were able to decrease the average buffeting amplitude from 32% to 47.5%. The airfoil shape control also decreased the frequency of the vibration by 34%. The airfoil shape control has presented a feasible engineering design solution where the piezoelectric shape control actuators are used to create favorable changes in lift characteristics.

4.3 Three-Dimensional Adaptive Wing

In order to eventually apply piezoelectric technology to real world adaptive wing applications, it is necessary to gain an understanding of how wings constructed with true aerodynamic shape would behave, and what construction methods would be appropriate for such a wing. To that end, a true three-dimensional wing with a symmetrical cross section was constructed, for comparison to the performance of the two dimensional wings used in previous studies. The wing was constructed like a conventional wing, with full leading and trailing edges connected by two skins. The wing had a constant cross section along its 0.6m span, a sweep angle of 0°, and a taper ratio of 1 for simplicity. Actuators and sensors were fully contained within the wing to preserve a clean aerodynamic profile (Figure 4.4).

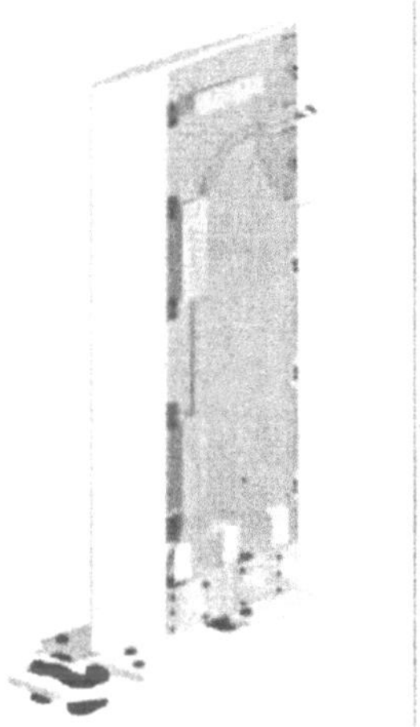

Figure 4.4 – 3-D wing test article

This wing was subjected to a range of aeroelastic phenomenon, including free-stream speeds where the natural modes of the wing were excited, as well as buffeting conditions. The high speed flutter behaviour of the wing was also to have be studied, but this was found to not be possible in the low speed wind tunnel used in the tests, and would also have almost ensured destruction of the wing. It was also desired to be able to compare the control effectiveness of piezoelectrics with that of conventional ailerons, and for that reason ailerons were built into the 3D wing. In addition, a motor pod was tested on the wing to examine the control effectiveness of the piezos when exited by mechanical vibration and to simulate the wing of an aircraft supporting an engine.

Wing Design

In order for the piezos to most effective and for the sake of simplicity, it was felt that they should be bonded directly to the skins of the wing. Moreover, in order to try and obtain flutter in the low speed wing tunnel, and to have the maximum possible level of vibration, the wing should be as made flexible as possible. The initial 3D wing was to built using two aluminum skins as the main elements, with a fully composite wing to be constructed in future projects. For these reason, the design options narrowed down to a type of stress skin construction, using the aluminum skins as the main structural elements. Leading and trailing edges would also be required to provide the proper aerodynamic shape. In order to minimize the structural stiffness of the wing structure, the leading and trailing edges were attached by brackets at five separate sections. The brackets also served as the shear web of the wing box.

The ideal placement of the actuators was determined from an analysis of the first mode shapes of the wing, obtained from a modal analysis of the wing finite element model. The first 15 mode shapes were requested, over the frequency range 0-100Hz, encompassing the frequency range of the aeroelastic response to be encountered in the wind tunnel experiments. The mode shapes of interest were the bending and torsion modes, since these modes were responsible for the vast majority of the deflection of the model. The coupling was clearly evident from the first five bending/torsion modes shown in Figure 5 below. The dashed lines represent the undeformed wing.

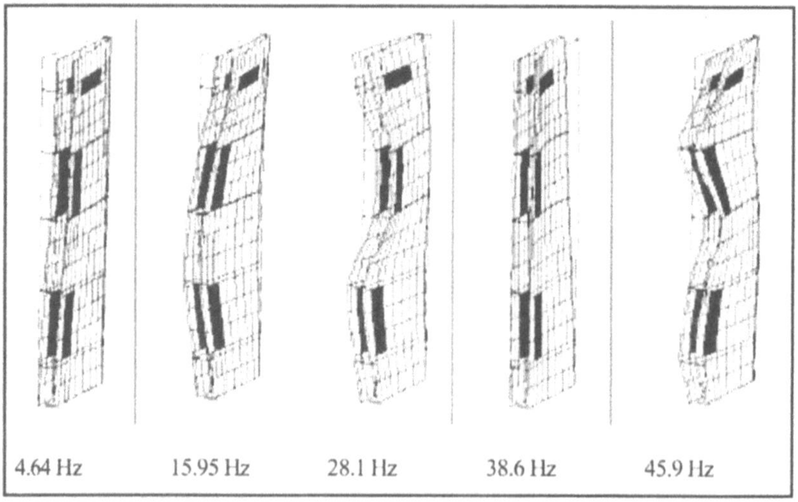

Figure 4.5 - First 5 bending/torsion mode shapes of wing

Based on the initial finite element model and the revised one including the actuators, it was determined that three pairs of actuators were appropriate. Two pairs were positioned vertically near the leading edge of the wing to control the bending of the wing, with another pair placed horizontally near the tip of the wing to control torsional modes, and it was hoped effect camber control. The actuators were all placed between the bracket attachment points in order to obtain the maximum possible control authority over the wing. Three pairs of smaller sensor piezos were placed along the root of the wing, just above the mounting block.

All of the actuator piezos were supplied by ACX™ with each patch containing two stacks of 0.254mm (0.010") piezo wafers that were encased in a polyimide coating and pre-wired for easy electrical connection. The vertical piezos were QP-40W's, while the horizontal ones were slightly smaller QP-40N's. The sensor piezos were raw 0.381mm (0.015") thick wafers supplied by Sensor Technology Limited. Both the sensors and actuators were bonded to the skins using epoxy supplied with the piezos.

Ailerons

In order to provide a comparison between the results obtained using piezo actuation to what could be obtain using conventional aerodynamic surfaces, the outboard two trailing edge sections were replaced with actuated ailerons. The ailerons were designed to be pivoted about their 30% chord point, in order to try and induce vibration in the ailerons themselves. Control surfaces pivoted about a point other than their leading edge tend to increase the level of vibration and reduce the energy needed for actuation. The aileron was divided into two sections as the trailing edge sections were, to preserve the flexibility of the wing.

Motor Pod

The final addition to the wing was a motor with propeller and its associated mounting bracket and fairing (Figure 4.6). The corresponding frequencies and mode shapes are shown in Figure 4.7. The motor was an electric one capable of drawing approximately 10A at 12V, and swinging a 25-15cm model aircraft prop with spinner. The motor was mounted on the wing using a steel bracket equipped with QP-40W actuators on either side of the main support plate and a piezoelectric sensor near its base. The mounting bracket was bolted into the steel bracket of the second leading edge section with four bolts and separated from the wing by an aluminum spacer block to prevent interference between the wing and motor mount. Wire feeds to the motor, sensor, and actuator, were passed though small gaps between the skin and leading edge block. The speed of the propeller was measured using an optical displacement probe connected to an oscilloscope to measure the frequency of the signal from the probe, and an accelerometer was also mounted on the bracket.

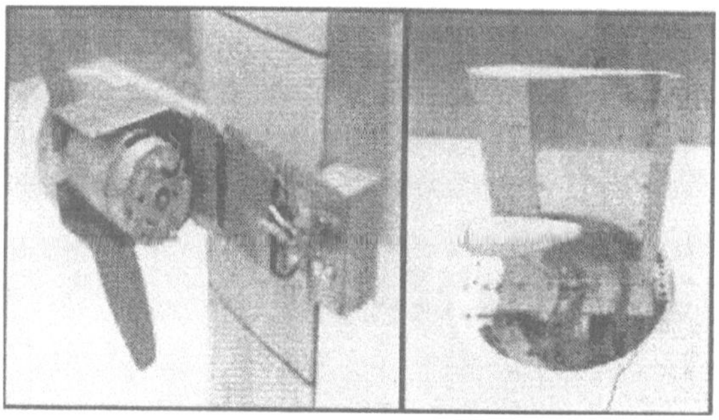

Figure 4.6 - Photographs of Motor Pod Installation

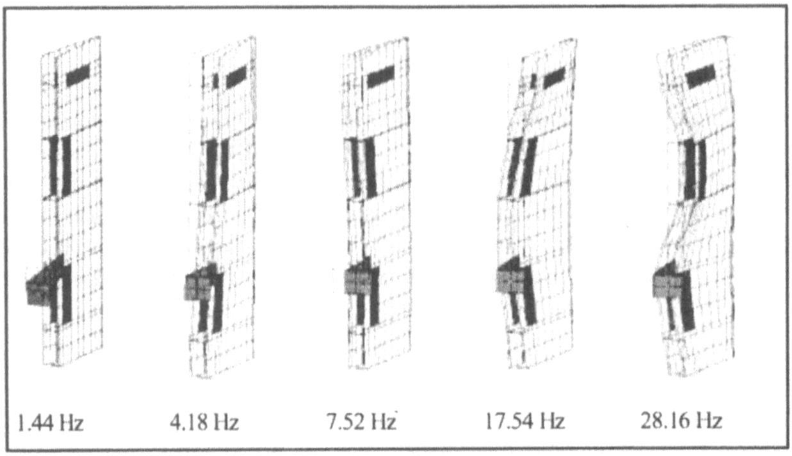

Figure 4.7 - First five Vibration Modes of Wing with Motor

Buffeting Wing

Mounting brackets for a lifting section wing were fabricated to allow buffeting tests to be performed on the 3D wing. The buffeting wing was to be positioned at the upstream end of the test section, roughly in line with the wing (Figure 4.8). The installation method allowed for lateral as well as rotational position adjustment of the wing, and consisted of two plywood plates bolted to the flange at the beginning of the test section. Two metal brackets bolted to the plywood had pins at their ends, which fit into an aluminum tube running through the length of the wing itself. A plate with a slot cut in it was glued to the top end of the wing and could be bolted to the top bracket to secure the wing at an angle of attack up to ±30°. A servo was also installed on the wing to control the aileron on the wing.

Figure 4.8 - Buffeting wing installation

4.4 Free-Stream Vibration

The aeroelastic response of the wing was tested at the three different Reynolds numbers for which the aerodynamic coefficients were determined, with both the aluminum and steel brackets installed. It was found that the signals from all three sensor pairs were essentially identical, with the pair near the leading edge producing the strongest signal. Consequently, that signal was used for the control laws in all of the subsequent experiments, and was also recorded as a measure of the wing's vibration. Sample periods extended up to 30s due to the transient nature of the vibrations. The wing was tested in both open and closed loop configurations, with various combinations of actuator pairs were utilized. In addition, the ailerons were tested as a method of aeroelastic control. The average signal amplitude from the piezo sensors was then compared between the controlled and uncontrolled states to determine the degree of vibration reduction. The results from this testing are summarized in Table 4.2. The experiments are catalogued by the gains used in the control laws, with the following categorization: (Gain Root Piezo Pair, Gain Mid-Wing Piezo Pair, Gain Wing-Tip Piezo Pair) (Gain Motor Mount Piezo Pair). When the gains were zero the corresponding entry was omitted. Single gain values indicate aileron-controlled experiments, as the 2 vertical piezo pairs were used together.

Table 4.2 - Free Stream Results

Test			Aileron	
	Aluminum Bracket Wing		Aluminum Bracket Wing	
	Gain	% Reduction	Gain	% Reduction
V = 11,4 m/s	(-3,5 3,5)	21,6	(-15)	12,5
Re = 2,6x10^5	(-3,5 3,5 -4)	37,5		
V = 18,3 m/s	(-5 5)	35,6	(-15)	15,4
Re = 4,2x10^5	(-5 5 -5)	42,7		
V = 24,6 m/s	(-5,5 5,5)	43,6	(-4)	8,5
Re = 5,7x10^5	(-5,5 5,5 -6)	51,8		

Several trends are apparent from this data, although more testing should be conducted to verify the following observations. The steel bracketed wing appears to be less controllable using piezos at higher Reynolds numbers (and hence larger oscillations), whereas the opposite is true for the aluminum bracket wing. It also appears that higher gains are required at higher Reynolds numbers, which is good from a power consumption perspective. Moreover, the degree of vibration reduction at lower vibration levels in a more flexible wing is less than that of a stiffer wing, while the reverse is true for larger oscillations. Finally, the aileron control is clearly less effective than piezo control, and was found to require lower gains at higher Reynolds numbers.

4.5 Wing Buffeting

The 3D wing was subjected to buffeting conditions in order to simulate the response of a conventional horizontal tail surface. The buffeting wing was adjusted to induce the maximum level of vibration in the 3D wing, which was found to occur with the leading edges of both wings roughly aligned, and the buffeting wing at an angle of attack of 30°. The wind tunnel was run so that the free-stream velocity was 12m/s. Again, tests were conducted using both the steel and aluminum bracketed wing under both piezo and aileron control (Table 4.3)

Table 4.3 - Buffeting Wing Results

Piezo Control			
Steel Bracket Wing		Aluminum Bracket Wing	
Gain	% Reduction	Gain	% Reduction
(-4 4)	27,9	(-4 4)	36,9
(-4 4 -4)	30,5	(-4 4 -5)	40,8

4.6 Motor Pod

The steel bracketed wing was tested with the motor pod mounted on it, in order to simulate the main wing of a multi engine aircraft. The motor was first tested at four power settings in static surrounding air, and then with the wind tunnel on. During the first two low power-setting tests, a nut was taped to the end of one of the propeller blades in order to induce vibration. The remaining tests did not use any intentional unbalancing of the propeller. The vibration levels were recorded and the results from the uncontrolled and controlled cases are tabulated in Table 4.4.

Table 4.4 - Motor Pod Results

Velocity	Rotor Speed	Gain	% Reduction
0 m/s	10 Hz	(-2 0) (-3)	77,9
	20 Hz	(-1,4) (-2,1)	0,5
	30 Hz	(-3 5 -7) (-2)	49,4
	60 Hz	(3) (0,8)	-29,1
25 m/s	130 Hz	(-4,5) (1,1) 5	48,2

4.7 Summary

The 3-D wing project was extremely successful, and has provided valuable insight for future adaptive wing projects. This study has produced a proof of concept model for a realistic wing structure model using stressed skins, and shown that piezoelectric control can be very effective.

References

1. McGowan, A. R., Heeg, J., and Lake, R. C., 1996, "Results of Wind-Tunnel Testing from the Piezoelectric Aeroelastic Response Tailoring Investigation," AIAA-96-1511-CP, NASA Langley Research Centre, Hampton Virginia

2. Active Control eXperts ACX™ Inc., "Quickpack Piezoelectric Actuators," 215 First Street, Cambridge, MA 02142-1227, 1997

3. Suleman, A., Moniz, P. A., Costa, A. P., 1999, "Experimental Flutter and Buffeting Suppression Using Piezoelectric Actuators and Sensors," ASME Adaptive Structures and Material Systems Conference, AD-Vol. 59/MD-Vol. 87, pp. 293-299.

5. An Adaptive Truss Structure

A. Suleman[1] and R. Sedaghati[2]

1. IDMEC-Instituto Superior Técnico, Lisbon, Portugal

2. University of Victoria, Victoria, B.C., Canada

Abstract A force method is proposed to analyze and optimize adaptive truss structures. It is observed that the structural strength is improved significantly using the adaptively optimized geometries while the computational effort required by the force method is found to be significantly lower than that of the displacement method.

5.1 Introduction

The concept of equilibrium of forces and compatibility of deformations is fundamental to analysis methods for solving problems in structural mechanics. The equilibrium equations need to be augmented by the compatibility conditions since the equilibrium equations are indeterminate by nature, and determinacy is achieved by adding the compatibility conditions. Generally, two analytical methods (displacement and force) are available to analyze determinate and indeterminate structures.

The application of the force method to analyze the topology optimization of adaptive truss structures under static loading has been reported by Sedaghati et al. [1]. In the present study, this method has been extended to analyze and optimize adaptive truss structures under dynamic loading. The equilibrium and compatibility equations are solved simultaneously. A direct method has been developed to generate the compatibility matrix for indeterminate truss structures. The method is based on the displacement-deformation relation and singular value decomposition (SVD) technique and there is no need to select consistent redundant members. For determinate truss structures the equilibrium matrix is generated automatically through the finite element analysis.

For optimum structural design, the design variables are selected so as to minimize/maximize the objective function while satisfying the required constraints. The constraints may include allowable stresses in the elements, limitation on geometric parameters, displacement limits at the joints, frequency specifications, system stability, etc. Depending on the nature of the applied loads, the structure and its geometry, one or more of the constraints can be active and control the design of the structure. For the geometry optimization of the adaptive truss structures, the structural strength is selected as the objective function and constraint equations are based on geometry parameters (such as angles) of the active members.

The application and efficiency of the proposed method is illustrated by optimizing the topology of an adaptive truss structure using active elements in order to maintain maximum structural strength under various impact loading conditions. The force method has proved to be perfectly natural and efficient because the primary variables in the topology optimization of adaptive truss structures are the member forces.

5.2 Geometry Optimization

An adaptive truss structure, which can change its configuration to maintain its structural strength by lengthening or shortening some of active member's length, is considered for geometry optimization in this study. When optimizing the configuration of the adaptive truss structure, the objective is to maximize the structural strength [2-4] , by changing the lengths of the active members. Here, it is assumed that the truss structure is composed of n members including n_a active members. The vector $\varphi = \{\varphi_1, \varphi_2, \ldots, \varphi_{n_a}\}$ includes the angles of the n_a active members and these are the design variables.

For a fixed position of the applied load $\boldsymbol{P}$, element force vector $\boldsymbol{F}$ depends on vectors $\boldsymbol{\varphi}$ and on the vector $\boldsymbol{\theta}$ which contains information on the directions of nonzero load components of $\boldsymbol{P}$. In linear problems, for a uniform applied load $\boldsymbol{P}$, $\boldsymbol{F}$ can be obtained from

$$\boldsymbol{F} = \mathbf{P}\,\boldsymbol{F}_u \tag{5.1}$$

where $\boldsymbol{F}_u$ is the element forces due to unit load in the position and direction of the applied load $\boldsymbol{P}$, the $(n \times n)$ matrix $\mathbf{P}$ includes the load factors corresponding to the relative elements. The allowable load factors corresponding to the relative elements can be obtained from

$$|\boldsymbol{F}_a| = \mathbf{P}_a |\boldsymbol{F}_u| \tag{5.2}$$

where $\boldsymbol{F}_a$ is the allowable element forces and diagonal matrix $\mathbf{P}_a$ includes the allowable load factors corresponding to the relative elements. The structural strength in static problems may be defined as the allowable load factor corresponding to whole structure and can be obtained from

$$p_a = \text{Min}\ \mathbf{P}_a \tag{5.3}$$

Finally, to optimize the geometry of the adaptive truss, requires finding the vector of geometrical design variables $\boldsymbol{\varphi}$ such that

$$\text{Max}\ p_a(\boldsymbol{\varphi}, \boldsymbol{\theta}) \tag{5.4}$$

subject to

$$G_i(\boldsymbol{\varphi}) \leq 0\ ,\ \ i = 1,2,\cdots,i_g \tag{5.5}$$

where i_g is the number of geometrical constraints.

In dynamic problems, the structural strength is obtained at each time step, $(p_a)_n$ $n = h, 2h, \ldots, Nh$, where $h = \Delta t$ is the time increments and N is the total number of time increments. The structural strength is defined as

$$P = \min\ (p_a)_n \qquad n = 1,2,\ldots, N \tag{5.6}$$

Finally, to optimize the geometry of the adaptive truss under dynamic loading, it is required to find the vector of geometrical design variables φ such that Eq. (6) is maximized subject to the geometrical constraints given by Eq. (5).

5.3 Illustrative Example

The twenty-four-bar adaptive truss shown in Figure 1 has been optimized topologically to obtain the maximum structural strength in the presence of various dynamic loading conditions [2]. It has four bays and every bay contains one active member. The truss consists of 24 members with active members 5, 9, 13 and 17. Angles $\varphi_1, \varphi_2, \varphi_3$ and φ_4 are taken as geometrical design variables, and these are related to the 4 active member's length. The ranges for those angles are specified as $0^\circ \le \varphi_1, \varphi_2, \varphi_3, \varphi_4 \le 90^\circ$. The material properties are: Young's modulus=7×10^{10} N/m^2 Yield stress $\sigma_y=10^8$ N/m^2. The geometrical parameters are: cross-sectional area A=10^{-4} m^2 and L=2 m.

The minimum and maximum natural frequency of the structure are $\omega_{min} = 82815$ Hz and $\omega_{max} = 1026.5$ Hz, respectively. An external impact load $P(t)$ is applied at node 1, and the following two cases are considered:

Case I: $$P(t) = \begin{cases} I_p / h(N) & 0 \le t \le h \\ 0 & t > h \end{cases}$$

Case II: $$P(t) = \begin{cases} I_p / 100h(N) & 0 \le t \le 100h \\ 0 & t > 100h \end{cases}$$

where, $I_p = \sum p(t)\,\Delta t$ is applied impact for each case. In each case, the time increment h=1.3×10^{-4} sec and a total numerical time of 1000×h=0.13 sec are considered.

In case I, the impact time on the structure is very short. It was observed that members 23 or 24 always produced the largest element forces, independently of the direction of the impact load $P(t)$, but if the external load is time-independent (static), member 18 in the bottom of the structure becomes the most critical. Members 23 and 24 are at the top of the structure where the direct impact load happens. The results reveal that for stability parameters $\beta = 0.25$ and $\gamma = 0.5$, the numerical solution displays spurious "beating" in which the amplitude of response repeatedly grows and decays. The Newmark method, with the selected stability parameters (average acceleration algorithm or trapezoidal rule) is unconditionally stable and it does not have artificial numerical damping which minimizes numerical noise. Taking γ>0.5 introduces artificial damping, which automatically dissipates the high-frequencies noises. In order to maximize the high-frequency dissipation for a given value of γ>0.5, we have taken $\beta = 0.25(\gamma + 0.5)^2$. Using γ =0.67 and $\beta = 0.25(\gamma + 0.5)^2$=0.34, the high-frequency noise is eliminated. For case I, no optimal geometries were found because the largest internal forces are always present in members 23 and 24. This means that for an impact load acting in a very short time, the structural strength (allowable applied impact) is not affected by the geometry of the structure. The required CPU time to generate the time-history of element forces through the force method was 1.7 sec in comparison to the 32 sec using the displacement method.

For case study II, the time taken for the impact load to act on the structure has been considerably increased. It is observed that high-frequency noise is present for the stability parameters of $\gamma = 0.5$ and $\beta = 0.25$, but the numerical noise is eliminated by using $\gamma > 0.5$ and $\beta > 0.25$. It is interesting to note that because the impact load acts over a relatively large time in comparison to the Case I, the effect of high-frequency noise is not as noticeable as in the Case I, especially for member No. 18. The structural strength for the fixed configuration $(\varphi_1 = \varphi_2 = \varphi_3 = \varphi_4)$ for both displacement and force methods are shown in the Figure 4 and the results from both methods totally match, demonstrating the accuracy of the proposed method. The structure becomes very rapidly weak for the impact loads applied in direction other than for $\theta = 0^{\circ}$. The CPU time for force and displacement methods are found to be 34 sec and 738 Sec, respectively. Thus the force method clearly provides a more efficient and computationally faster solution. The optimal results for the various values of the external impact load direction θ are given in the Table 1. The optimal geometries of the adaptive structure are shown in the Figure 3, corresponding to the different load directions. Figure 4 illustrates the changes in the structural strength for the adaptive truss structure in comparison to the fixed structure for various direction of θ and it can also be inferred that the structural strength is improved significantly as the structure adaptively changes its geometry to accommodate the changes in loading direction.

5.4 Conclusions

The integrated force method has been implemented to analyze and optimize the geometry of adaptive truss structures under dynamic loading. The compatibility matrix is derived directly using the displacement-deformation relation and the SVD technique.

The application and efficiency of the proposed method was illustrated by optimizing the topology of an adaptive truss structure using active elements in order to maintain structural strength under various loading conditions. The method has proved to be perfectly natural and efficient because the primary variables in topology optimization of adaptive truss structures are the member forces. It has been shown that the structural strength can be improved significantly by optimally changing the geometric configuration to counteract the changes in the direction of the applied load. The computational time has been reduced drastically when using the force method to analyze adaptive structures under dynamic loads when compared to the numerical solution using the displacement method.

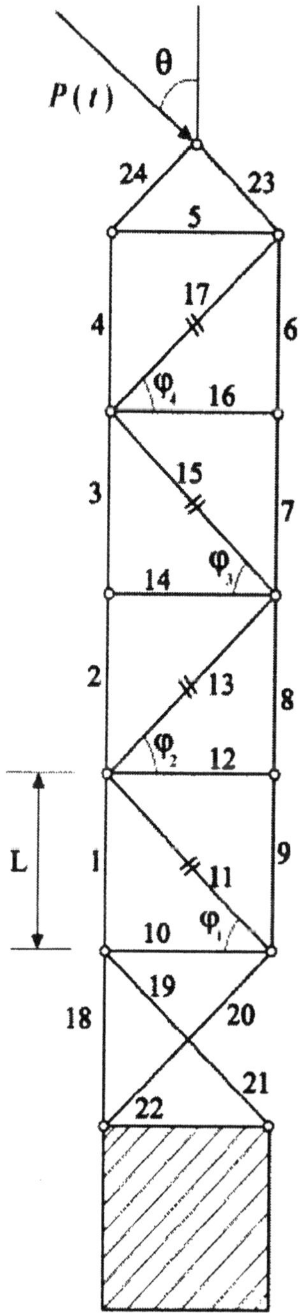

Figure 5.1 The 24-bar plane adaptive truss structure with active members 11, 13, 15 and 1

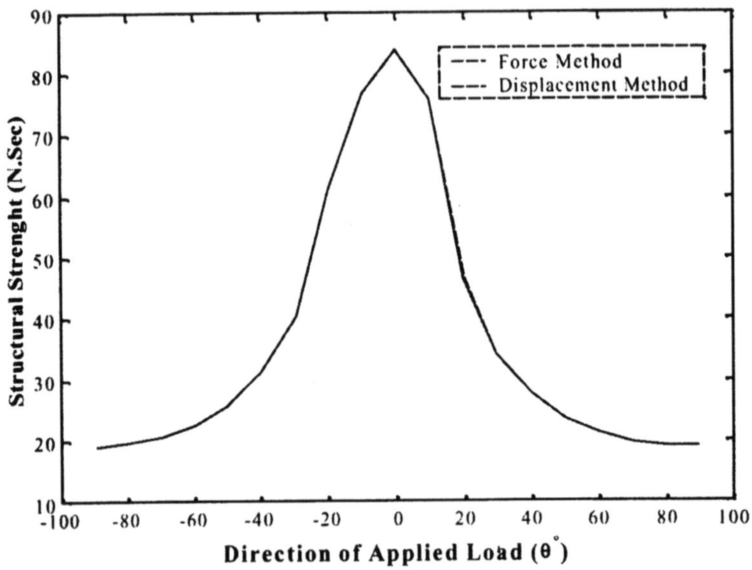

Figure 5.2 Structural strength versus the direction of the applied load for the fixed structure.

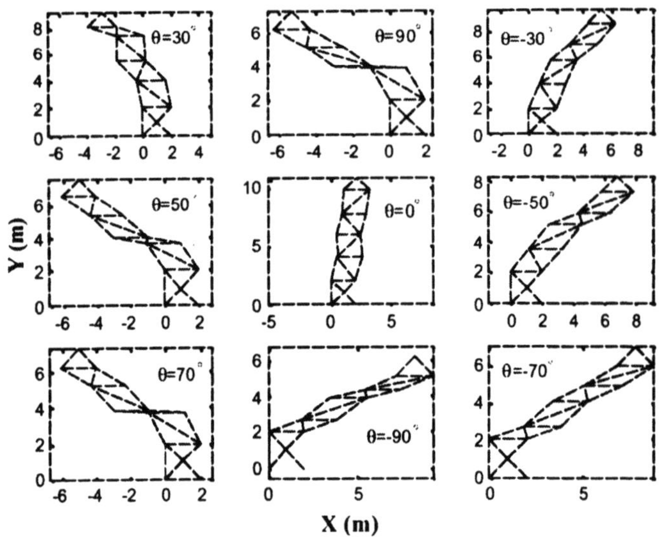

Figure 5.3 Optimal adaptive shapes of the truss structure

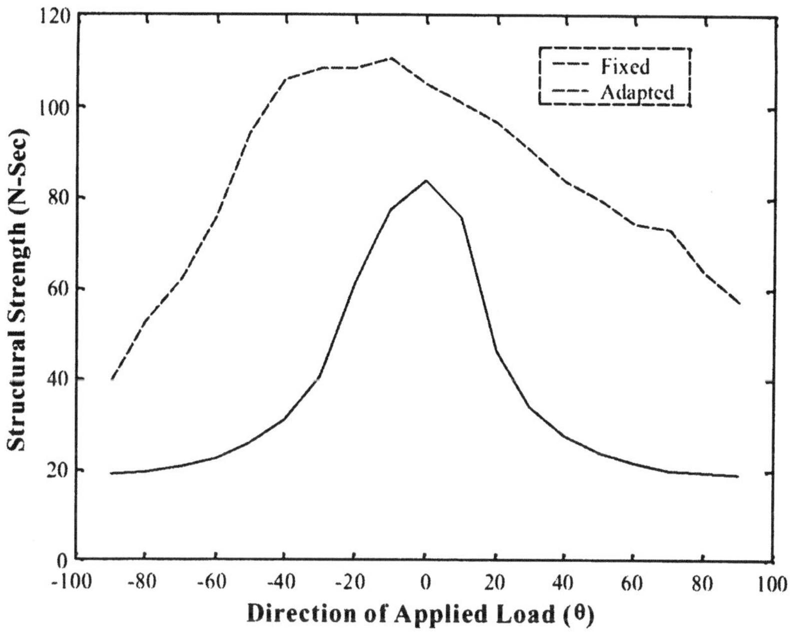

Figure 5.4 Optimal adaptive and fixed structural strength versus the direction of the applied load.

References

1. Sedaghati, R., Tabarrok, B., Suleman, A. and Dost, S., "Optimization of Adaptive Truss Structures using Finite Element Force Method based on Complementary Energy," to be appeared in the CSME, Nov. 1999.

2. Murotsu, Y. and Shao, S., " Some Approaches to the Optimal Adaptive Geometries of Intelligent Truss Structures,", First Joiny US/Japan Conference on Adaptive Structures, Nov. 13-15 1990, Hawaii US, pp. 743-771.

3. Murotsu, Y. and Shao, S., "Optimum Adaptation of Intelligent Truss Structures.", J. of Intell, Mater. Syst. And Struct., Vol. 1, No. 2, April 1990, pp. 175-199.

4. Murotsu, Y. and Shao, S., "Optimal Adaptive Geometry of an Intelligent Truss Structure", AIAA-90-1093.

CHAPTER II

ACTIVE VIBRATION CONTROL

L. Gaul
Professor
Institute A of Mechanics , University of Stuttgart
Pfaffenwaldring 9
70550 Stuttgart - GERMANY
Tel: 49-711-685 6278/6277
Fax: 49-711-685 6282
e-mail: gaul@mecha.uni-stuttgart.de

1. Adaptive Structure Projects at the Institute A of Mechanics imbedded in a Collaborative Research Center at the University of Stuttgart

L. Gaul, H. Albrecht, M. Kögl, R. Nitsche, U. Stöbener, J. Wirnitzer

Institute A of Mechanics, University of Stuttgart
Pfaffenwaldring 9, D-70550 Stuttgart
EMail: gaul@mecha.uni-stuttgart.de

Abstract. A brief survey about research projects on adaptive structures at the Institute A of Mechanics is presented. These projects have been initiated with the preparation of the collaborative research center "Adaptive Aerospace and Lightweight Structures" (Sonderforschungsbereich - SFB 409). As two projects are part of SFB 409 and the others have strong ties, the idea of the collaborative research center funded by the Deutsche Forschungsgemeinschaft (DFG) is outlined according to a presentation given by the chairman of SFB 409 Prof. B. Kröplin at the SPIE Conference in San Diego in 1998 Kröplin et al. (1998). The current SFB program incorporates 16 research projects. With review periods of 3 years it can run three or four periods. The covered fields range from smart materials and their integration into structures and composites to shape adaptation of aircraft wings, active vibration suppression, and noise control. The project is initiated by the university, supported by the DFG and designed in coordination to related industrial projects.

1 Introduction

Passive measures to reduce noise and vibrations or to ensure optimal performance even in off-design states of flexible aerospace and lightweight structures have reached certain limits. This is why adaptive or smart structures have gained importance. The integration of actuators, sensors, and controls with a material or structural component define what comprises a smart structure. Its goal or objective is to create a material system with intelligence and life features integrated in the microstructure to reduce mass and energy and produce adaptive functionality Rogers (1993). By its nature, the field of smart structures depends on interdisciplinary research since numerous disciplines are involved to set up smart structure system solutions. Developing such novel system solutions not only necessitates advanced knowledge in each isolated discipline (e.g. control theory) but, particularly, also fruitful cooperation among them. Therefore, advances in smart structure research strongly depend on a well-balanced cooperative research environment.

Special collaborative research centers (Sonderforschungsbereich – SFB) have been established by the German Research Foundation (Deutsche Forschungsgemeinschaft – DFG) about 20 years ago. These SFBs are long term programs (9-12 years) which concentrate and coordinate research activities of various institutes and schools of a university in order to solve complex problems. As the experience of about 400 SFBs in Germany has shown, an important benefit of the interdisciplinary structure of these programs is the improvement of the quality and applicability of results by stimulating the communication and cooperation of the scientists involved.

The University of Stuttgart consists of 13 schools with the focus on natural sciences and engineering, particularly aerospace engineering and engineering cybernetics. Twelve institutes (see Table 1) of 5 different schools cooperate in 16 projects forming together the SFB "Adaptive Aerospace and Lightweight Structures". The involved institutes range from materials science, control theory, engineering mechanics, fluid and structural dynamics to aircraft design.

Table 1. Participating Institutes

Institute	Project Manager	E-mail address	Project
Institute of Aerodynamics and Gasdynamics	S. Wagner U. Rist	wagner@iag.uni-stuttgart.de rist@iag.uni-stuttgart.de	A4,A5,B4 A4
Institute of Aircraft Construction	F.J. Arendts R. Voit-Nitschmann	arendts@ifb.uni-stuttgart.de rvn@ifb.uni-stuttgart.de	C8 B4,C8
Institute of Control Technology of Machine Tools and Manufacturing Systems	G. Pritschow K.-H. Wurst	xkl@isw.uni-stuttgart.de	C6
Institute of Flight Mechanics and Control	K.H. Well	well@ifr.uni-stuttgart.de	A2,B6
Institute A of Mechanics	L. Gaul A. Kistner K. Willner	gaul@mecha.uni-stuttgart.de kistner@mecha.uni-stuttgart.de willner@mecha.uni-stuttgart.de	B2 A1,B2 B2
Institute B of Mechanics	W. Schiehlen	wos@mechb.uni-stuttgart.de	B1
Institute of Space Systems	E.W. Messerschmid U.M. Schöttle	messerschmid@irs.uni-stuttgart.de schottle@irs.uni-stuttgart.de	B3 B3
Institute of Statics and Dynamics of Aerospace Structures	B.-H. Kröplin S. Rudolph	kroeplin@isd.uni-stuttgart.de rudolph@isd.uni-stuttgart.de	A3,A5,C4 A1
Institute of Synthetics Analysis and Synthetics Science	G. Busse	busse@ikpiris4. verfahrenstechnik.uni-stuttgart.de	C5
State Material Testing Institute	F. Aldinger	sigmund@aldix.mpi-stuttgart.mpg.de	C1
Institute of Physiology. University of Tübingen	R.W. Gülch	rw.guelch@uni-tuebingen.de	C3
DASA/Dornier Satellite Systems GmbH	F. Döngi	doengi@sd.dofn.de	A3,B6

2 Central idea

The aim of this research program is not to produce singular, independent research results but to develop integrated system solutions for smart aerospace structures, see Fig. 1.

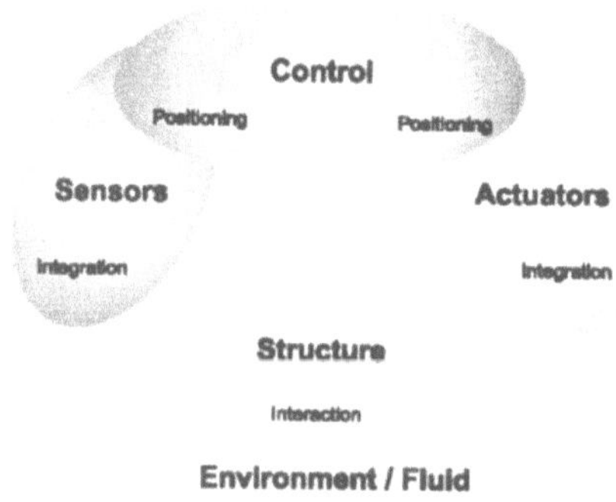

Figure 1. Integrated System Solution

Obviously, researchers cannot be familiar with all disciplines involved, rather they will be specialists in one. The common strategy of specialized researchers is to advance their own discipline by investigating some detailed problem, and trying to fill gaps of knowledge by new findings. Now, the central idea of the present SFB is to guide the detailed investigations of each "specialized" researcher to fill these gaps in a way such that the autonomous disciplines grow together forming a smart system solution. Besides, an important benefit of the SFB's structure is to build an environment for interdisciplinary postgraduate education.

Setting up novel smart system solutions first requires the understanding of the complete spectrum of disciplines. Therefore, in such a university research program all disciplines must be involved, from material science to aircraft design. The orientation of researchers working on independent projects towards the common idea of smart structures is guaranteed by the definition of the following three fields of application:

- shape adaptation,
- active vibration suppression, and
- noise control.

Having these fields of application in mind, the SFB's smart structure research proceeds as shown in Fig. 2. First, the physical phenomena involved in aerospace vehicles and lightweight structures are investigated, such as the analysis of the complex helicopter rotor aerodynamics by high-dimensional numerical models. Based on this physical understanding of the problem which has been accumulated at the University of Stuttgart during past research activities, outlines of possible conceptual designs are developed according to the needs of the cooperating industry.

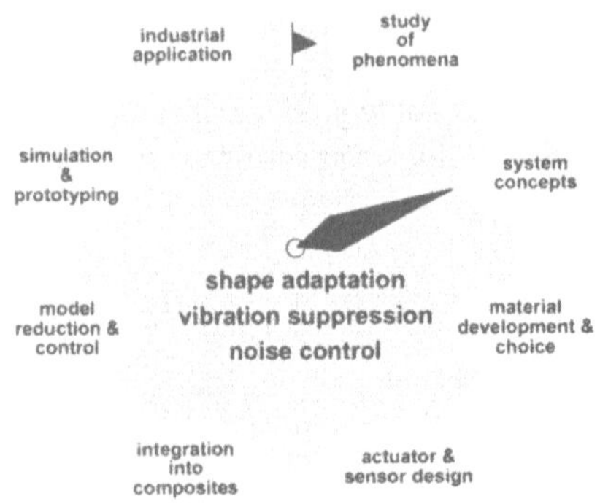

Figure 2. The SFB's Central Idea

The material scientists, apart from developing novel smart materials and investigating the manufacturing process, support actuator and sensor designers with information on the potential and limits of existing smart materials. Furthermore, problems arising from the integration of smart materials into composite structures are investigated by means of non-destructive damage diagnostics such as ultrasonics, X-rays, interferometry, and laser-vibrometry. Besides, the question of effectiveness of embedded smart materials in a wide range of frequencies is addressed.

The complex high-order mathematical models used to investigate the physical phenomena are not necessarily suitable for model-based controller design. For this purpose, model reduction techniques are applied in order to generate compact controller design models. As an alternative to model-based control system synthesis, neural control strategies are developed with a special focus on dimensionally homogeneous similarity networks based on the principles of similarity mechanics.

With growing degree of maturity of the research program, prototypes are tested in experiments as well as in numerical simulations in order to demonstrate and prepare industrial application.

3 Projects

The SFB consists of 16 research projects which are combined into three project groups, as shown in Table 2. Additional projects of the Institute A of Mechanics are associated to these project groups.

Group A "Fluid-Structure Control" combines problems in the medium to high frequency range dealing with structural acoustics and aeroservoelasticity.

The aim in structural acoustics is the reduction of noise by controlling the acoustic medium with an adaptive boundary. This is motivated by the necessity of isolating the payload of the Ariane 5 launch vehicle from external noise as well as by the desire to increase modern passenger aircraft comfort, see Fig. 3.

Table 2. The SFB projects

Fluid-Structure Control	
A 1	Neural network based intelligent structures
A 2	Controller design for coupled fluid-structure systems
A 3	Active structure acoustics control (ASAC)
A 4	Drag reduction by active fluid control
A 5	Smart helicopter rotor blades
Active Vibration Damping and Shape Adaptation	
B 1	Integrated optimization of structures and control systems
B 2	Adaptive control of joints in lightweight structures
B 3	Adaptive tethers for orbital systems
B 4	Adaptive wing
B 6	Smart structures for high precision satellites and instruments
Materials and Actuators	
C 1	Development and characterization of ferroelectric ceramics
C 3	Stimulated polymer gels for pseudomuscular actuators
C 4	Modeling and simulation of stimulated polymer gels
C 5	Non-destructive characterization of shape adaptive structures
C 6	Actuator concepts for smart structures
C 8	Integration of piezoceramics into composites

In the field of aeroservoelasticity, problems include the reduction of aerodynamic drag by the active control of laminar-turbulent transition and the vibration and noise control of helicopter rotor blades. These problems necessitate high order simulation models, low order controller design models, and model reduction techniques. Control algorithms applied to this class of problems are designed using both model-based and neural network techniques.

For the investigation of active vibration control (AVC) and active structural acoustic control (ASAC) simulations and experiments with plates are carried out. The plates are equipped with piezoelectric patches and sheets in order to achieve actuatoric and sensoric features. The design and the placement of the sensors and actuators is important for the success of the control. Therefore a design and placement strategy for structures with distributed parameters is developed. Since the overall reduction of vibration and radiated sound requires a high authority control, the concept of modal state space control is applied. The real time estimation of modal displacements and modal velocities is obtained by using modal filters or observers. For experiments the modal controller is implemented on a digital controller board and the results of these experiments show a significant reduction of the vibration amplitude and the related sound radiation for harmonic and stochastic excitation signals Stöbener and Gaul (1998).

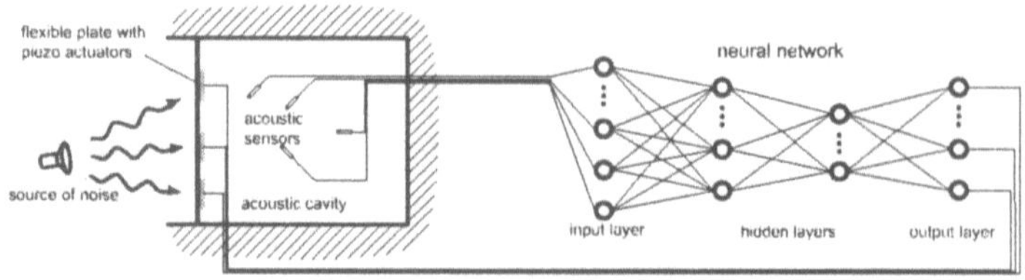

Figure 3. Neural Network Based Intelligent Structures

Advanced studies are carried out with a car body. The necessary modal data, such as eigenfrequencies, modal damping ratios and shape functions, are determined by an experimental modal analysis. With the experiences of the plate control, the piezoelectric sensors and actuators are designed and positioned on the floor panel of the car body. By controlling four modes the vibration amplitude and the radiated sound is reduced within a frequency range from 200 to 500 Hz. These studies are published in Stöbener and Gaul (Feb 7 10 2000).

Other works pursue the objective to combine passive and active damping layers, such as foam mats and piezoelectric devices. The advantage of the combination is the broadband vibration reduction which is obtained by the energy dissipation of the passive damping layer for high frequencies and the active control by the piezoelectric actuators for low frequencies.

Group B "Active Vibration Damping and Shape Adaptation" involves active vibration damping for high precision space structures as well as low frequency shape adaptation.

For high precision tracking of large space structures with low structural and aerodynamic damping, artificial damping has to be introduced into the system using active control strategies.

As an example of shape adaptation, an adaptive wing concept is investigated, see Fig. 4. In order to minimize drag in transsonic flow, airfoil contours are usually optimized with respect to a single flight condition. In order to avoid the occurrence of shocks over a wide range of flight conditions, local shape adaption of the airfoil is explored.

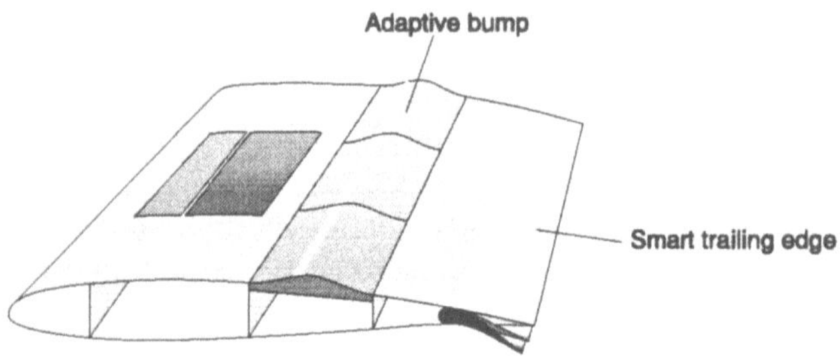

Figure 4. Adaptive Wing

The low structural damping of large space structures often causes a problem when high positioning accuracies are required in orbit. Therefore effective measures for vibration suppression are required. Lightweight truss structures often consist of flexible struts which are assembled by joint connections. The energy dissipation due to the interfacial slip in the joint connections turned out to be a major source of damping. Experiments on a two-dimensional truss structure with bolted joints are performed in order to analyse the influence of dry friction on the vibration characteristics, and the experimental data are used for validation of the numerical simulations Albrecht et al. (1999). Based on the numerical model of the structure, issues such as optimal placement of the adaptive joint connections within the truss structure as well as design of control strategies are treated.

The energy dissipation in the bolted joint can be controlled by varying the normal pressure in the contact using a piezoelectric stack disc as washer. For a truss structure with several adaptive joint connections either a decentralised control strategy with a local controller for each adaptive joint or a global control approach using one single controller can be applied.

An appropriate design method for a local controller is the use of Lyapunov function technique. The control objective is to maximize energy dissipation in an instantaneous and local sense. In general a state observer is required for the implementation of such a 'damping controller'. As the system behaviour is nonlinear due to friction, nonlinear controller and observer design methods have to be used Gaul and Nitsche; Gaul and Nitsche; Gaul et al. (; 1999; Mar 11 13 1998). The control concept is also termed semi-active because it uses active control theory to vary the properties of a passive element.

An alternative control approach uses neural networks which are capable of learning an appropriate nonlinear control law to achieve a given control objective. Neural control seems to be a promising concept for the design of a global controller.

Group C "Materials and Actuators" includes development, characterization, modeling and testing of smart materials, actuator design as well as integration into composites.

Basic material development and mathematical modeling is done with emphasis on ferroelectric ceramics and polymer gels. A classification of smart materials in the frequency-displacement domain is depicted in Fig. 5. Problems arising from the integration of smart materials into composites are investigated by means of non-destructive damage diagnostics such as ultrasonics, X-rays, interferometry, and laser-vibrometry.

Because smart structures are gaining increasing importance, it is necessary to have numerical methods at hand with which accurate and reliable solutions for complex piezoelectric problems can be obtained. Besides Finite Element Methods, Boundary Element Methods are a powerful tool for the numerical solution of field problems of mathematical physics. They have gained increasing popularity in the last two decades, due to some advantages which they have over Finite Element Methods. These include for example the reduction of the spatial dimension of the problem by one, which greatly simplifies the meshing process. Furthermore, the use of fundamental solutions and the mixed approach, in which both displacements and tractions are approximated, result in a high accuray in stress computations.

A Boundary Element Method is developed for the solution of three-dimensional problems of dynamic piezoelectricity. This method uses the fundamental solution of the static piezoelectric operator, and the resulting domain integral containing the inertia term is transformed to the boundary using the Dual Reciprocity Method. The procedure is described in detail

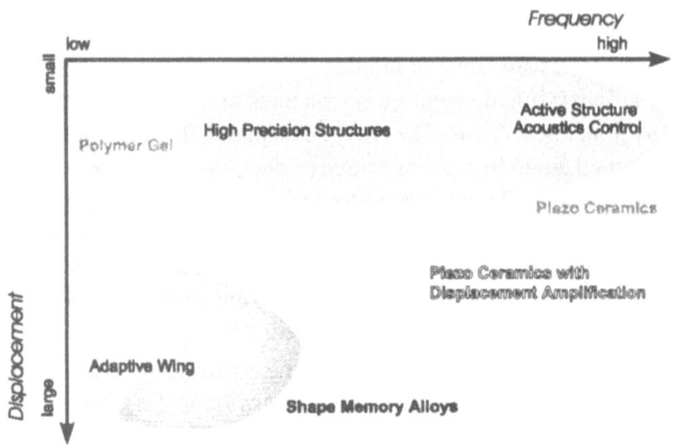

Figure 5. Classification of Smart Materials

in Kögl and Gaul; Kögl and Gaul (; 1999) and provides an appropriate tool for simulating the performance of actuator or sensor designs.

4 Conclusion

The collaborative research center SFB 409 "Adaptive Aerospace and Lightweight Structures" at the University of Stuttgart has been outlined as a frame which incorporates adaptive structures research at the Institute A of Mechanics. SFB 409 started in January 1998, it consists of 16 projects involving 12 institutes of 5 different schools. The program is expected to run 3 to 4 funding periods of 3 years each, giving about 100 doctoral students the opportunity to do research in an interdisciplinary environment.

Additional to the projects on neural network based intelligent structures (A1) and adaptive control of joints in lightweight structures (B2), the Institute A of Mechanics is involved in projects on acoustic control of individual panels or those assembled in a car body, on control concepts for semi-active damping, and the development of boundary and domain discretization methods for piezoelectric members. For more details, the reader is referred to the cited references.

Acknowledgments

Support by the Deutsche Forschungsgemeinschaft (DFG) for SFB 409 and Graduate Collegium GKKS is gratefully acknowledged.

References

Albrecht, H., Wirnitzer, J., and Gaul, L. (1999). Vibration suppression for lightweight space structures using adaptive joints. In *Proceedings of the Intern. Conference on Adaptive Structures and Technologies (ICAST)*, 111–118.

Gaul, L., and Nitsche, R. Friction control for vibration suppression. *Mechanical Systems and Signal Processing (accepted for publication)*.

Gaul, L., and Nitsche, R. (1999). Contact pressure control in bolted joint connections. In Gaul, L., and Brebbia, C., eds., *Computational Methods in Contact Mechanics IV*, 369–378. Southampton, Boston: WIT Press.

Gaul, L., Nitsche, R., and Sachau, D. (Mar 11-13, 1998). Semi-active vibration control of flexible structures. In *Proc. Euromech 373 Colloquium: Modelling and control of adaptive mechanical structures*. Innovationskolleg ADAMES, 217–226.

Kögl, M., and Gaul, L. A Boundary Element Method for transient piezoelectric analysis. *Engineering Analysis with Boundary Elements (submitted)*.

Kögl, M., and Gaul, L. (1999). Dual reciprocity boundary element method for three-dimensional problems of dynamic piezoelectricity. In *Boundary Elements XXI*, 537–548. Southampton: Computational Mechanics Publications.

Kröplin, B., Grohmann, B., and Konstanzer, P. (1998). Smart structure research at the University of Stuttgart. In *Smart Structures and Materials 1998: Mathematics and Control in Smart Structures*, volume 3323, 2–6. SPIE.

Rogers, C. A. (1993). Intelligent material systems - the dawn of a new materials age. *J. Intell. Mater. Syst. Struct.* 4:4–12.

Stöbener, U., and Gaul, L. (1998). Active control of plate vibration by discrete PVDF actuator and sensor segments. In Hagood IV, N. W., and Atalla, M. J., eds., *Ninth International Conference on Adaptive Structures and Technologies*, 349–358.

Stöbener, U., and Gaul, L. (Feb. 7-10, 2000). Active structural modal control for the floor panel of a car body. In *Proceedings of IMAC-XVIII: A Conference on Structural Dynamics, San Antonio*, volume 1, 153–159. Society of Experimental Mechanics.

2. Piezoelectric Stack Actuator: FE-Modeling and Application for Vibration Isolation

Uwe Stöbener and Lothar Gaul

Institute A of Mechanics, University of Stuttgart
Pfaffenwaldring 9, D-70550 Stuttgart

Abstract The piezoelectric stack actuator is one of the most established actuator device for active systems. In this paper a finite element (FE) formulation for a piezoelectric rod is derived. The FE formulation provides a model of stack actuators by which the dynamic behaviour is calculated. To validate the FE formulation and to evaluate the performance of a vibration isolation system with stack actuator an experimental setup is designed and tested. In a first set of experiments a feedforward control concept is used and the results are compared with the FE simulation. Feedforward control requires a reference signal which is not always available in real applications. Therefore a second set of experiments is performed with a proportional feedback controller.

1 Introduction

Piezoelectric stack actuators are manufactured by several companies and used in a variety of technical applications, such as for motion control and nano positioning. They are available in different sizes, for different voltages (HV=High Voltage 0-1000V, LV=Low Voltage 0-100V) and for different load levels. Nevertheless the design principle, which is shown in Fig. 1, is identical for all stack actuators. The stack consists of PZT layers and electrodes. The displacement of the complete stack is given by the sum of the displacements of all layers. The typical layer thickness is $0.5mm$ up to $1.0mm$ and the typical actuator displacement is $10\mu m$ up to $150\mu m$. The electrodes are alternating connected to the voltage supply and to the ground. Due to the polarization of the PZT material the piezoelectric effect is sensitive to the orientation of the electric field. To achieve a maximum actuator displacement all PZT layers are build into the stack with an alternating polarization direction. Therefore all layers change their lenght in the same direction.

By neglecting nonlinear material and hysteresis effects the stack can be replaced by a single PZT rod, whose lenght is identical to the thickness of the sum of all layers. The driving voltage U_{act} has to be replaced by a 'virtual' voltage U^*_{act}.

$$U^*_{act} = n_{PZT} U_{act} \tag{1}$$

The voltage U^*_{act} is calculated by the product of the number of PZT layers n_{PZT} and the driving voltage U_{act}.

2 FE formulation for a piezoelectric rod

Earlier work of FE formulations for piezoelectric material was published by Allik and Hughes; Naillon et al. (1970; 1983). In the nineties a lot of publications have been written about FE models of piezoelectric structures, e.g. Gaudenzi and Bathe; Ghandi and Hagood; Shieh (1995; ; 1993) In the next chapters a FE formulation for a piezoelectric actuator in stacked design is derived. As mentioned above the stack is modeled by a rod.

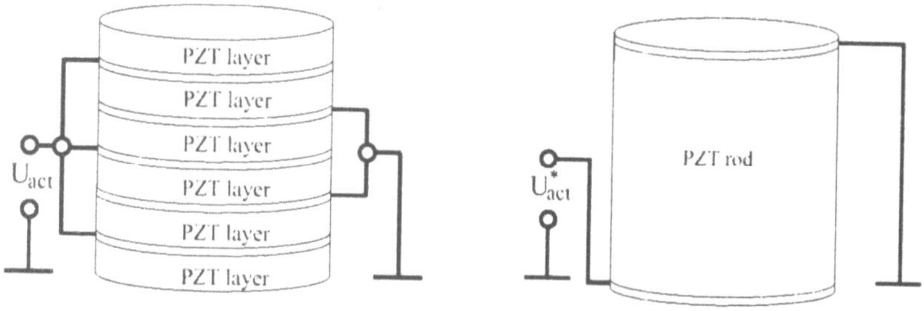

Figure1. Piezoelectric stack and equivalent rod

2.1 Finite rod element

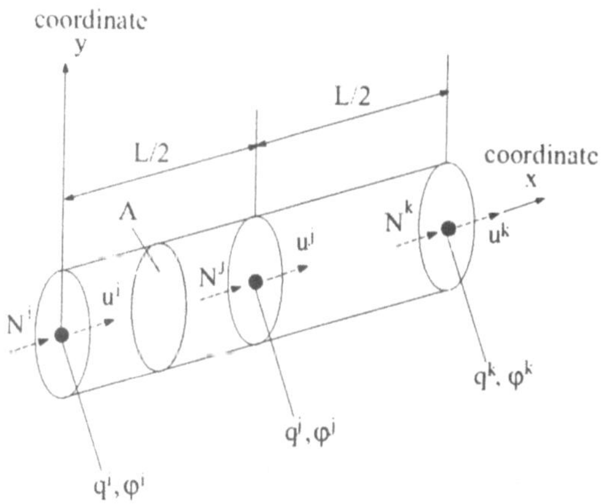

Figure2. Three node finite element of a piezoelectric rod

Fig. 2 illustrates a 3-node finite element for a piezoelectric rod. The rod length is L and its cross section is A. The nodal forces are $\mathbf{N}^i$, $\mathbf{N}^j$, $\mathbf{N}^k$, the nodal displacements are $\mathbf{u}^i$, $\mathbf{u}^j$, $\mathbf{u}^k$, the nodal electric charges are q^i, q^j, q^k and the nodal electric potentials are φ^i, φ^j, φ^j. The superscripts i,j,k refer to the node numbers. The forces and displacements are vector quantities with components in cartesian coordinates (e.g. $\mathbf{u}^i = \mathbf{u}^i_x + \mathbf{u}^i_y + \mathbf{u}^i_z$) whereas the charges and potentials are scalar quantities.

2.2 Virtual work

The equation of motion is derived from D'Alembert's generalization of principle of virtual displacements

$$\int_V (\rho\delta u_i\ddot{u}_i + \delta\varepsilon_{ij}\sigma_{ij})\, dV = \int_V \delta u_i b_i dV + \int_A \delta u_i t_i dA, \tag{2}$$

with the volume V and the cross section A of the rod. The mass density is expressed by ρ, δu_i is the virtual displacement, $\ddot{u}_i$ is the acceleration, $\delta\varepsilon_{ij}$ is the virtual strain, σ_{ij} is the stress, b_i is the volume force and t_i is the traction. The electric flux conservation is described by

$$\int_V \delta E_i D_i dV + \int_V \delta\varphi q_V dV + \int_A \delta\varphi q_A dA = 0, \tag{3}$$

with D_i as the electric flux or electric displacement. δE_i is the virtual electric field, q_V is the charge per volume, q_A is the charge per area and $\delta\varphi$ is the virtual electric potential. Superposition of these equations leads to

$$\int_V (\rho\delta u_i\ddot{u}_i + \delta\varepsilon_{ij}\sigma_{ij})\, dV - \int_V \delta E_i D_i dV =$$

$$\int_V \delta u_i b_i dV + \int_A \delta u_i t_i dA + \int_V \delta\varphi q_V dV + \int_A \delta\varphi q_A dA. \tag{4}$$

2.3 Constitutive equations

In the following step the constitutive equations for the piezoelectric material are introduced. The nonlinear behaviour and hysteresis effects of piezoelectric materials are neglected in this paper. The assumption of linear material behaviour is valid for small electric field strenght and small mechanical stresses since significant hysteresis effects are related to high voltages and large mechanical loads. Under these conditions the following linear set of equations is used

$$\sigma_{ij} = C^E_{ijkl}\varepsilon_{kl} - e_{ijm}E_m, \tag{5}$$

$$D_m = e_{mij}\varepsilon ij + \xi^\sigma_{mn}E_n. \tag{6}$$

In these equations σ_{ij} are the stress coordinates ($[N/m^2]$), C^E_{ijkl} are the elasticity stiffness constants ($[N/m^2]$) measured for a constant electric field. ε_{kl} are the strain coordinates, E_m are the electric field coordinates ($[N/C]$ or $[V/m]$), e_{ijm} are the piezoelectric field stress coefficients ($[C/m^2]$ or $[N/(V\ m)]$), D_m is the electric flux density ($[C/m^2]$) and ξ^σ_{mn} is the permittivity coefficient measured for constant stress ($[C^2/(N\ m^2)]$ or $[F/m]$). Inserting the constitutive Eqs. (5,6) to Eq. (4) gives

$$\int_V \rho\delta u_i\ddot{u}_i dV + \int_V \delta\varepsilon_{ij}C^E_{ijkl}\varepsilon_{kl}dV - \int_V \delta\varepsilon_{ij}e_{ijm}E_m dV$$

$$-\int_V \delta E_m e_{mij}\varepsilon_{ij}\,dV - \int_V \delta E_m \xi^{\sigma}_{mn} E_n\,dV =$$

$$\int_V \delta u_i b_i\,dV + \int_A \delta u_i t_i\,dA + \int_V \delta\varphi q_V\,dV + \int_A \delta\varphi q_A\,dA. \quad (7)$$

Eq. 7 is the coupled electromechanical virtual work principle.

2.4 Finite element approximation

The finite element method approximates the displacement field by nodal weighted shape functions. Often Hermitian polynomials are used for this approximation. They are formulated in natural coordinates, as depicted in Fig. 3. For a three node rod element the quadratic polynomials and their derivatives with respect to the natural coordinates are given by

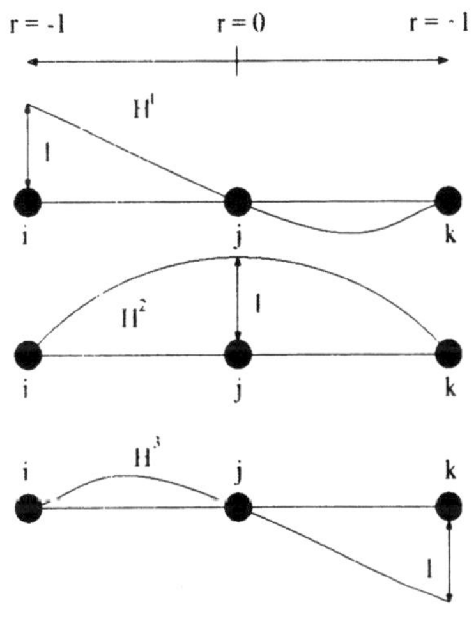

Figure3. Hermitian polynomials

$$H^1 = \tfrac{1}{2}\left(r^2 - r\right), \; B^1 = r - \frac{1}{2}, \quad (8)$$

$$H^2 = 1 - r^2, \quad B^2 = -2r, \quad (9)$$

$$H^3 = \tfrac{1}{2}\left(r^2 + r\right), \; B^3 = r + \frac{1}{2}. \quad (10)$$

The polynomials are assembled in the following matrices

$$\mathbf{H} = \begin{bmatrix} \mathbf{H}_m & \mathbf{0} \\ \mathbf{0} & \mathbf{H}_{el} \end{bmatrix} = \begin{bmatrix} H^1 H^2 H^3 & 0 & 0 & 0 \\ 0 & 0 & 0 & H^1 H^2 H^3 \end{bmatrix}, \quad (11)$$

$$\mathbf{B}=\begin{bmatrix}\mathbf{B}_m & \mathbf{0}\\ \mathbf{0} & -\mathbf{B}_{el}\end{bmatrix}=\begin{bmatrix}B^1 B^2 B^3 & 0 & 0 & 0\\ 0\ 0\ 0 & -B^1 & -B^2 & -B^3\end{bmatrix}, \tag{12}$$

where the index m stands for mechanical quantities and the index el refers to electrical quantities.

The continuous displacement field $\mathbf{u}$ and the continuous electric potential φ are approximated by the matrix of Eq. (11)

$$\mathbf{p}=\begin{bmatrix}\mathbf{u}\\ \varphi\end{bmatrix}=\begin{bmatrix}\mathbf{H}_m & \mathbf{0}\\ \mathbf{0} & \mathbf{H}_{el}\end{bmatrix}\begin{bmatrix}\mathbf{u}_e\\ \varphi_e\end{bmatrix}=\mathbf{H}\mathbf{p}_e. \tag{13}$$

The vector $\mathbf{p}$ in Eq. (13) contains the continuous displacement and potential field whereas $\mathbf{p}_e$ contains nodal values. The coordinates in the element state vector $\mathbf{p}_e$ are given by

$$\mathbf{p}_e=\begin{bmatrix}\mathbf{u}_e\\ \varphi_e\end{bmatrix}=\begin{bmatrix}\mathbf{u}^i\\ \mathbf{u}^j\\ \mathbf{u}^k\\ \varphi^i\\ \varphi^j\\ \varphi^k\end{bmatrix}. \tag{14}$$

The strain ε_m and the electric field $\mathbf{E}_{el}$ are approximated by the matrix of Eq. (12)

$$\begin{bmatrix}\varepsilon_m\\ \mathbf{E}_{el}\end{bmatrix}=\begin{bmatrix}\mathbf{B}_m & \mathbf{0}\\ \mathbf{0} & -\mathbf{B}_{el}\end{bmatrix}\begin{bmatrix}\mathbf{u}_e\\ \varphi_e\end{bmatrix}. \tag{15}$$

2.5 Equation of motion

The equation of motion is obtained by implementing the finite element approximation in the coupled mechanical and electrical virtual work principle Eq. (7). This results in

$$\begin{bmatrix}\delta\mathbf{u}_e\\ \delta\varphi_e\end{bmatrix}^T\int_{V_e}\rho\begin{bmatrix}\mathbf{H}_m^T\mathbf{H}_m & \mathbf{0}\\ \mathbf{0} & \mathbf{0}\end{bmatrix}dV\begin{bmatrix}\ddot{\mathbf{u}}_e\\ \ddot{\varphi}_e\end{bmatrix}+\begin{bmatrix}\delta\mathbf{u}_e\\ \delta\varphi_e\end{bmatrix}^T\int_{V_e}\begin{bmatrix}\mathbf{B}_m^T E\mathbf{B}_m & \mathbf{B}_m^T e\mathbf{B}_{el}\\ \mathbf{B}_{el}^T e\mathbf{B}_m & -\mathbf{B}_{el}^T\xi\mathbf{B}_{el}\end{bmatrix}dV\begin{bmatrix}\mathbf{u}_e\\ \varphi_e\end{bmatrix}$$

$$=\begin{bmatrix}\delta\mathbf{u}_e\\ \delta\varphi_e\end{bmatrix}^T\int_{A_e}\begin{bmatrix}\mathbf{H}_m^T\\ \mathbf{H}_{el}^T\end{bmatrix}\begin{bmatrix}\mathbf{t}\\ q_s\end{bmatrix}dA. \tag{16}$$

The first term of Eq. (16) describes the forces of inertia, the second term is related to the stiffness of the rod and the term on the right hand side represents the external load. Electrical effects of inertia are neglected since these effects do not occur in the dynamic range of a piezoelectric actuator. Therefore the mass matrix has just one element. The stiffness matrix is fully populated. The term $\mathbf{B}_m^T E\mathbf{B}_m$ expresses the mechanical stiffness, $\mathbf{B}_m^T e\mathbf{B}_{el}$ and $\mathbf{B}_{el}^T e\mathbf{B}_m$ are electro-mechanical coupling terms and $-\mathbf{B}_{el}^T\xi\mathbf{B}_{el}$ represents the electrical stiffness. Eq. (16) can be rewritten in a compact notation

$$\delta\mathbf{p}_e^T\left(\mathbf{M}_e\ddot{\mathbf{p}}_e+\mathbf{K}_e\mathbf{p}_e-\mathbf{F}_e\right)=0, \tag{17}$$

with the element mass matrix $\mathbf{M}_e$, the element stiffness matrix $\mathbf{K}_e$ and the element load vector $\mathbf{F}_e$. The load vector includes normal forces and electric charges at both ends of the rod. Volume forces, e.g. the gravity force, are not considered. Since the virtual state vector $\delta \mathbf{p}_e$ is arbitrary the term in parenthesis has to vanish and the equation of motion is deduced from Eq. (17)

$$\mathbf{M}_e \ddot{\mathbf{p}}_e + \mathbf{K}_e \mathbf{p}_e = \mathbf{F}_e. \tag{18}$$

By comparing Eq. (16) with Eq. (17) the physical meaning of the integrals becomes obvious. The first integral is the consistent mass matrix $\mathbf{M}_e$, the second is the stiffness matrix $\mathbf{K}_e$ and the third is the load vector $\mathbf{F}_e$. Using the shape functions (8,9.10) the matrices can be calculated as

$$\mathbf{M}_e = \frac{\rho A L}{30}\begin{bmatrix} 4 & 2 & -1 & 0 & 0 & 0 \\ 2 & 16 & 2 & 0 & 0 & 0 \\ -1 & 2 & 4 & 0 & 0 & 0 \\ 0 & 0 & 0 & 0 & 0 & 0 \\ 0 & 0 & 0 & 0 & 0 & 0 \\ 0 & 0 & 0 & 0 & 0 & 0 \end{bmatrix}, \mathbf{K}_e = \frac{A}{12L}\begin{bmatrix} 7E & -8E & 1E & 7e & -8e & 1e \\ -8E & 16E & -8E & -8e & 16e & -8e \\ 1E & -8E & 7E & 1e & -8e & 7e \\ 7e & -8e & 1e & -7\xi & 8\xi & -1\xi \\ -8e & 16e & -8e & 8\xi & -16\xi & 8\xi \\ 1e & -8e & 7e & -1\xi & 8\xi & -7\xi \end{bmatrix}, \mathbf{F}_e = \begin{bmatrix} N^i \\ N^j \\ N^k \\ q^i \\ q^j \\ q^k \end{bmatrix}. \tag{19}$$

This FE formulation is used in the following chapter.

3 Active vibration isolation

Figure4. Electric engine and mounting

A lot of technical systems are equipped with engines, pumps and other aggregates which generate vibrations. Due to mechanical connections the vibration propagates into the supporting structure. The concept of Active Vibration Isolation (AVI), which is described in Fuller et al. (1996), is used to cut the path of structure borne sound between the vibration source and the receiving structure. In Fig. 3 the vibration source is an electric motor. The passive isolation by resilient mounts is changed into an Active Vibration Isolation system by inserting a piezoelectric stack.

3.1 SDOF oscillator

For the presented investigations a simple test structure, a single-degrees-of-freedom (SDOF) oscillator, is used. The oscillator is shown in Fig. 5. Since the vibrating mass 5 is supported by the stack actuator 1 the dynamic reaction force on the foundation 6 can be controlled. This force is measured with the force transducer 4 whereas the acceleration of the mass 5 is sensed with an acceleration pick-up 3. The mass 5 is excited by using a second actuator 2.

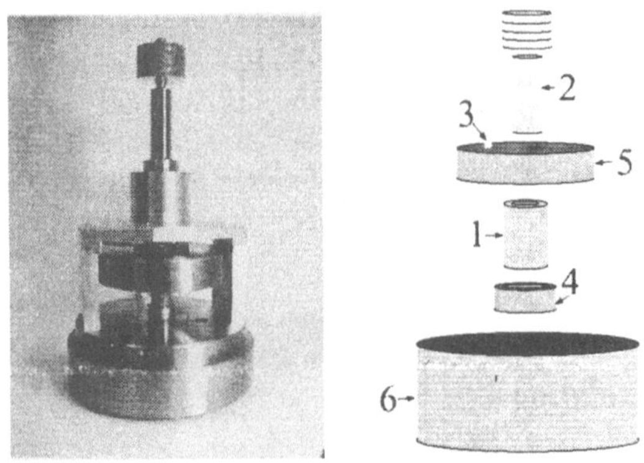

Figure5. SDOF oscillator: Photo and components

On the left hand side of Fig. 6 the actuator 1 is depicted as a three node finite element. By using the derived FE formulation the following equation of motion is obtained

$$\frac{\rho AL}{30}\begin{bmatrix} 4 & 2 & -1 \\ 2 & 16 & 2 \\ -1 & 2 & 4+\tilde{M} \end{bmatrix}\begin{bmatrix} 0 \\ \ddot{u}_2 \\ \ddot{u}_3 \end{bmatrix}+\frac{AE}{12L}\begin{bmatrix} 7 & -8 & 1 & 7d & -8d & 1d \\ -8 & 16 & -8 & -8d & 16d & -8d \\ 1 & -8 & 7 & 1d & -8d & 7d \end{bmatrix}\begin{bmatrix} 0 \\ \ddot{u}_2 \\ \ddot{u}_3 \\ \phi_1 \\ \phi_2 \\ \phi_3 \end{bmatrix}=\begin{bmatrix} F_1 \\ 0 \\ F_{exc} \end{bmatrix}, \tag{20}$$

where the part of the equation containing the electric charge is neglected. The piezoelectric field stress coefficient e is replaced by the piezoelectric field strain coefficient d ($[m/V]$ or $[C/N]$) and the lumped mass is $\tilde{M} = 30M/(\rho AL)$.

In order to calculate the required voltage level for the AVI application the FE model is interpreted in terms of a linear spring and an active componente (see right hand side of Fig. 6). The reaction force for the foundation equals the sum of the spring force F_s and the actuator force F_{act}. To cancel the reaction force the actuator force F_{act} has to be equal to the negative spring force $-F_s$.

By neglecting the mass of the actuator the equation of motion is given as a special case of Eqn. (20)

$$M\ddot{u} + K_{act}u + K_{act}d_{33}U^*_{act} = \hat{F}_{exc}e^{i\Omega t}, \tag{21}$$

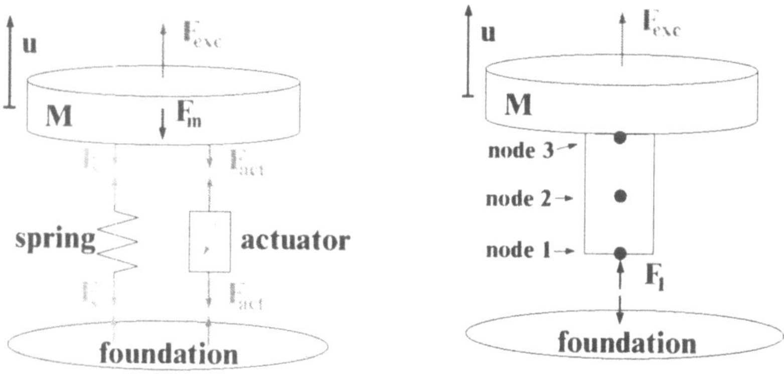

Figure6. FE model and free body diagram

where M is the mass of the oscillator, K_{act} is the stiffness of the actuator. d_{33} is the piezoelectric field strain coefficient, U^*_{act} is the 'virtual' voltage applied to the actuator, $\hat{F}_{exc}$ is the excitation force amplitude and Ω is the excitation frequency. The reaction force vanishes if the voltage is set to

$$\hat{U}^*_{act} = \frac{1}{\Omega^2 M d_{33}} \hat{F}_{exc} \rightarrow \hat{U}_{act} = \frac{1}{n_{PZT} \Omega^2 M d_{33}} \hat{F}_{exc}. \tag{22}$$

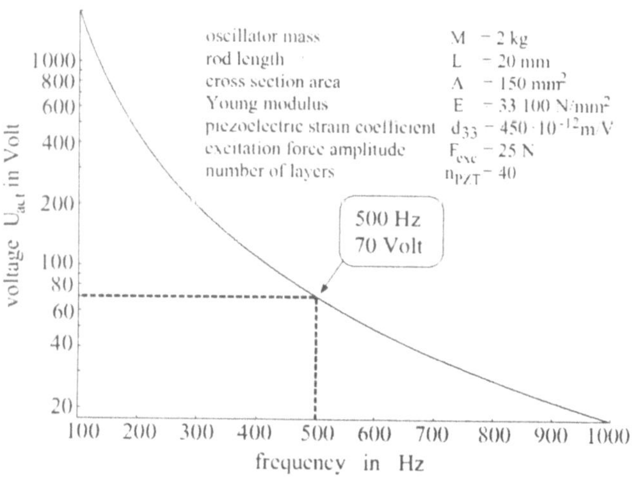

Figure7. Benchmark example

In Fig. 3.1 the data for a benchmark example are listed. The graph of Fig. 3.1 shows the actuator voltage, which is needed for AVI, as a function of the frequency. The dashed lines mark the frequency which is used for the FE calculation and the experiment. The actuator voltage increases rapidly for low frequencies. Therefore a lower limit for the practical application of AVI exists. An upper frequency limit is given by the resonance frequency of the actuator.

3.2 FE simulation

With the values of the benchmark example and the finite element represented by Eq. (20) and Fig. 6 left hand side a finite element simulation is performed. The time response of the force and the displacement are calculated and plotted in Fig. 8. Initially the actuator voltage is set to zero. Without control the reaction force is approximately equal to the excitation force. At $t = 0.05$ s the actuator voltage is set to the precalculated value (see Fig. 3.1). Therefore the reaction force drops from 25 N to 3 N and the associated displacement increases from $0.5\mu m$ to $1.2\mu m$.

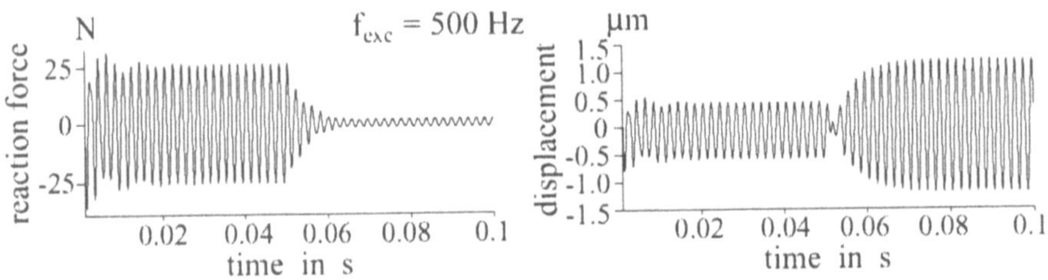

Figure8. Results of FE-calculation

3.3 Experiment with feedforward control

In Fig. 9 the experimental setup for the feedforward control concept is illustrated. In the center of Fig. 9 the test specimen (parts 1 to 6), which has been described above, is shown. The excitation signal is generated by the function generator 7 and then feed in the analog multiplier 8. In the multiplier the excitation signal is multiplied with two constant voltages in the range of ±1 Volt. The constant voltages can be adjusted by two potentiometers. As a result of the multiplication the two output signals have a phase relation of 0^0 or 180^0 degrees. The display module 9 is used to observe the voltages and the two channel high voltage (HV) amplifier 10 generates the necessary voltages of a maximum of 1000 Volt. The outputs of the HV amplifiers are connected to the PZT actuators 1 and 2. The signals of the acceleration pick up 3 and the force transducer 4 are amplified by the charge amplifiers 11 and displayed by the oscilloscope 12. In order to determine the displacement of the mass 5 the acceleration signal is numerically integrated with respect to time.

The test procedure starts with the adjustment of the excitation force by tuning the potentiometer and observing the display of the oscilloscope. Then the second potentiometer is used to activate the compensation actuator 1. Due to the activation of actuator 1 the measured reaction force changes. An appropriate setting is found if the reaction force goes to zero. For an

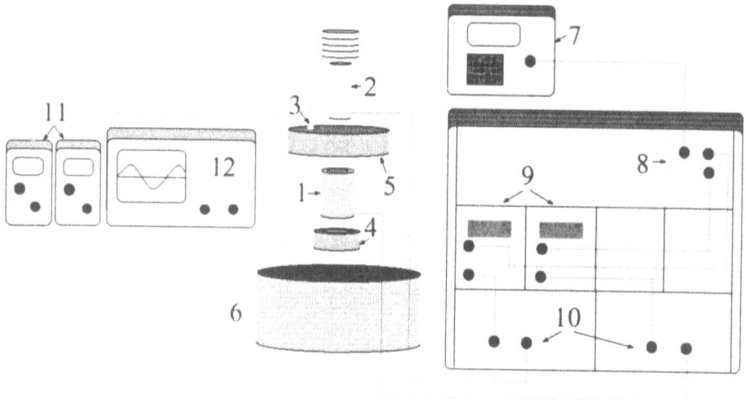

Figure9. Experimental setup for feedforward control

excitation force of 25 N the experimentally determined voltage U_{act} equals 72 Volt. This value confirms the calculation presented above (see Fig. 3.1). In order to compare the calculated dynamical behaviour of the oscillator (see Fig. 8) with an experimental result the transient force and acceleration signal is measured. These results are plotted in Fig. 10 and it can be seen that they validate the FE calculation. There is a slight difference between the calculated and the measured signals after the activation of the compensation during the transient periode. This difference is a result of the dynamics of the used HV amplifier which is not a part of the FE model.

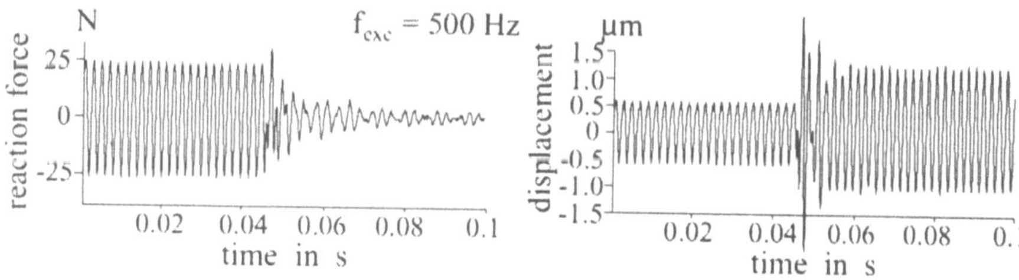

Figure10. Results of the experiment with feedforward control

3.4 Experiment with Feedback Control

Since feedforward control requires a reference signal of the excitation its application is limited to technical systems where such a reference signal is available. If no reference signal is available the feedforward control concept has to be replaced by a feedback controller. The experimental setup depicted in Fig. 9 can be easily converted into a setup for feedback control. Therefore the function generator 7 is directly connected to the display module and the HV amplifier. The excitation signal is used to drive the excitation actuator 2 only. In contrast to the previous setup the measured signal of the reaction force, which is amplified within the charge amplifier 11, is feed in the analog multiplier 8. The potentiometer of the multiplier is used to adjust the proportional feedback factor and the output of the multiplier is connected with the display module and the second channel of the HV amplifier. With this setup a closed loop (force transducer → charge amplifier → multiplier → display module → HV amplifier → compensation actuator) is built.

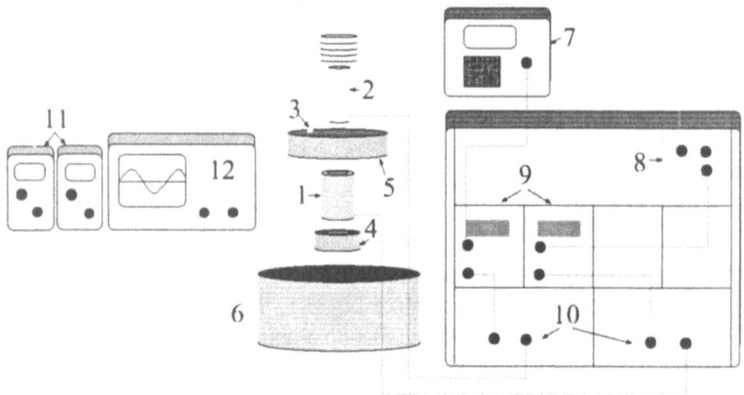

Figure11. Experimental setup for feedback control

Because of the closed loop the feedback controller is not always stable. Therefore the feedback factor is limited. In this experiment the factor is evaluated by increasing the factor till the limit of stability is reached. The results are plotted in Fig. 12. It can be seen that the feedback controller reduces the reaction force from 25 N to 13 N but a complete cancelation can not be achieved. The compensation of the remaining reaction force by the use of an PI controller is not possible because the dynamics of the disturbance is higher than the dynamics of the system and the controller.

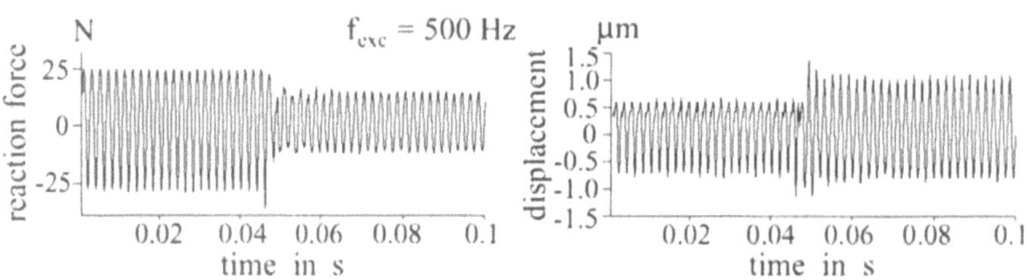

Figure12. Results of the experiment with feedback control

4 Summary and Conclusions

A finite element formulation for piezoelectric rods has been derived. The formulation includes the electro mechanical coupling of the piezoelectric material but it neglects nonlinear and hysteretic effects which occur for high voltages and high stresses. To validate the FE formulation the concept of AVI has been presented and the calculation of a SDOF oscillator equipped with a piezoelectric stack actuator has been performed. The results of the calculation have been compared with experimental results and it has been shown that a good agreement between the calculation and the experiment is obtained. The experimental investigations are divided into feedforward and feedback control strategies. Since feedforward controllers require a reference signal feedback control allows a larger variety of applications but feedforward control leads to a complete cancelation of the reaction force whereas in proportional feedback control a steady state error remains.

References

Allik, H., and Hughes, T. (1970). Finite element method for piezoelectric vibration. *International Journal of Numerical Methods in Engineering* 2:151–158.

Fuller, C., Elliott, S., and Nelson, P. (1996). *Active Control of Vibration*. Academic Press, London.

Gaudenzi, P., and Bathe, K. (1995). An iterative finite element procedure for the analysis of piezoelectric continua. *Journal of Intelligent Material Systems and Structures* 2(6):266–273.

Ghandi, K., and Hagood, N. Nonlinear finite element modeling of phase transitions in electro-mechanically coupled material. In *Proceedings of SPIE*, volume 2715, 121–140. SPIE.

Naillon, M., Coursant, R., and Besnier, F. (1983). Analysis of piezoelectric structures by the finite element method. *ACTA Electronica* 4(25):341–361.

Shieh, R. (1993). Finite element formulation for dynamic response analysis of multiaxial active 3-d piezoelectric beam element structures. In *Collection of Technical Papers Proceedings of the AIAA/ASME Structures, Structural Dynamics and Materials Conference*, volume 6, 3250–3258. AIAA/ASME.

3. Damping of Structural Vibrations Using Adaptive Joint Connections and Neural Control

Hans Albrecht. Jan Wirnitzer and Lothar Gaul.

Institute A of Mechanics, University of Stuttgart
Pfaffenwaldring 9, D-70550 Stuttgart
EMail: {albrecht, wirnitzer, gaul }@mecha.uni-stuttgart.de

Abstract In the present paper, a concept for vibration suppression of flexible structures based on controlled energy dissipation in adaptive joint connections is presented. First, the damping effect of dry friction on the dynamical behaviour of a two-dimensional truss structure consisting of five struts and one bolted joint connection is investigated experimentally and numerically. Experiments are carried out with varied normal force in the bolted joint in order to analyse the influence on the vibration characteristics of the flexible structure. For the experimental structure a FE formulation is presented including a model of the frictional joint connection. Finally, a neural control approach for the adaptive joint connections is described. An application example illustrates this so-called semi-active vibration damping concept.
The present paper shows that this concept significantly enhances the vibration suppression of a flexible lightweight structure. Based on these results, an adaptive three-dimensional large space structure with adaptive joints is in reach in the near future.

1 Introduction

The low structural damping of large space structures often causes a problem when high positioning accuracies are required in orbit. Therefore effective measures for vibration reduction are required. Several approaches for active vibration suppression of lightweight flexible structures has been presented. In Breitbach et al. (1994) piezo-stacks embedded in several struts of the truss structure are used as actuators for a force feedback control. An alternative are tendons whose length can be controlled by piezoelectric actuators Preumont and Bossens (1999). Additionally, friction in joint connections of the structure has been considered as a significant source of damping. Since truss structures often consist of flexible struts which are connected by joint connections, this approach can easily be employed without increasing the structural weight. Several researchers investigated the dynamic behaviour of flexible structures with frictional joint connections Beards and Woowat (1985), Gaul and Lenz (1997) as well as the application to vibration control Ferri et al. (1992), Gaul et al. (1998), Nitsche and Gaul (September 12 15 1999).

A new concept of adaptive joints (patented by L. Gaul under DE 19702518) is based on the controlled energy dissipation in bolted joint connections. The energy dissipation due to interfacial slip in the joint can be controlled by varying the normal pressure in the contact area using a piezo-disc actuator. The first part of this paper demonstrates the applicability of friction damping for vibration suppression of a two-dimensional truss structure. The experiments are confined to a pin-type joint connection enabling relative rotation. However, the results can also be applied to joint connections allowing translational relative motion in addition. For the experimental structure a

FE formulation is set up including a model of the frictional joint connection and the simulation results are validated by comparison with measurements. In the second part, a neural control approach is used in order to optimize the energy dissipation and thus the vibration suppression. Using this approach the design of a neural controller can be carried out either model-based by using a numerical plant model or model-free by incorporating the experimental structure in a hardware-in-the-loop optimization.

2 Truss structure with bolted joints

The two-dimensional truss structure investigated in the present paper can be viewed as a segment of a three-dimensional space structure. Its geometry is presented in Figure 1. It consists of five steel struts which are mounted to the four corner-parts by pin-type joints (Figure 2). Approximation of a two-dimensional free-free boundary condition is achieved by suspending the structure horizontally at its four corners using steel wires of sufficient length.

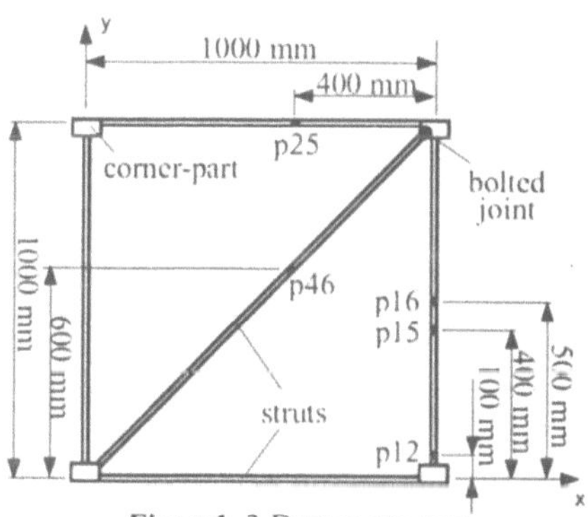

Figure1. 2-D truss structure

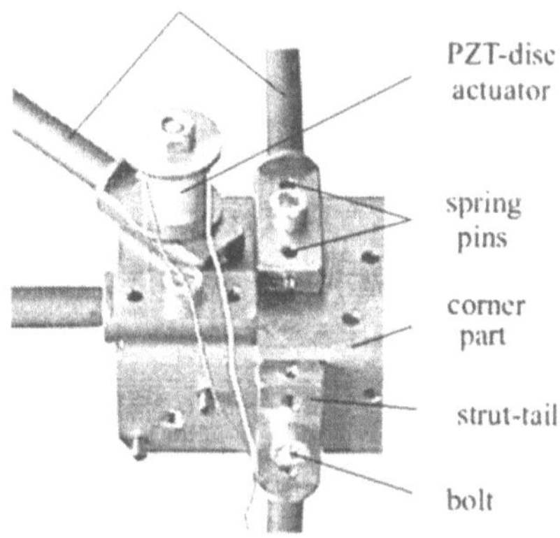

Figure2. Bolted Joint Connections

Except one bolted joint of the diagonal strut, all other connections are fixed absolutely tight preventing any kind of slip motion. The bolted joint only enables rotation about its brass bolt. The joint is manufactured with a precise fit which eliminates the possibility of clearance. The effects caused by impacts in the joint (see Folkman et al. (1996)) are not considered in this paper. Investigations incorporating experiments with a thinner pin and a PTFE-jack have shown that the friction occuring in the pin joint is negligible. The main source of friction damping stems from the slip between the plane interfaces of the strut-tail and the corner-part.

For the control of the normal force in the contact a piezoelectric(PZT)-disc actuator will be used which is placed under the bolt (Figure 2). However, for the experimental analysis presented in this paper a constant normal force is required. Therefore, a stack of plate springs is used which allows to vary the normal force in the range between 0 and 800 N simply by modifying the height of the stack which has a well known spring characteristic.

3 Experiments with a passive joint connection

In the following, the influence of friction on the dynamic behaviour of the truss structure is investigated. The transfer behaviour of a friction joint is highly nonlinear and the response depends on the amplitude of the excitation. The interpretation of fundamental harmonic FRF-data obtained from such a nonlinear system has been treated in Ewins and Vakakis (1992). The nonlinearity raises restrictions on the possible interpretations of the measured data due to the fact that the system cannot be described by uncoupled modes.

In order to characterize the friction damping qualitatively, the response of the truss structure is measured for different normal pressures in sliding contact. For each adjusted normal force, the FRFs of the truss structure are recorded at points p46 (in y-direction), p15 (in x-direction) and p25 (in y-direction) which are marked in Figure 1. The structure is excited by an impulse load applied at point p16 in x-direction. The sum of the three FRFs is plotted in Figure 3 for different normal forces. The modes listed in Table 1 and Table 2 which correspond to peaks in Figure 3 are denoted by their mode number.

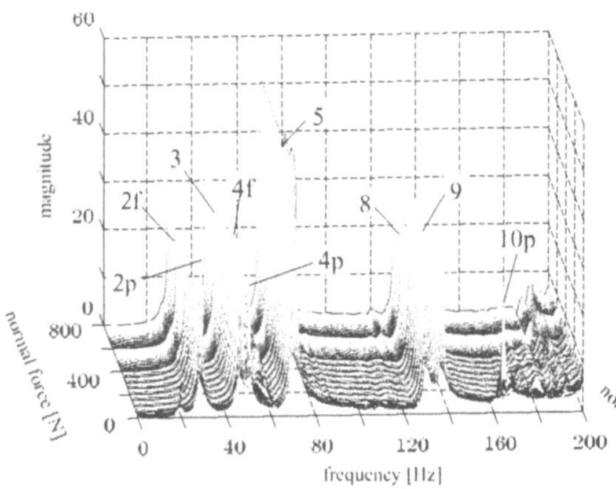

Figure3. FRFs for varying normal forces

Figure4. Zoom of the plot in Figure 3

As can be seen in Figure 3, the modes 3, 5, 8 and 9 are almost uneffected by the variation of the normal force. This can be explained by considering the mode shapes. As an example, Figure 5 shows the mode shapes of the mode 3 and mode 8 obtained from an Experimental Modal Analysis (EMA). All of them have a common feature, there is no change in angle between the diagonal strut and the corner-part. And therefore, no sliding can occur in the contact. The mode shapes at these frequencies are very similar for the structure with a fixed joint as well as for the structure with a frictionless pin joint. And, as can be seen in Table 1 and Table 2, the eigenfrequencies of the uneffected modes are almost unchanged. There is no chance to influence these modes by a frictional joint at this position within the structure. The consequences for an optimal placement strategy of several joint connections are evident.

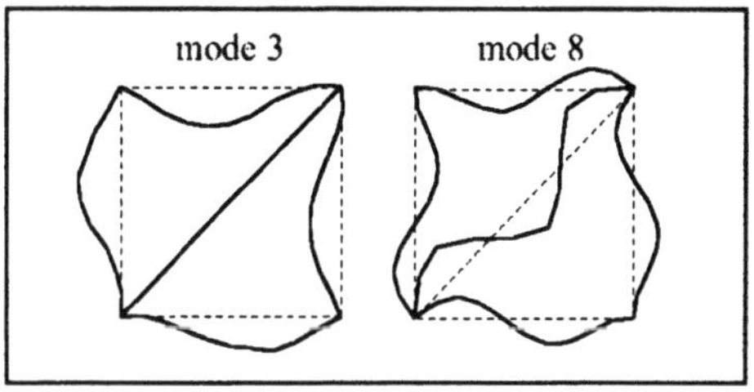

Figure5. Mode shapes of uneffected modes

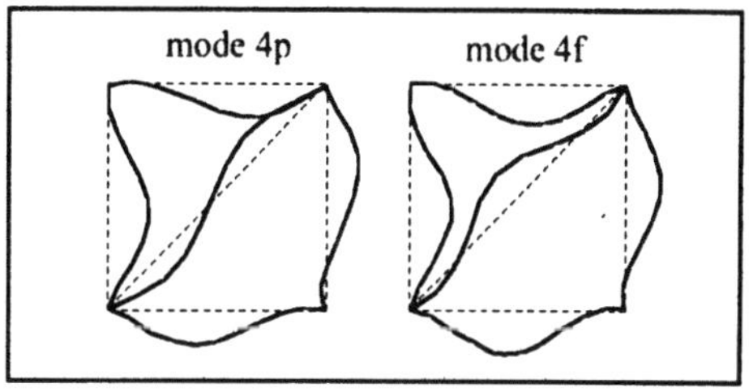

Figure6. Mode shapes of effected modes

The zoom plotted in Figure 4 shows the damping effect of the bolted joint in the frequency range from 50 to 60 Hz. By increasing the normal force mode 4p (pin joint) gradually changes to mode 4f (fixed joint) with a high reduction of the resonance amplitude inbetween. The corresponding mode shapes are depicted in Figure 6. Comparable experiments with more than one passive joint connection at different positions within the structure can be performed.

4 FE Formulation

Further investigations such as an optimal placement strategy of adaptive joints require a numerical model of the truss structure including the bolted joint connections. Several well known lumped parameter joint models used in the past to describe the nonlinear behaviour of isolated joints are capable of describing global states of slip and stick only. However, the actual normal contact pressure distribution in a dynamically loaded joint is not uniform in the interface. Depending on the transmitted load, the contact interface is divided into stick and slip regions. Before gross slip behaviour occurs, the so called macroslip, only local slip exists, the so called microslip. An overview of static and dynamic friction models is given in Gaul and Nitsche (1999).

A quite powerfull dynamic friction model which is especially suitable for control purposes is the LuGre model Canudas de Wit et al. (1995). However, for a FE formulation the Valanis model Valanis (1991) adopted from the theory of plasticity is more suitable. Its capability to approximate a hysteresis loop caused by friction in the bolted joint already has been shown in Gaul and Lenz (1997). The change of the friction moment in time $\dot{M}_{\mathrm{Val}}$ is given by the evolution equation

$$\dot{M}_{\mathrm{Val}} = f\left(\theta, \dot{\theta}, M_{\mathrm{Val}}\right) = \frac{E_0\,\dot{\theta}\left[1 + \frac{\lambda}{E_0}\,\mathrm{sgn}(\dot{\theta})\,(E_t\,\theta - M_{\mathrm{Val}})\right]}{1 + \kappa\frac{\lambda}{E_0}\,\mathrm{sgn}(\dot{\theta})\,(E_t\,\theta - M_{\mathrm{Val}})}\,, \tag{1}$$

where M_{Val} is the transmitted moment and θ is the relative rotation about the axis of the pin. θ can be expressed by means of the rotational degrees of freedom (dof) m and n ($m, n \in [1, N]$) which are coupled by the joint connection

$$\theta = (x_m - x_n) = \mathbf{b}^{\mathrm{T}}\mathbf{x} \quad \text{mit} \quad \mathbf{b} = [\underset{m\text{-th}}{\dots\ 01}\ 0\ \dots\ \underset{n\text{-th column}}{0\text{-}1}\ 0\ \dots\ 0]^{\mathrm{T}}_{(1,N)} \tag{2}$$

The stiffness modulus of the sticking condition is denoted by E_0, while the tangent modulus E_t describes the slope of slip motion. The parameter κ controls the influence of microslip and λ is obtained from the stick limit equivalent to the yield stress defined by $\sigma_0 = \frac{E_0}{\lambda\left(1-\kappa\frac{E_t}{E_0}\right)}$.

Neglecting material damping, the system of equations defining the structure with a bolted joint connection becomes

$$\mathbf{M}\ddot{\mathbf{x}} + \{\mathbf{K} + \mathbf{K}_{\text{Val}}(\mathbf{x}, \dot{\mathbf{x}})\}\,\mathbf{x} = \mathbf{f} \tag{3}$$

with the nonlinear stiffness matrix

$$k_{\text{Val}_{ij}} = c_{\text{Val}}\, b_i\, b_j\,, \qquad \text{where} \qquad c_{\text{Val}} = \frac{E_0\left[1 + \frac{\lambda}{E_0}\,\text{sgn}\,(\mathbf{b}^{\mathrm{T}}\dot{\mathbf{x}})\left\{E_t\,(\mathbf{b}^{\mathrm{T}}\mathbf{x}) - M_{\text{Val}}\right\}\right]}{1 + \kappa\frac{\lambda}{E_0}\,\text{sgn}\,(\mathbf{b}^{\mathrm{T}}\dot{\mathbf{x}})\left\{E_t\,(\mathbf{b}^{\mathrm{T}}\mathbf{x}) - M_{\text{Val}}\right\}} \tag{4}$$

is the tangent stiffness obtained from equation (1) which is defined with respect to the rotational dof. The remaining dofs of the joint connection are coupled by kinematic constrains preventing any relative motion. Due to their linear nature they can be inserted into the stiffness matrix $\mathbf{K}$. For a structure with N_k frictional joint connections, $\mathbf{K}_{\text{Val}}$ is obtained by

$$\mathbf{K}_{\text{Val}} = \sum_{k=1}^{N_k} \mathbf{K}^k_{\text{Val}}\,. \tag{5}$$

The time integration of equation (3) is carried out using the Newmark method. Due to the nonlinear system, the displacement solution ${}^{n+1}\mathbf{x}^{(i)}$ at time step t_{n+1} is calculated in an iterative manner by using the Newton-Raphson method Bathe (1982) of the form

$${}^{n+1}\mathbf{K}^{(i-1)}\,\delta\mathbf{x}^{(i)} = {}^{n+1}\mathbf{f} - {}^{n+1}\mathbf{r}^{(i-1)} \quad \text{and} \quad {}^{n+1}\mathbf{x}^{(i)} = {}^{n+1}\mathbf{x}^{(i-1)} + \delta\mathbf{x}^{(i)}\,, \tag{6}$$

where ${}^{n}\mathbf{x}$ is the converged solution in the preceding time step t_n, ${}^{n+1}\mathbf{r}$ is the internal nodal-load vector corresponding to the internal stresses and ${}^{n+1}\mathbf{f}$ contains the external nodal-loads which are assumed to be independent of $\mathbf{x}$. In every iteration i the contribution ${}^{n+1}\mathbf{K}^{(i-1)}_{\text{Val}}$ to the tangent-stiffness matrix ${}^{n+1}\mathbf{K}^{(i-1)}$ using equation (5) and ${}^{n+1}\mathbf{r}^{(i-1)}_{\text{Val}}$ to the internal load vector ${}^{n+1}\mathbf{r}^{(i-1)}$ accordning to

$${}^{n+1}\mathbf{r}^{(i-1)}_{\text{Val}} = \sum_{k=1}^{N_k} {}^{n+1}\mathbf{r}^{k,(i-1)}_{\text{Val}} \qquad \text{with} \qquad \mathbf{r}^k_{\text{Val}} = M^k_{\text{Val}}\,\mathbf{b} \tag{7}$$

must be calculated, respectively. Therefore, the differential equation (1) is integrated for the time step $\Delta t_{n+1} = t_{n+1} - t_n$ by using, e. g., the Runge-Kutta or Adams-Bashforth-Moulton method.

4.1 Simulation results

Table 1 provides a comparison between calculated and measured eigenfrequencies of the structure with a fixed joint. The Modal Assurance Criterion (MAC) value used as a measure for the correlation between the calculated and the measured mode shapes is listed as well. A MAC-value of 1.0 indicates that the compared mode shapes are identical. Table 2 contains the corresponding data for a frictionless pin joint. The worse correlation in Table 2 is mostly attributed to the weak nonlinearity caused by slight friction between the pin and borehole.

No.	EMA [Hz]	FEM [Hz]	Error	MAC
1f	26.4	26.4	0.0%	0.87
2f	29.3	29.3	0.0%	1.00
3f	46.5	45.9	1.3%	0.96
4f	57.4	57.3	0.2%	0.98
5f	70.5	71.1	0.9%	0.94
6f	77.6	76.3	1.7%	0.98
7f	121.4	119.1	1.9%	0.99
8f	129.7	129.4	0.2%	0.99
9f	136.8	137.0	0.1%	0.93
10f	192.4	194.6	1.1%	0.79

Table1. Fixed joint (sticking condition)

No.	EMA [Hz]	FEM [Hz]	Error	MAC
1p	19.1	19.6	2.6%	0.97
2p	27.8	28.7	3.2%	0.99
3p	46.1	45.9	0.4%	0.89
4p	49.5	48.7	1.6%	0.97
5p	70.6	71.1	0.7%	0.94
6p	72.2	71.4	1.1%	0.88
7p	106.8	103.7	2.9%	0.98
8p	129.9	129.3	0.5%	0.98
9p	136.7	137.0	0.2%	0.91
10p	164.5	165.2	0.4%	0.99

Table2. Pin joint (no friction)

In order to simulate the dynamic behaviour of a particular bolted joint, the corresponding parameters of the Valanis model have to be determined. They can be identified from a measured hysteresis loop of an isolated bolted joint in a torsional resonator Gaul and Lenz (1997). However, to circumvent the effort associated with the design of a torsional resonator, the accelerations of a single strut which is clamped at its left end and supported by a bolted joint at its right end are recorded. By fitting the calculated responses to the measured data, the Valanis parameters are obtained for different constant normal forces. As a result of the identification process the dependency of the parameters E_0 and σ_0 on the normal force F_N turns out to be approximately linear, whereby κ has only very little influence and E_t is negligibly small.

The truss structure with one bolted joint is subjected to an impulse load at point p12 (Figure 1). The accelerations are recorded at point p46 in y-direction for a fixed joint and for a joint with a normal force $F_N = 200$ N, respectively. As can be seen from Figure 7, even one frictional joint in the structure can significantly reduce the vibration. Approximately after 0.5 seconds, sticking in the joint occurs and there is no further reduction in amplitude. To achieve further reduction the normal force would have to be decreased. Therefore, a controller is required which adjusts an optimal normal force. Figure 8 shows the comparison between measurement and simulation.

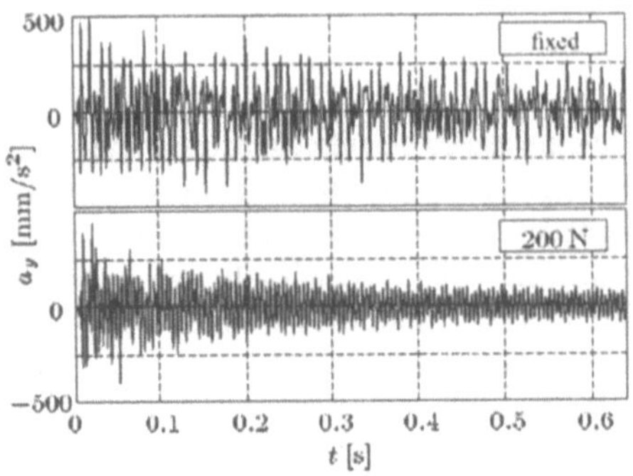

Figure7. Response of the undamped and damped truss structure

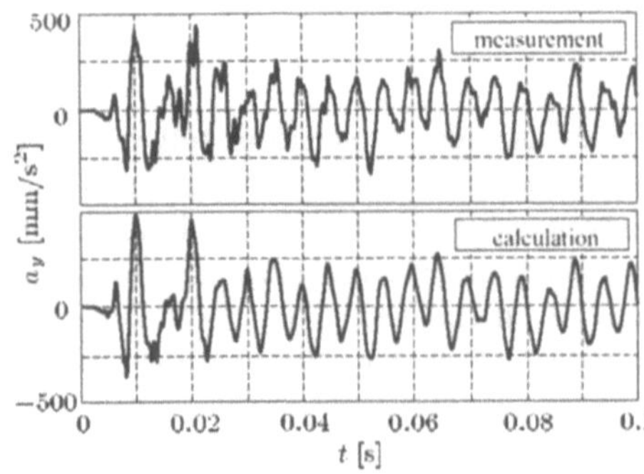

Figure8. Response of the damped structure (F_N = 200 N)

5 Semi-active control

The truss structure investigated so far included passive damping elements. In order to enhance the energy dissipation the normal force of each bolted joint connection has to be optimally adjusted. This is achieved by controlling the voltage applied to a piezoelectric(PZT)-disc actuator placed under the bolt. This approach is called semi-active damping since the properties of a passive element are changed by active control.

There are two control concepts to optimize the vibration suppression of a truss structure. Either the control task is achieved by several local controllers, one for each adaptive joint connection, or by just a single global controller. Especially, for the design of a global controller neural networks are a promising alternative to methods of nonlinear control theory.

5.1 Neural control

Neural Networks (NN) have already been successfully used for pattern recognition and signal processing and their application to system idetification and control has been increasingly reported in the recent past Hunt et al. (1992). Due to the nonlinear nature of NN they are capable of representing any arbitrary nonlinear mapping and moreover, they are able to learn this mapping by examples. Classical adaptive control theory also provides solutions to some of these tasks, such as Self-Tuning Regulator (STR) or Model-Reference-Adaptive-Control (MRAC). However, these control methods only provide the optimization of coefficients of predefined control rules which must be a priori known. The controller design methods of neural control do not have this limitation and the design procedure yields an optimal control rule, as well. Further advantages of NN for control purposes are the fast overall processing due to their inherent parallel structure and the robustness due to their ability to handle unknown inputs (generalization). A drawback of neural control is the lack of profound concepts for stability theory. However, stability is not an important issue in semi-active control. Since no energy can be put into the system by the actuator, the closed-loop system remains stable if the open-loop system guarantees stability.

The present approach makes use of static multi-layer feedforward networks (FFN) consisting of K layers with connections from neurons of the layer k to neurons of the next layer $k+1$. The ouput $\mathbf{a}^k(n)$ of the s^k neurons in the k-th layer is given by

$$\mathbf{a}^k(n) = f_{act}^k\left(\mathbf{net}^k(n) + \mathbf{b}^k\right) , \quad \text{with} \quad \mathbf{net}^k(n) = \mathbf{W}^k\,\mathbf{a}^{k-1}(n) , \tag{8}$$

where $f_{i,act}^k$ $(i = 1, \ldots, s^k)$ is the activation function of the i-th neuron and $\mathbf{net}^k$ is the weighted sum of the output $\mathbf{a}^{k-1}$ of the preceding layer $k-1$. The element w_{ij} of matrix $\mathbf{W}^k$ represents the weight of the connection from neuron i (of the layer $k-1$) to neuron j (of the layer k). b_i^k is the bias (threshold) of the neuron i. Equation (8) is initialized by assigning the pattern $\mathbf{x}(n)$ to the input layer $k = 0$, i. e., $\mathbf{a}^0 = \mathbf{x}(n)$. Despite their simple structure, FFN are capable of approximating an arbitrary nonlinear function.

A method for training a neural controller to learn a functional mapping of the plant state (or the measured variables) to action variables that would be optimal with regard to an arbitrary quality criterion is the so called Reinforcement-Learning algorithm Werbos (1990). Unlike the mostly used training methods, such as Internal Model Control (IMC) or MRAC, the Reinforcement-Learning algorithm does not require a neural identification model of the plant and thus, it can be used not only for model-based Hrycej (1994) but also for model-free training. The training of the neural controller is performed by optimization of a given cost function with respect to the weights $\mathbf{W}^k$ and biases $\mathbf{b}^k$ which are the controller parameters. The cost function is evaluated from the measured response of the closed-loop system as shown in Figure 5.1. It should be noted that the optimization may be performed by any optimization method (gradient-based or stochastic).

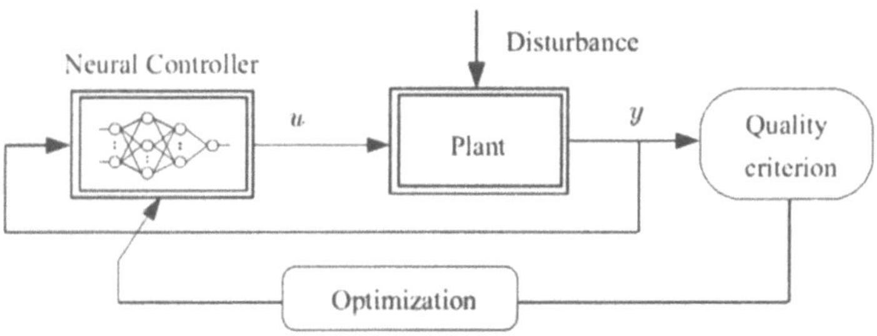

Figure9. Reinforcement-Learning

5.2 Example

Figure 10 shows two Euler-Bernoulli beams coupled by a semi-active bolted joint connection. The system can be seen as an elementary part of the truss structure under investigation.

Neglecting material damping, the system can be described in modal space by

$$\mathbf{M}\ddot{\mathbf{x}} + \mathbf{K}\mathbf{x} + \mathbf{b}\,M_r = \mathbf{f} , \qquad \mathbf{x}(0) = \mathbf{x}_0 \tag{9}$$

with the modal mass matrix $\mathbf{M}$, the modal stiffness matrix $\mathbf{K}$, the load vector $\mathbf{f}$ and the vector $\mathbf{x}$ containing $N = N_1 + N_2$ modal coordinates w_{ij}^* of both beams ($i = 1, 2$ and $j = 1, \ldots, N_i$). The

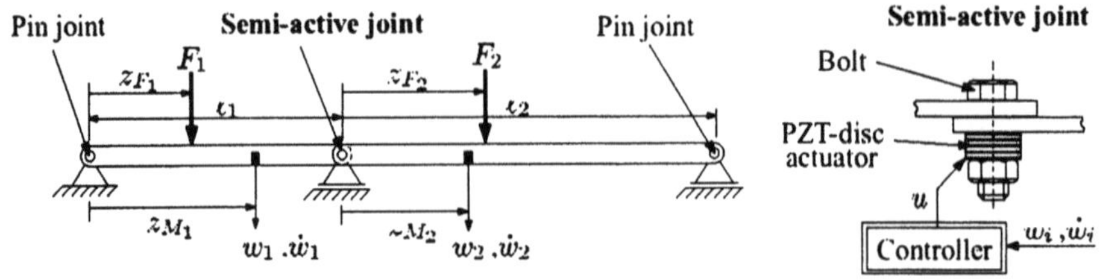

Figure10. Adaptive beam structure

beam deflections at z_{M_i} can be calculated by means of the modal shape vectors $\phi_{ij}(z_{M_i})$ from $w_i(z_{M_i},t) = \sum_{j=1}^{N_i} \phi_{ij}(z_{M_i})\, w_{ij}^*(t)$. The friction moment

$$M_r = r\, F_N\, \mu\left(\varphi, \dot{\varphi}, \dot{\theta}\right) \tag{10}$$

is defined by the LuGre model Canudas de Wit et al. (1995) as a function of the normal force F_N and the dynamic friction coefficient μ which depends on the relative angular velocity $\dot{\theta} = \mathbf{b}^{\mathrm{T}}\dot{\mathbf{x}}$ in the joint and on the inner variable φ. For a detailed derivation of equation (9) and a definition of the model parameters the reader is refered to Nitsche and Gaul (September 12 15 1999). In the following, the applied voltage which corresponds to the control output u is assumed to be proportional to the normal force F_N, i. e., $u \sim F_N$.

Two sensors placed at $z_{M_1} = 0.75\ell_1$ and $z_{M_2} = 0.25\ell_2$ provide the beam deflections w_i and their time derivatives $\dot{w}_i$ as system output which is directly used as input to the controller. For the local controller a two-layer ($K = 2$) FFN is chosen with $s^0 = 4$ input neurons, $s^1 = 6$ neurons in the hidden layer ($k=1$) and $s^2 = 1$ neuron in the output layer ($k = 2$). The activation function of the output layer is a bounded axis-symetric function in order to guarantee a bounded control output $u = [u_{min}, u_{max}]$ and the hidden layer uses the sigmoid function. The optimal network parameters $\hat{\mathbf{W}}^k$ (including the weights $\mathbf{W}^k$ and biases $\mathbf{b}^k$) are obtained by solving the optimization problem

$$\min_{\hat{\mathbf{W}}^k} \quad \mathcal{J}_C = \sum_{i=1}^{2} \frac{1}{2} \int_{t_0=0}^{T} \mathbf{w}_i^{\mathrm{T}} \mathbf{w}_i \, dt \tag{11}$$

using a gradient descent algorithm with a finite-difference approximation of the gradients. The choice of an appropriate excitation for the training procedure is not a critical issue. However, it is advantageous to define an excitation that causes a significant vibration level of both beams.

Figure 11 shows the response of the controlled and passive ($u = u_{min} = \text{const}$) system calculated at the sensor locations z_{M_i}. The forces F_i of the test excitation are applied at $z_{F_i} = 0.41\ell_i$. The excitation includes impulses as well as harmonic loads with a frequency equal to the first eigenfrequency of the left and right beam, respectively.

In order to check the performance of the neural controller, a reference controller is needed. Here, the Lyapunov controller presented in Nitsche and Gaul (September 12 15 1999) is used (except that no observer is involved). From the undistinguishable closed-loop responses in Figure 11, it can be suggested that the control strategies of both controllers are very similar. The

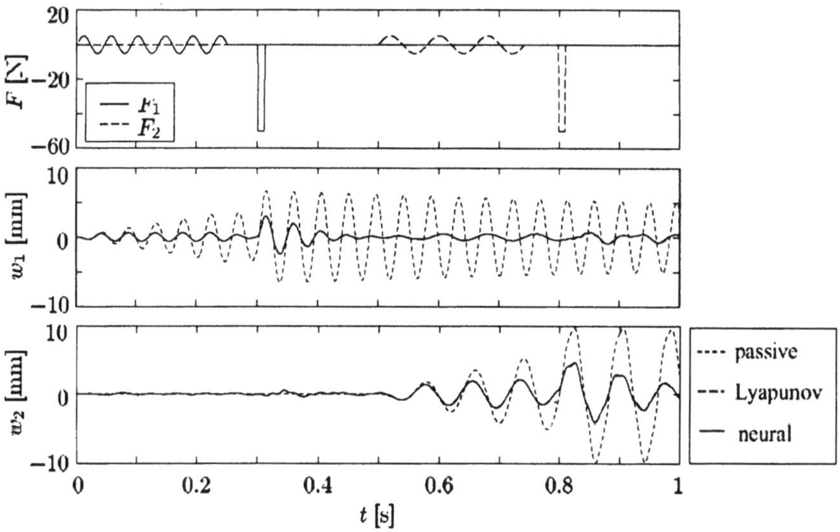

Figure11. Test excitation $F_i(t)$ and system response $w_i(z_{M_i}, t)$

confirmation is given in Figure 12 where the control signal u and the transmitted friction moment M_r for both controllers are plotted.

An alternative approach employs an additional neural network which estimates the joint state, i. e., the relative angle θ and angular velocity $\dot{\theta}$, from the sensor output. The controller then uses the joint state as input. Since the joint state cannot be measured, a plant model is required in order to generate training data for the supervised learning of the neural estimator by Backpropagation Hunt et al. (1992). Therefore, this approach can only be utilized for model-based design of the controller.

Simulations have confirmed that both approaches provide good robustness with respect to deviations of the system parameters used for training and testing. The robustness analysis focused on alteration of the sensor positions, addition of noise to the sensor signals and increase in the number of modal coordinates N. The robustness of the approach using a neural estimator turned out to be slightly better.

6 Conclusion

In the present paper a new approach for vibration suppression of lightweight truss structures has been presented. The concept utilizes the energy dissipation caused by friction in bolted joint connections. The applicability of friction damping has been demonstrated on a two-dimensional truss structure with a passive bolted joint. Furthermore, an approach for semi-active vibration damping using neural networks has been introduced. As a next step, the neural control approach has to be applied to the two-dimensional structure and the model-free training of the controller must be tackled.

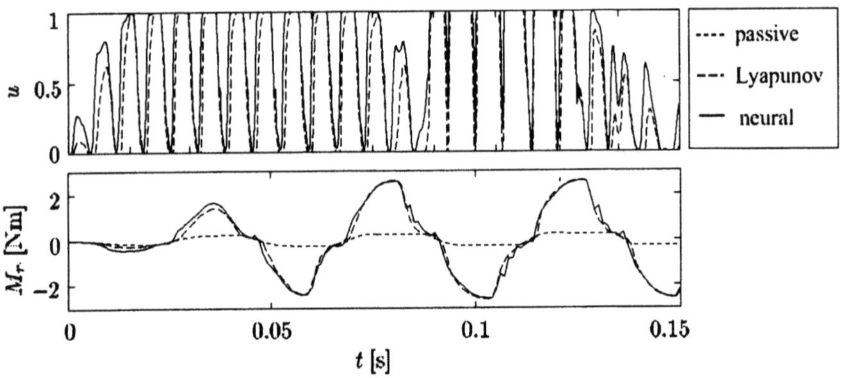

Figure 12. Control output u and friction moment M_r

References

Bathe, K.-J. (1982). *Finite Elemente Procedures in Engineering Analysis*. Englewood Cliffs/NJ: Prentice Hall.

Beards, C. F., and Woowat, A. (1985). The control of frame vibration by friction damping in joints. *Journal of Vibration, Acoustics, Stress, and Reliability in Design* 107:26–32.

Breitbach, E. J., Lammering, R., Melcher, J., and Nitzsche, F. (1994). Smart structures research in aerospace engineering. In *European Conference on Smart Structures and Materials*, 11–18.

Canudas de Wit, C., Olsson, H., Åström, K. J., and Lischinsky, P. (1995). A new model for control of systems with friction. *IEEE Transactions on Automatic Control* 40(3):419–425.

Ewins, D. J., and Vakakis, A. F. (1992). Effects of weak nonlinearities on modal analysis. In *Proceedings of the Intern. Modal Analysis Conference (IMAC)*, 72–78.

Ferri, A. A., Heck, B. S., and Lane, J. S. (1992). Vibration control using semi-active friction damping. In R. Ibrahim, A. S., ed., *Friction-Induced Vibration, Chatter, Squeal, and Chaos. ASME-Publication*, volume DE-Vol. 49. 165–171.

Folkman, S., Ferney, B., Bingham, J., and Dutson, J. (1996). Friction and impact damping in a truss using pinned joints. In Guran, A., Pfeiffer, F., and K.Popp., eds., *Dynamics with Friction: Modeling, Analysis and Experiment. Series on Stability, Vibration and Control of Systems, Series B*, volume 7. World Scientific Publishing Co. 137–168.

Gaul, L., and Lenz, J. (1997). Nonlinear dynamics of structures assembled by bolted joints. *Acta Mechanica* 125:169–181.

Gaul, L., and Nitsche, R. (1999). Nonlinear dynamics of structures with joint connections. In D. J. Ewins, D. J. I., ed., *Structural Dynamics 2000, Series Engineering Dynamics*, volume DE-Vol. 49. 165–171. to appear.

Gaul, L., Lenz, J., and Sachau, D. (1998). Active damping of space structures by contact pressure control in joints. *Mechanics of Structures and Machines* 26(31):81–100.

Hrycej, T. (1994). Model-based training of control sequences in neural controllers. *Neural Network World* 2:173–188.

Hunt, K. J., Sbarbaro, D., Zbikowski, R., and Gawthrop, P. J. (1992). Neural networks for control systems – a survey. *Automatica* 28(6):1083–1112.

Nitsche, R., and Gaul, L. (September 12-15, 1999). Friction control for vibration suppression. In *Proceedings of the ASME Design Engineering Technical Conference*, volume DECT99/VIB-8191.

Preumont, A., and Bossens, F. (1999). Active tendon control of a truss structure. In *International Conference on Adaptive Structures and Technologies*, 305–312. Paris: Technomic Publishing.

Valanis, K. C. (1991). Fundamental consequences of a new intrinsic time measure. plasticity as a limit of the endochronic theory. *Archive of Mechanics* 32:171–191.

Werbos, P. J. (1990). A menu of designs for reinforcement learning over time. In Miller, T. W., Sutton, R. S., and Werbos, P. J., eds., *Neural Networks for Control*. MIT Press. chapter 1, 67–96.

4. Sensor and actuator design methods in active vibration control for distributed parameter structures

Hans Albrecht, Uwe Stöbener and Lothar Gaul

Institute A of Mechanics, University of Stuttgart
Pfaffenwaldring 9, D-70550 Stuttgart

Abstract The efficiency of active structures is determined by their design. An important part of the design process is the placement and shaping of sensors and actuators necessary for the control. This paper focuses on placement and shaping methods for distributed parameter structures, such as plates and shells. The methods are based on a modal representation of the structural vibration. According to the standards of control theory and signal processing the state space formulation is used to collect modal data into matrices and to describe the dynamics of the system. For the normalization of these matrices the H_2 and H_∞ norms are used. The actuators are considered to be distributed piezoelectric sheets which are bonded to the surface of the structure. The efficiency of the actuators is expressed in scalar performance indices. The calculation of these indices requires the normalization of the input matrix and the generation of a so-called placement matrix. An example is presented for a plate to demonstrate the design process for distributed actuators. For the detection of modal displacements and velocities the concept of point sensors is applied which requires a positioning of sensors into an array. To obtain an appropriate sensor layout the placement of each sensor with respect to the vibration amplitude and the correlation of each sensor within the array has to be considered. The sensor performance is also expressed by scalar indices, which are calculated by the same formulas as for the actuators. The sensor correlation is investigated by the condition of the output matrix. A comparison of different point sensor arrays is carried out for the mentioned plate example.

1 Modal state space formulation

1.1 System equation

The motion of structures with distributed parameters, such as rods, beams and plates can be described by the following partial differential equation (PDEq.)

$$M\ddot{w}(x,y,t) + D\Delta^2\dot{w}(x,y,t) + B\Delta^2 w(x,y,t) = \sum_{n=1}^{N} \bar{u}_n(x,y)u_n(t), \tag{1}$$

with the 2-D Laplace operator

$$\Delta = \left(\frac{\partial^2}{\partial x^2} + \frac{\partial^2}{\partial y^2}\right). \tag{2}$$

The PDEq. (1) describes the time variant and space dependent displacement $w(x,y,t)$ of a plate in response to a number of N applied control inputs $u_n(x,y,t)$ created by distributed segmented actuators. Here M is the mass per unit area, D is the viscous damping constant and B is the bending stiffness.

To transform the PDEq. (1) into a modal formulation the displacement $w(x, y, t)$ is expressed by a superposition of modal coordinates $q_k(t)$. The space dependency of the displacement is taken into consideration by the modal shape functions $\phi_k(x, y)$

$$w(x, y, t) = \sum_{k=1}^{\infty} \phi_k(x, y) q_k(t). \tag{3}$$

The evaluation of the modal shape functions requires the solution of the eigenvalue problem

$$\Delta^2 \phi_k(x, y) = \lambda_k M \phi_k(x, y), \tag{4}$$

which is derived from the PDEq. (1) for the undamped structure ($D = 0$). Thus the eigenvalues λ_k are related to the eigenfrequencies ω_k of the undamped structure by

$$\lambda_k = \omega_k^2. \tag{5}$$

Finally the equation of motion can be written in modal coordinates of mode k as

$$\ddot{q}_k(t) + 2\zeta_k \omega_k \dot{q}_k(t) + \omega_k^2 q_k(t) = \sum_{n=1}^{N} b_{kn} u_n(t), \tag{6}$$

where ζ_k is the modal damping ratio and b_{kn} is the modal input factor of the n-th actuator defined by

$$b_{kn} = \frac{\int\limits_a \int\limits_b \bar{u}_n(x, y) \phi_k(x, y) dx dy}{M \int\limits_a \int\limits_b \phi_k^2(x, y) dx dy}. \tag{7}$$

The integral limits a and b characterize the dimensions of the structure (lenght and width). The state space representation of the system for the k-th mode is obtained by transforming the second order DEq. (6) into a set of two first order DEqs.

$$\begin{Bmatrix} \dot{q}_k(t) \\ \ddot{q}_k(t) \end{Bmatrix} = \begin{bmatrix} 0 & 1 \\ -\omega_k^2 & -2\zeta_k \omega_k \end{bmatrix} \begin{Bmatrix} q_k(t) \\ \dot{q}_k(t) \end{Bmatrix} + \begin{bmatrix} 0 & \cdots & 0 \\ b_{k1} & \cdots & b_{kN} \end{bmatrix} \begin{Bmatrix} u_1(t) \\ \vdots \\ u_N(t) \end{Bmatrix}. \tag{8}$$

In this formulation the modal displacements q_k and modal velocities $\dot{q}_k$ are collected in the modal state space vector $[x_k]$ and the eigenfrequencies and modal damping ratios are put into the system matrix $[A_k]$. The inputs are represented by the input matrix $[B_k]$ and the input vector $\{u_k\}$. The input matrix $[B_k]$ contains the modal transformation coefficients b_{kn} which are defined by Eq. (7).

Since Eq. (8) describes the dynamic of the structure for just one mode the complete system Eq. is obtained by superposition of all modes. In practice the number of considered modes is limited to a selected number K. The system equation for K modes can be written as

$$\{\dot{x}\} = [A]\{x\} + [B]\{u\}. \tag{9}$$

There are different types of compositions of the modal state vector presented in the literature, e.g Gawronski (1998). In this paper the modal state vector is arranged as follows

$$\{x\}^T = \{\, q_1 \cdots q_K \; \dot{q}_1 \cdots \dot{q}_K \,\}^T , \tag{10}$$

which leads to a real valued system matrix.

1.2 Output equation

Active structures are equipped with sensors. Therefore the system Eq. (9) has to be supplemented by an output Eq. which describes the sensor outputs due to the current state of the structure. The sensor signals are collected into the output vector $\{y\}$ and are related to the modal state vector $\{x\}$ by the output matrix $[C]$

$$\{y\} = [C]\,\{x\}. \tag{11}$$

For a number of N sensors and K modes the size of the output matrix is $2K$xN what can be seen from the following Eq.

$$\begin{Bmatrix} y_1(t) \\ \vdots \\ y_N(t) \end{Bmatrix} = \begin{bmatrix} c_{q11} & \cdots & c_{q1K} & c_{v11} & \cdots & c_{v1K} \\ \vdots & \ddots & \vdots & \vdots & \ddots & \vdots \\ c_{qN1} & \cdots & c_{qNK} & c_{vN1} & \cdots & c_{vNK} \end{bmatrix} \begin{Bmatrix} q_1(t) \\ \vdots \\ q_K(t) \\ \dot{q}_1(t) \\ \vdots \\ \dot{q}_K(t) \end{Bmatrix}. \tag{12}$$

The indices q and v used for the elements of the output matrix indicate their relation to modal displacements and accordingly to modal velocities. Sometimes it is advantageous to formulate the output Eq. in form of a sum

$$\{y\} = [C_q]\,\{x_q\} + [C_v]\,\{x_v\}. \tag{13}$$

This formulation requires a separation of the state space vector into displacements and velocities. Due to this separation the output matrix is devided into the matrices $[C_q]$ and $[C_v]$.

$$\begin{Bmatrix} y_1(t) \\ \vdots \\ y_N(t) \end{Bmatrix} = \begin{bmatrix} c_{q11} & \cdots & c_{q1K} \\ \vdots & \ddots & \vdots \\ c_{qN1} & \cdots & c_{qNK} \end{bmatrix} \begin{Bmatrix} q_1(t) \\ \vdots \\ q_K(t) \end{Bmatrix} + \begin{bmatrix} c_{v11} & \cdots & c_{v1K} \\ \vdots & \ddots & \vdots \\ c_{vN1} & \cdots & c_{vNK} \end{bmatrix} \begin{Bmatrix} \dot{q}_1(t) \\ \vdots \\ \dot{q}_K(t) \end{Bmatrix} \tag{14}$$

2 Norms

For the investigation of the sensor placement and the actuator layout a normalization of the input matrix $[B]$ and the output matrix $[C]$ is required. In this paper the H_2 and the H_∞ norms are used to normalize the matrices.

Since the dynamic of the structure is described by the matrices $[A]$, $[B]$ and $[C]$ the transfer function of the structure $[G(\omega)]$ is given as

$$[G(\omega)] = [C]\,(j\omega\,[I] - [A])^{-1}\,[B] \tag{15}$$

and is used to define the H_2 norm of the structure

$$\|\,[G]\,\|_2 = \sqrt{\frac{1}{2\pi}\int_{-\infty}^{\infty} tr\left([G^*(\omega)]\,[G(\omega)]\right)d\omega} \tag{16}$$

and the H_∞ norm of the structure

$$\|\,[G]\,\|_\infty = \max\,\sigma_{max}\left([G(\omega)]\right), \tag{17}$$

where σ_{max} is the largest singular value of $[G(\omega)]$. According to Eq. (16) and (17) $\|\,[G]\,\|_2$ can be interpreted as the area enclosed by the transfer function and $\|\,[G]\,\|_\infty$ is the value of largest peak of the transfer function. A convenient way to determine the numerical values of the both norms is given by

$$\|\,[G]\,\|_2 \cong \sqrt{\textstyle\sum_{k=1}^{K}\|\,[G_k]\,\|_2^2}, \quad \text{with} \quad \|\,[G_k]\,\|_2 \cong \frac{\|\,[B_k]\,\|_2\;\;\|\,[C_k]\,\|_2}{2\sqrt{\zeta_k\omega_k}}, \tag{18}$$

$$\|\,[G]\,\|_\infty \cong \max_k \|\,[G_k]\,\|_\infty \quad \text{with} \quad \|\,[G_k]\,\|_\infty \cong \frac{\|\,[B_k]\,\|_2\;\;\|\,[C_k]\,\|_2}{2\zeta_k\omega_k}, \tag{19}$$

where $\|\,[G_k]\,\|_2$ and $\|\,[G_k]\,\|_\infty$ are the modal norms of the mode k.

3 Design for distributed actuators

3.1 Input matrix

For the design process of distributed PVDF actuators the input matrix $[B]$ is needed. The n-th actuator, which is bonded to the surface of the structure, generates the bending moments m_{nx} and m_{ny} along the x- and y-axis

$$m_{nx} = BC_{nx}\left(H(x - x_{n1}) - H(x - x_{n2})\right)\left(H(y - y_{n1}) - H(y - y_{n2})\right), \tag{20}$$

$$m_{ny} = BC_{ny}\left(H(x - x_{n1}) - H(x - x_{n2})\right)\left(H(y - y_{n1}) - H(y - y_{n2})\right). \tag{21}$$

The constants C_{nx} and C_{ny} represent the gradient of the actuator induced strain. The calculation of these constants is described in Fuller et al. (1996) and will not be discussed in this paper. $H(\cdot)$ is the Heaviside step function and $x_{n1}, x_{n2}, y_{n1}, y_{n2}$ are the cartesian coordinates of the actuator boundaries. In order to replace the line moments by equivalent loads the following relation is used

$$\bar{u}_n(x, y) = -\left(\frac{\partial^2 m_{nx}(x, y)}{\partial x^2} + \frac{\partial^2 m_{ny}(x, y)}{\partial y^2}\right). \tag{22}$$

The physical input $\bar{u}_n(x, y)$ is transformed into a modal input by applying Eq. (7). The evaluation of the integral of the numerator of Eq. (7) results to

$$\int_a \int_b \bar{u}_{n(x,y)} \phi_{k(x,y)} dx dy =$$

$$-B\left(C_{nx} \int_{y_{n1}}^{y_{n2}} \phi_{2k}(y) dy \left[\frac{d\phi_{1k}(x)}{dx}\right]_{x_{n1}}^{x_{n2}} + C_{ny} \int_{x_{n1}}^{x_{n2}} \phi_{1k}(x) dx \left[\frac{d\phi_{2k}(y)}{dy}\right]_{y_{n1}}^{y_{n2}}\right) \tag{23}$$

Finally the input matrix is created by calculating the modal input factors of each actuator and each mode. The input factors of an actuator are collected into a column and the input factors of a row are related to a mode. It has to be mentioned that the required mode shapes can be determined analytically or numerically Stöbener and Gaul (1999) or by an experimental modal analysis Stöbener and Gaul (2000).

3.2 Actuator placement indices matrix

The placement index indicates the controllability of the n-th actuator with respect to the k-th mode and it is defined as

$$\sigma_{2_{nk}} = \frac{\| [G_{kn}] \|_2}{\| [G] \|_2} \tag{24}$$

by using the H_2 norm and

$$\sigma_{\infty_{nk}} = \frac{\| [G_{kn}] \|_\infty}{\| [G] \|_\infty} \tag{25}$$

for the H_∞ norm Gawronski (1998). The actuator norms $\| [G_{kn}] \|_2$ and $\| [G_{kn}] \|_\infty$ are calculated by the following formulas

$$\| [G_{kn}] \|_2 \cong \frac{\| [B_{kn}] \|_2 \ \| [C_k] \|_2}{2\sqrt{\zeta_k \omega_k}}, \tag{26}$$

$$\| [G_{kn}] \|_\infty \cong \frac{\| [B_{kn}] \|_2 \ \| [C_k] \|_2}{2\zeta_k \omega_k}. \tag{27}$$

The modal norms defined in Eq. (18), (19) are the root-mean-square (rms) sums of the individual actuator norms. The placement indices are arranged into the placement indices matrices $[\Sigma]_2$ and $[\Sigma]_\infty$.

$$[\Sigma]_2 = \begin{bmatrix} \sigma_{2_{11}} & \cdots & \sigma_{2_{1N}} \\ \vdots & \ddots & \vdots \\ \sigma_{2_{K1}} & \cdots & \sigma_{2_{KN}} \end{bmatrix}, \quad [\Sigma]_\infty = \begin{bmatrix} \sigma_{\infty_{11}} & \cdots & \sigma_{\infty_{1N}} \\ \vdots & \ddots & \vdots \\ \sigma_{\infty_{K1}} & \cdots & \sigma_{\infty_{KN}} \end{bmatrix}. \tag{28}$$

Each row of these matrices is related to a mode and each column corresponds to an actuator. Since the output matrix is included in the calculation of the modal norms (see Eq. (26)) the actuator placement indices are dependent from the sensors. Therefore the actuator performance is estimated with respect to the sensor performance. If the performance of the actuators is needed without including the sensors, then the input matrix has to be replaced by the unity matrix. The unity matrix does not influence the actuator norm calculated by Eq. (26) and emulates an ideal sensor array which enables the observation of all modes.

3.3 Control performance indices

The actuator placement indices are evaluated for each actuator and each mode. In order to determine the performance of all actuators together a control performance index is introduced. The calculation of the performance index starts with the summation of all N actuator placement indices for the k-th mode in the way of root mean square (rms) values for σ_{a2_k} or by choosing the largest actuator placement index related to the k-th mode for $\sigma_{a\infty_k}$

$$\sigma_{a2_k} = \sqrt{\sum_{n=1}^{N} \sigma_{2_{kn}}^2} \,, \tag{29}$$

$$\sigma_{a\infty_k} = \max_k (\sigma_{\infty_{kn}}) \,. \tag{30}$$

These values indicate the control performance of a selected number of actuators with respect to a selected mode and they are collected into the modal control performance matrices

$$[\Sigma]_{a2}^T = \left[\sigma_{a2_1} \cdots \sigma_{a2_K}\right]^T \,, \tag{31}$$

$$[\Sigma]_{a\infty}^T = \left[\sigma_{a\infty_1} \cdots \sigma_{a\infty_K}\right]^T \,. \tag{32}$$

To obtain a numerical expression for the control performance for all modes the values σ_{a2_k} or $\sigma_{a\infty_k}$ are summarized by the formulas

$$\alpha_{a2} = \frac{1}{K} \sum_{k=1}^{K} \sigma_{a2_k} \,, \tag{33}$$

$$\alpha_{a\infty} = \frac{1}{K} \sum_{k=1}^{K} \sigma_{a\infty_k} \,, \tag{34}$$

$$\beta_{a2} = \sqrt[K]{\prod_{k=1}^{K} \sigma_{a2_k}} \,, \tag{35}$$

$$\beta_{a\infty} = \sqrt[K]{\prod_{k=1}^{K} \sigma_{a\infty_k}} \,. \tag{36}$$

α_{a2} and $\alpha_{a\infty}$ are the so-called arithmetic mean (am) control performance indices of an actuator array and β_{a2}, $\beta_{a\infty}$ are the geometric mean (gm) control performance indices.

3.4 Design example

The design procedure is determined by the following steps:
1. Select the modes which shall be controlled.
2. Create an actuator array according to the shape of the lowest mode.

3. Calculate the control performance indices β_{a2} and $\beta_{a\infty}$.
4. If β_{a2} and $\beta_{a\infty}$ are equal to zero, then create an actuator array according to the shape of the next higher mode.
5. Calculate the control performance indices α_{a2} and $\alpha_{a\infty}$.
6. If α_{a2} and $\alpha_{a\infty}$ are lower than a certain limit, then create an actuator array according to the shape of the next higher mode and return to step 3.
7. Stop the design procedure if α_{a2} and $\alpha_{a\infty}$ are higher than the limit.

The mentioned limit for the control performance indices α_{a2} and $\alpha_{a\infty}$ has to be chosen from practical experience. Limits above 0.2 are recommended.

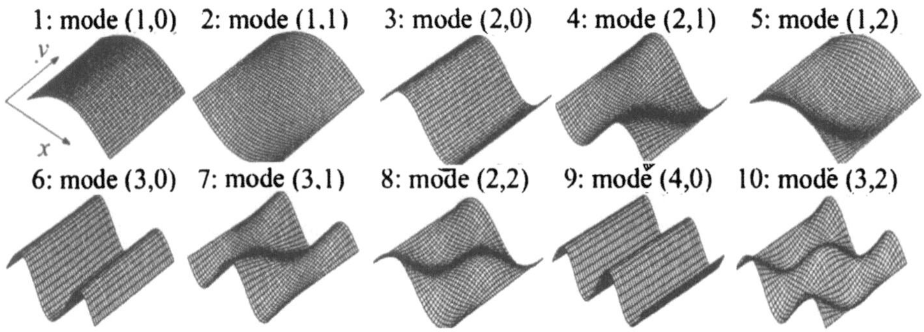

Figure1. Mode shapes of a rectangular plate

As a design example a rectangular plate with two parallel free edges and two parallel simply supported edges is used. The first ten mode shapes of this plate are depicted in Fig. 1. The objective of the vibration control is the reduction of the amplitude for these modes. Since PVDF actuators have to be implemented in large sheets the index calculation of various actuator arrays starts from the assumption that the total plate surface is covered by PVDF actuators. As depicted in Tab. 1 the segmentation of the actuator areas corresponds to the nodal lines of mode (1,0), (2,0), (1,1), (2,1), (3,0), (1,2) and (3,2). The control performance indices α_{a2}, $\alpha_{a\infty}$ and β_{a2}, $\beta_{a\infty}$ clarify that only the actuator configuration with nine segments leads to a controllability for all ten modes ($\beta_{a2} \neq 0$, $\beta_{a\infty} \neq 0$) and to the highest values for the arithmetic mean performance indices.

4 Placement of PVDF sensors

The optimal sensor placement depends in twofold respect on the output matrix $[C]$. On the one hand the magnitudes of the single elements indicate the sensitivity of the single sensors, on the other hand the condition indicates their correlation.

actuator segments	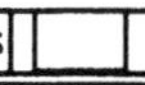	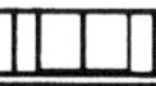	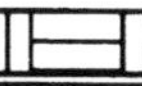	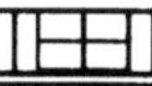		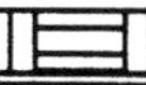	
α_{a2}	0.1492	0.1680	0.1982	0.2314	0.1968	0.2353	0.2823
$\alpha_{a\infty}$	0.1668	0.1269	0.2092	0.1933	0.2080	0.2737	0.2777
β_{a2}	0.0000	0.0000	0.0000	0.0000	0.0000	0.0000	0.2397
$\beta_{a\infty}$	0.0000	0.0000	0.0000	0.0000	0.0000	0.0000	0.2500

Table1. Control performance indices for selected actuator arrays

4.1 Calculation of the output matrix

Since this paper focuses on the concept of point sensors it is assumed that small quadratic PVDF sensor foils are used to measure velocity-proportional signals. After the determination of the mode shapes for the controlled modes the respective elements of the $[C]$-matrix can be calculated as follows

$$c_{v_{nk}} = -c_{ns} \frac{t_s}{\epsilon_{33}^S} \frac{(t_p + t_s)}{2A_s} \left(e_{31} \int_{x_{n1}}^{x_{n2}} \int_{y_{n1}}^{y_{n2}} \frac{\partial^2 \phi_k}{\partial x^2} dy dx + e_{32} \int_{x_{n1}}^{x_{n2}} \int_{y_{n1}}^{y_{n2}} \frac{\partial^2 \phi_k}{\partial y^2} dy dx \right), \quad (37)$$

where c_{ns} is the sensor amplification constant of the n-th sensor, t_s is the sensor thickness, t_p is the thickness of the structure A_s is the sensor area, ϵ_{33}^S is the dielectric constant, e_{31}, e_{32} are the piezoelectric charge constants and x_{n1}, x_{n2}, y_{n1} and y_{n2} are the coordinates of the sensor boundaries.

4.2 Sensitivity of single sensors

The first objective is to get preferably high elements of $[C]$. In Straßberger (1997) this is obtained by vivid placement of the sensors on the antinodes of the controlled modes. However this approach is limited to simple structures and a small number of modes. In more complicated cases a sensitivity performance index for each sensor arrangement has to be calculated. This can be carried out according to the control performance indices derived in Eq. (24) to (36). Where Eq. (26) and (27) change to

$$\| [G_{nk}] \|_2 \cong \frac{\|[B_n]\|_2 \ \|[C_{kn}]\|_2}{2\sqrt{\zeta_k \omega_k}} \quad (38)$$

$$\| [G_{nk}] \|_\infty \cong \frac{\|[B_n]\|_2 \ \|[C_{kn}]\|_2}{2\zeta_k \omega_k} \quad (39)$$

and for $[B_n]$ the unity matrix has to be inserted. The sensitivity indices α_{s2}, $\alpha_{s\infty}$, β_{s2} and $\beta_{s\infty}$ for the sensors are determined like the respective actuator performance indices.

4.3 Condition of the output matrix

The solid computational solubility of the equation system determined by $[C]$ depends primary on its so-called condition. As one uses usually the same kind of sensors and the same measurement circuit for each sensor and measures either the displacement or the velocity the $2K \times N$ – $[C]$–Matrix becomes a $N \times N$ quadratic $[C_q]$ or $[C_v]$ matrix as described in eq. (13) and (14). As in this paper it is assumed that the control input is the modal vector of velocities the $[C_v]$-matrix is used. The easiest way to solve the respective system of linear equations is given when the describing matrix is orthogonal. In the opposite case the matrix gets singular. If $[C_v]$ is near to the singularity the calculation of the desired modes will be impossible because the error caused by the measurement noise and calculation errors due to digitizing the analog measurement signals will lead to very unprecise results. Therefore a scalar quantity to classify the status of $[C_v]$ between orthogonality and singularity is needed. This quantity can be found in the so-called condition number of a matrix Zurmühl and Falk (1984). It can be defined as

$$cond([C_v]) = ||[C_v]|| \cdot ||[C_v]^{-1}||. \tag{40}$$

Where $||[C_v]||$ means one of the norms of the matrix explained in Zurmühl and Falk (1984). As problems can occur with the calculation of the inverse matrix as needed in (40) according to Mackiewicz et al. (1996) the condition number can alternatively be calculated as

$$cond([C_v]) = \frac{\sigma_1([C_v])}{\sigma_N([C_v])}, \tag{41}$$

where σ indicates in this case a singular value of the matrix $[C_v]$. The singular values are the eigenvalues of the matrix given by $[C_v]^T][C_v]$ listed by their size, i. e. the index 1 indicates the highest and the index N the lowest singular value. An orthogonal matrix has the condition number 1, an arbitrary matrix a number greater than 1, in case of singularity it gets infinite. From this it gets visible that an appropriate sensor design minimizes the condition number of the sensor matrix $[C_v]$.

In the presented example a measurability index structure has been chosen that includes the condition number according to (41) and one of the developed sensitivity indices α or β. The measurability index of the sensor array called i_s is

$$i_s = \frac{mag_s}{cond([C])}, \tag{42}$$

where mag_s is the magnitude index α or β. The index i_s ranges between 1 for the best possiple configuration and 0 for a defective one. The use of commercial available optimization routines is possible with the recipocal value of i_s, because then the optimization problem gets a minimization task of i_s in dependency of the sensor coordinates.

4.4 Sensor placement example

As an example the earlier mentioned rectangular plate is used. The objective will be to control the modes $(1,2)$, $(3,0)$, $(3,1)$, $(2,2)$, $(4,0)$, $(3,2)$, $(3,2)$ and $(4,1)$. Therefore the plate is equipped with nine PVDF-sensor patches. The patch size is $15 \times 15mm$ and the sensor thickness $110\mu m$. As a first placement approach three sensors at a time are placed on the antinode lines of the mode $(3,0)$ because these locations promise quite good signals from the most of the desired modes. As only seven modes have to be detected it is possible to combine seven sensors respectivly and thus compare different sensor combinations. In table (2) the results are listed in descendent order, i.

e. the first combination of sensors resulted in the best accordance between measurement and the analytical calculated expectancy (judged by the subjective impression as a rough criterion) and the last the worst. In this table the sensor placement index i_s is calculated with α_{s2} according

combination	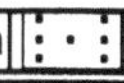	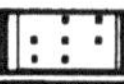				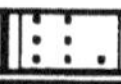		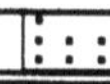
α_{s2}	0.337	0.338	0.334	0.336	0.335	0.336	0.332	0.338
$\alpha_{s\infty}$	0.274	0.274	0.269	0.274	0.269	0.274	0.250	0.274
β_{s2}	0.272	0.278	0.253	0.266	0.258	0.269	0.255	0.276
$\beta_{s\infty}$	0.233	0.233	0.211	0.233	0.211	0.233	0.204	0.233
$cond([C])$	10.64	9.242	40.71	44.32	39.17	44.14	$1.7 \cdot 10^{16}$	$2.5 \cdot 10^{5}$
i_s	0.032	0.037	0.008	0.008	0.009	0.008	$2.0 \cdot 10^{-17}$	$1.4 \cdot 10^{-6}$

Table2. Sensitivity indices for different sensor combinations

(42). It turns out that the sensitivity indices show admissible values while the influence of the condition number was not predictable at the first approach. The combination of both is necessary to indicate the performance of the array.

A procedure for the design of applicative sensor arrays contains the following steps:

1. State the mode shapes of the structure and the surface bending.
2. State the modes that have to be controlled. Their number equals the number of sensors.
3. Place the sensors near the bending antinodes of the mode shapes.
4. Calculate the output matrix and the measurability index.
5. If the index is to low go through condition number and magnitude indices.
6. If one of the β_s-indices is to low, one sensor is placed awkward. Go to the line of the output matrix with the lowest values and replace the referring sensor.
7. If the condition number is to high, go to the column with the lowest values and move a sensor to one of the antinodes of the referring mode, if in all lines and columns at least one tolerably high value occurs change one sensor arbitrarily.
8. Go with the new placement combination to step 4 until the measurability index fits the requirements.

5 Conclusions

In the present paper an approach to optimize the design of smart structures is shown. For better understanding it is clarified with an example of a plane plate. However this does not restrict the practicability of the methods to simple problems. Once the modal shape functions of a shell structure are known, no matter if from analytical, finite element or experimental way, the presented optimization methods can be applied.

References

Fuller, C., Elliott, S., and Nelson, P. (1996). *Active Control of Vibration*. London: Academic Press.

Gawronski, W. (1998). *Dynamics and Control of Structures*. Springer New York, Inc.

Mackiewicz, A., Holnicki-Szulc, J., and López-Almansa, F. (1996). Optimal sensor location in active control of flexible structures. *AIAA* 34(4).

Stöbener, U., and Gaul, L. (1999). Active control of plate vibration by discrete pvdf actuator and sensor segments. In Hagood, N., and Atalla, M., eds., *Ninth International Conference on Adaptive Structures and Technologies*, 349–358. Technomic Co., Inc.. Lancaster, PA 17601.

Stöbener, U., and Gaul, L. (2000). Active structural modal control for the floor panel of a car body. In *International Modal Analysis Conference-IMAC XVIII San Antonio*. Society of Experimental Mechanics. (to appear).

Straßberger, M. (1997). *Aktive Schallreduktion durch digitale Zustandregelung der Strukturschwingungen mit Hilfe piezo-keramischer Aktoren*. Ph.D. Dissertation, Ruhr-Universität Bochum.

Zurmühl, R., and Falk, S. (1984). *Matrizen und ihre Anwendungen Teil 1 und 2*. Springer Verlag.

5. Controller design for friction driven systems

R. Nitsche and L. Gaul

Institute A of Mechanics, University of Stuttgart
Pfaffenwaldring 9, D-70550 Stuttgart
e-mail: {nitsche, gaul}@mecha.uni-stuttgart.de

Abstract The design of a 'damping controller' for friction driven systems is introduced. A state variable bristle model, the *Lund-Grenoble* (LuGre) model is used to describe the friction behavior between the two sliding surfaces in the friction interface. The controller design is based on Lyapunov technique in order to maximize the energy dissipated in a friction interface in an instantaneous and local sense. The introduced control design method will be applied to a smart structure, namely an active friction joint. As the control law requires the knowledge of unmeasurable states, the design of an extended Kalman filter is introduced.

1 Introduction

There are many examples of friction-driven systems, where the transfer/dissipation of power between to systemcomponents is by a friction interface. Contact friction in tires is one well known example. The power is generated by the combustion driving the vehicle engine wheel axis. The tire/road contact friction transmitts part of the motor power to the vehicle body producing the requested motion.

In this paper friction dampers are introduced as an application, where it is desirable to maximize the energy dissipation in the friction interface.

Active control theory is used to design a controller which maximizes energy dissipation of the friction damper in an instantaneous and local sense. The concept is termed *semi-active control* because active control theory is used to vary the property of a passive element. The semi-active control design method is attributable to Karnopp and co-workers who first proposed varying the force of a viscous damper by controlling its orifice area Karnopp et al. (1974). The appeal of semi-active control is that performance levels rivaling fully active control can be achieved with a fraction of input power required of active control. Closed loop semi-active systems are guaranteed to be stable since the control system is 'passive'. As energy is always dissipated, the semi-active concept is more stable than a fully active one and is insensitive with respect to the spillover problem and sometimes more effective than a purely passive approach.

Several approaches has been taken to derive semi-active control laws which maximizes damper energy dissipation. One is for example to write a Lyapunov function representing the system energy. By inspection of its time derivative, one may arrive at a bang-bang control law which maximizes the damping contribution of the controlled term Nitsche and Gaul; Dupont and Stokes (1999; 1995). In this paper, a control design method based on a Lyapunov technique which is known as Jurdjevic-Quinn control Jurdjevic and Quinn (1978) is presented, where the design task can be solved in a systematic way. If only a partial measurement of the considered state is avaliable, as will often be the case, a state estimator is required for the implementation of the proposed state feedback. Therefore the design of an extended Kalman filter for state estimation is introduced. With an appropriate computer-algebra system (e. g. MATHEMATICA) the controller can easily be calculated and implemented on a real-time processor board. As an application of the

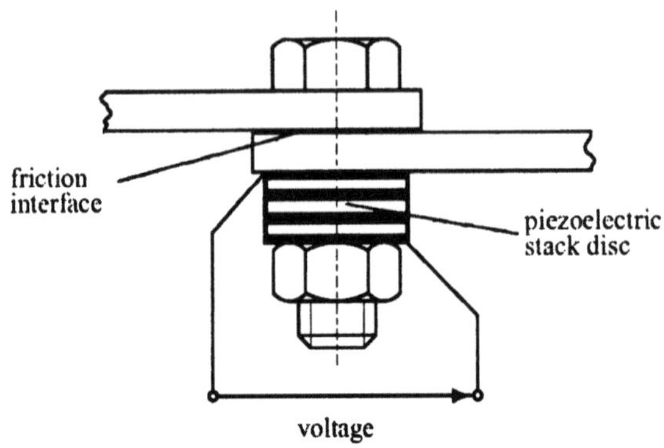

Figure1: Active friction joint.

introduced concept a simple bolted joint connection is turned into a semi-active friction damper using a piezoelectric stack disc as a washer (see Fig. 1). To test the efficiency of the proposed control design method, an experimental setup (Fig. 3) of the active joint connection will be introduced. It can be shown that the semi-actively controlled joint improve the damping performance significantly.

2 Models

The system dynamics is assumed to be linear with ℓ degrees of freedom while the friction in the m dampers are nonlinear. In what follows, the nonlinear friction model under consideration and the linear system dynamics will be introduced.

2.1 Friction modeling

Friction has often been modeled by an algebraic equation relating velocity and normal force to friction force. It is well known, however, that friction possesses dynamics associated with varying velocity.

A rather complete description of friction modeling and its impact on control appears in Armstrong-Hélouvry et al. (1994). The friction model under consideration is a modification of the so-called LuGre model Canudas de Wit and Olsson (1998), where the friction interface is thought of as a contact between bristles (Fig. 2). This model was designed to reproduce all observed friction phenomena over a wide range of operating conditions and was already used to describe the nonlinear transfer behavior of an active joint connection Gaul and Nitsche; Gaul and Nitsche (; 1999). The governing equations are

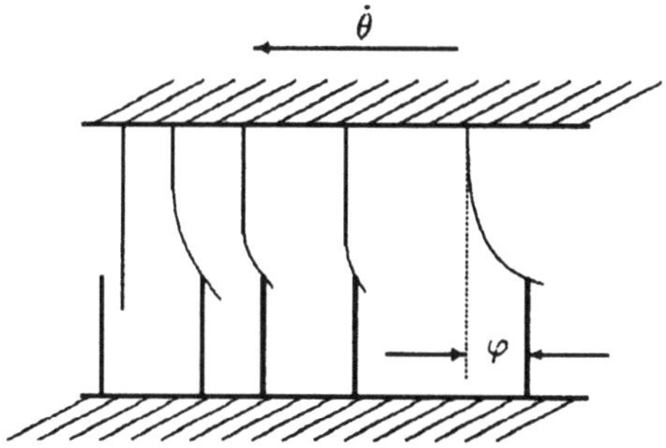

Figure2: Model of the friction interface.

$$\tau_f = F_N r(\sigma_0 \varphi + \sigma_1 \dot{\varphi} + \sigma_2 \dot{\theta}) =: F_N r \mu(\varphi, \dot{\varphi}, \dot{\theta}) \tag{1a}$$

$$S_\varphi : \quad \dot{\varphi} = \dot{\theta} - \frac{|\dot{\theta}|}{g(\dot{\theta})} \varphi, \quad \varphi(0) = \varphi_0 \tag{1b}$$

$$\sigma_0 \, g(\dot{\theta}) = M_c + M_\Delta \exp\left(-\left(\dot{\theta}/\dot{\theta}_s\right)^2\right) \tag{1c}$$

where the frictional moment is τ_f, and the relative sliding velocity at the friction interface is $\dot{\theta}$. The internal friction state φ represents the average deflection of the bristles. The dynamic of the friction internal state φ is represented by the subsystem S_φ, i. e., (1b), which describes the (relative) motion of the surfaces in contact during the stiction phase. The function ($\infty > M_c + M_\Delta \geq g(\dot{\theta}) \geq M_c > 0$) describes the part of the 'steady-state' characteristic of the model for constant velocity motions, including Stribeck velocity $\dot{\theta}_s$ [rad/s], static friction ($M_c + M_\Delta$) [Nm], and Coulomb friction M_c. The steady-state friction characteristics are given by

$$F_{ss} = \left[M_c + M_\Delta \exp\left(-(\dot{\theta}/\dot{\theta}_s)^2\right)\right] \operatorname{sgn}(\dot{\theta}) + \sigma_2 \dot{\theta} \tag{2}$$

where σ_2 [Nm s/rad] represents the viscous friction. The complete friction model is thus characterized by four 'static' parameters M_c, M_Δ, σ_2, $\dot{\theta}_s$, and two 'dynamic' parameters σ_0 and σ_1. The parameter σ_0 [Nm rad] can be understood as being a stiffness coefficient of the microscopic deformations of φ [rad] during the presliding displacement (microslip) and σ_1 [Nm s/rad] as being the damping coefficient associated with $\dot{\varphi}$. The variable μ, which is defined in (1a), can be interpreted as a state dependent friction coefficient. The LuGre friction model (1) allows analytical stydies (stability and existence of solution), generally not possible with discontinuous friction models. Moreover the model has several properties (c. f. Canudas de Wit (1998)) which are very suitable for control systems. A very usefull property which will be used later is the following:

Property 2.1 (Boundedness) *An intuitive property of the bristle model in Fig. 2 is that the deflection of the bristles should be finite:*

Assume that $0 < g(\dot\theta) \le a$. If $|\varphi(0)| \le a$ then $|\varphi(t)| \le a \quad \forall t \ge 0$.

Proof: Canudas de Wit and Olsson (1998)
Let $V = \varphi^2/2$, then the time derivative of V evaluated along the solution of S_φ, e. g. (1b), also called the *Lie derivative*, is

$$L_{S_\varphi} V = \frac{\mathrm{d}V}{\mathrm{d}t} = \varphi(\dot\theta - \frac{|\dot\theta|}{g(\dot\theta)}\varphi) = -|\dot\theta||\varphi|\left(\frac{|\varphi|}{g(\dot\theta)} - \mathrm{sgn}(\dot\theta)\,\mathrm{sgn}(\varphi)\right). \tag{3}$$

The Lie derivative $L_{S_\varphi}V$ is negative when $|\varphi| > g(\dot\theta)$. Since $g(\dot\theta)$ is strictly positive and bounded by a, it is easy to see that the set $\Omega = \{\varphi : |\varphi| \le a\}$ is an invariant set fot the solutions of (1b), i. e., all the solutions of $\varphi(t)$ starting in Ω remain there. □

Remark 2.2 (Lie Derivative) This paper does not require the formalism of differential geometry and employs the Lie derivative only for notational convenience: If $\boldsymbol{f} : \mathbb{R}^n \to \mathbb{R}^n$ is a vector field and $V : \mathbb{R}^n \to \mathbb{R}$ is a scalar function, the notation $L_f V$ is used for $\frac{\partial V}{\partial \boldsymbol{x}} \cdot \boldsymbol{f}(\boldsymbol{x})$. It is recursively extended to

$$L_f^k V(\boldsymbol{x}) = L_f(L_f^{k-1}V(\boldsymbol{x})), \;\; k > 1, \quad L_f^0 V(\boldsymbol{x}) = V(\boldsymbol{x}).$$

2.2 Mechanical models

The mechanical system is assumed to be linear while the friction dampers are nonlinear:

$$M\ddot{\boldsymbol{q}} + D\dot{\boldsymbol{q}} + K\boldsymbol{q} + \sum_{i=1}^{m} \tau_{\mathrm{f}i}\boldsymbol{c}_i = \boldsymbol{\tau}_{\mathrm{ext}}, \qquad \boldsymbol{q} \in \mathbb{R}^\ell \tag{4}$$

with mass matrix M, damping matrix D and stiffness matrix K. The frictional moment of the m friction dampers will be described by (1), and $\boldsymbol{\tau}_{\mathrm{ext}}$ is the vector of the disturbances. The vector $\boldsymbol{c}_i$ is defined such that $\boldsymbol{c}_i^\top \boldsymbol{q}$ is the relative displacement θ_i in the friction interface of the ith friction damper.

With the new state vector

$$\boldsymbol{x} = [\boldsymbol{q}, \dot{\boldsymbol{q}}, \boldsymbol{\varphi}]^\top, \quad \boldsymbol{x} \in \mathbb{R}^{(2\ell+m)} \tag{5}$$

and ignoring the forcing term, the system (1), (4) can be written as a control-affine system

$$\dot{\boldsymbol{x}} = \boldsymbol{f}(\boldsymbol{x}) + \sum_{i=1}^{m} u_i g_i(\boldsymbol{x}), \quad \boldsymbol{x} \in \mathbb{R}^n \tag{6}$$

while u_i is the normal force $F_{\mathrm{N},i}$ in the i-th friction interface and is the control input.

3 Controller design

Our intention is to design a 'damping controller' which additionally dissipates system energy due to friction in friction dampers (e. g. an active joint). In many applications the equilibrium $\boldsymbol{x} = \boldsymbol{0}$ of the uncontrolled part

$$\dot{\boldsymbol{x}} = \boldsymbol{f}(\boldsymbol{x}), \quad \boldsymbol{f}(\boldsymbol{0}) = \boldsymbol{0} \tag{7}$$

of the system (6) is stable and the task of the control is to provide additional damping which will render $\boldsymbol{x} = \mathbf{0}$ asymptotically stable. If a radially unbounded Lyapunov function $V(\boldsymbol{x})$ is known such that

$$L_f V(\boldsymbol{x}) \leq 0 \quad \forall \boldsymbol{x} \in I\!R^n \tag{8}$$

then it is tempting to employ $V(\boldsymbol{x})$ as a Lyapunov function for the whole system (6).

In view of $L_f V(\boldsymbol{x}) \leq 0$, the time-derivative of $V(\boldsymbol{x})$ for (6) satisfies

$$\dot{V}(\boldsymbol{x}) \leq L_{g_i} V(\boldsymbol{x}) u_i, \quad i = 1, \ldots, m. \tag{9}$$

This shows that $\dot{V}(\boldsymbol{x})$ can be made more negative with the control law

$$\boldsymbol{u}(\boldsymbol{x}) = -\boldsymbol{\kappa} \left(L_{g_1} V(\boldsymbol{x}), \ldots, L_{g_m} V(\boldsymbol{x}) \right)^{\mathsf{T}} \quad \forall \kappa_i > 0, \quad i =, \ldots, m. \tag{10}$$

Remark 3.1 *A proof that the feedback law (10) globally asymptotically stabilizes the sytem (6) can be found in Sontag (1998). The proof is a simple application of the LaSalle invariance principle.*

Remark 3.2 *The introduced controller design requires the calculation of the Lie derivative of the vector field (6) which can yield large expressions. For this, a* MATHEMATICA *toolbox for symbolic nonlinear controller and observer design is developed. With this toolbox it is easy to make the required analysis (e. g. (8)) and the calculation of the nonlinear controller (10). Furthermore, with the toolbox it is possible to generate software code which can be directly down loaded on a controller board for real-time simulations or controller implementation.*

In general only partial measurement of the required states for the feedback (10) are available. In the next section an extended Kalman filter will be introduce to estimate the states for the implementation of the state feedback.

4 Observer design: extended Kalman filter

The system under consideration will now be extended to a nonlinear *stochastic* differential equation:

$$\begin{aligned} \dot{\boldsymbol{x}} = \boldsymbol{f}(\boldsymbol{x}, \hat{\boldsymbol{u}}) + \boldsymbol{r}_1(t), \quad & \boldsymbol{x} \in I\!R^n, \hat{\boldsymbol{u}} = \boldsymbol{u}(\hat{\boldsymbol{x}}) \in I\!R^m, \\ & \hat{\boldsymbol{x}}_0 = E[\boldsymbol{x}_0], \boldsymbol{r}_1(t) \sim N(\mathbf{0}, R_1(t)) \end{aligned} \tag{11a}$$

$$\boldsymbol{\eta} = \boldsymbol{h}(\boldsymbol{x}) + \boldsymbol{r}_2(t), \quad \boldsymbol{\eta} \in I\!R^q, \boldsymbol{r}_2(t) \sim N(\mathbf{0}, R_2(t)) \tag{11b}$$

The function $\boldsymbol{r}_1(t)$ and $\boldsymbol{r}_2(t)$ describes the system and measurement noise respectively and are assumed to be unbiased, gaussian (normal) random vectors with covariance $R_1(t)$ and $R_2(t)$ respectively. The covariance matrices have the following property

$$E\left[(\hat{\boldsymbol{x}}(t) - \boldsymbol{x}(t))(\hat{\boldsymbol{x}}(t) - \boldsymbol{x}(t))^{\mathsf{T}} \right] = P_0 \tag{12a}$$

$$E[\boldsymbol{r}_1(t) \boldsymbol{r}_2^{\mathsf{T}}(t)] = 0 \tag{12b}$$

where $E[\boldsymbol{\xi}]$ is the expection of $\boldsymbol{\xi}$.

4.1 Extended Kalman filter

For the system (11), (12) an *extended Kalman filter* with time varying gains can be designed c. f. Gelb (1974):

$$\dot{\hat{\boldsymbol{x}}} = \boldsymbol{f}(\hat{\boldsymbol{x}}, \hat{\boldsymbol{u}}) + K(t)\,(\boldsymbol{\eta} - \boldsymbol{h}(\hat{\boldsymbol{x}}))\,, \quad \hat{\boldsymbol{x}}(0) = \hat{\boldsymbol{x}}_0 \tag{13a}$$

$$\dot{P}(t) = A(t)P(t) + P(t)A^{\top}(t) + R_1(t) - P(t)C^{\top}(t)R_2^{-1}(t)C(t)P(t), \quad P(0) = P_0 \tag{13b}$$

$$K(t) = P(t)C^{\top}(t)R_2^{-1}(t). \tag{13c}$$

The matrices $A(t)$, $C(t)$ are calculated linearizing about the reconstructed trajectory $\hat{\boldsymbol{x}}(t)$:

$$A(t) = \left.\frac{\partial \boldsymbol{f}(\boldsymbol{x}, \hat{\boldsymbol{u}})}{\partial \boldsymbol{x}}\right|_{\hat{\boldsymbol{x}}(t)}, \quad C(t) = \left.\frac{\partial \boldsymbol{h}(\boldsymbol{x})}{\partial \boldsymbol{x}}\right|_{\hat{\boldsymbol{x}}(t)}. \tag{14}$$

Remark 4.1 *For the implementation of the proposed Kalman filter the matrix Riccati differential equation (13b) has to be solved on-line. If the system is situated in the neighborhood of a stationary solution, a* stationary *extended Kalman filter algorithm can be used, which will be outlined in the following.*

4.2 Stationary extended Kalman filter

For the stationary extended Kalman filter the gain matrix K is constant and can be calculated be solving a stationary matrix Riccati equation. The filter equations are:

$$\dot{\hat{\boldsymbol{x}}} = \boldsymbol{f}(\hat{\boldsymbol{x}}, \hat{\boldsymbol{u}}) + K\,(\boldsymbol{\eta} - \boldsymbol{h}(\hat{\boldsymbol{x}}))\,, \quad \hat{\boldsymbol{x}}(0) = \hat{\boldsymbol{x}}_0 \tag{15a}$$

$$\boldsymbol{0} = AP + PA^{\top} + R_1 - PC^{\top}R_2^{-1}CP, \tag{15b}$$

$$A = \left.\frac{\partial \boldsymbol{f}(\boldsymbol{x}, \boldsymbol{u})}{\partial \boldsymbol{x}}\right|_{\boldsymbol{x}_e, \boldsymbol{u}_e}, \quad C = \left.\frac{\partial \boldsymbol{h}(\boldsymbol{x})}{\partial \boldsymbol{x}}\right|_{\boldsymbol{x}_e} \tag{15c}$$

$$K = PC^{\top}R_2^{-1}. \tag{15d}$$

Remark 4.2 *The solution of the stationary matrix Riccati equation (15b) can only be solved numericaly. Therefore an interface from the* MATHEMATICA *toolbox to the numerical tool* MATLAB *is designed to solve (15b) in the observer design process. The constant filter gains (15d) can be precalculated i. e., off-line, in opposite to the time varying gains (13c) which has to be calculated on-line. Thus the stationary Kalman filter is more suitable for a real-time application.*

5 Application to a Smart Structure: An Active Joint Connection

To test the efficiency of the proposed control law experimentally, an active joint was designed as shown in Fig. 3. The friction interface comes up in the contact area between the inner ① and the outer ② part of the joint. The inner part is attached to a shaker which excites the joint with an external force f_{ext}. The normal force at the friciton interface is actively controlled using a piezoelectric stack actuator (PZT). The displacement is measured very closely to

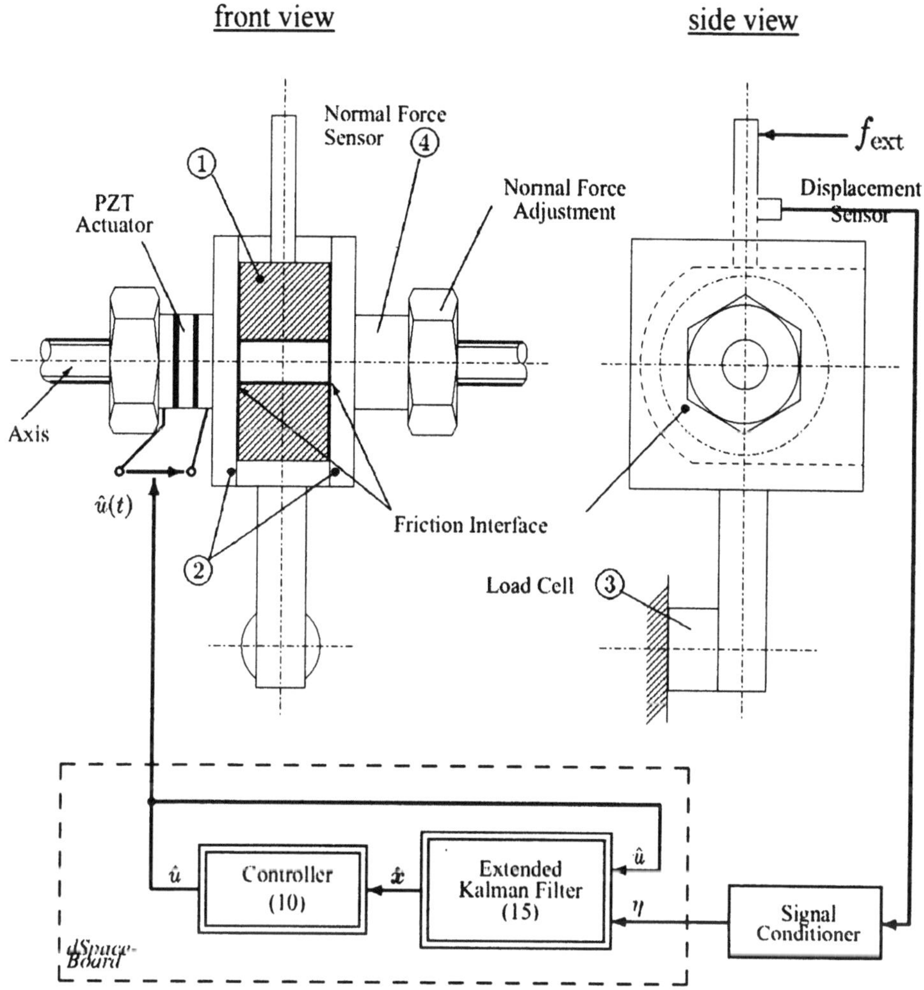

Figure3: Scheme of the experimental setup of the active joint.

the friction interface, therefore most elastic deformations are excluded. By regarding only small displacements and due to the boundedness of the britstles (Property 2.1) the requirements for a *stationary* extended Kalman Filter are met.

The normal force controller (10) and the stationary extended Kalman Filter (15) are realized on a DSPACE digital signal processor board. A nominal normal force can be applied to the joint using the normal force adjustment mechanism (Fig. 3). The active rotary joint can be modelled with a single degree of freedom system:

$$J\ddot{\theta} + d\dot{\theta} + c\theta + \overbrace{\tau_{\rm f}}^{=(1)} = r f_{\rm ext} \tag{16a}$$

with bounded input $F_{\rm N}^{\rm min} \leq u \leq F_{\rm N}^{\rm max}$ and the measurement

$$\eta = \theta. \tag{16b}$$

Determining the parameters of the model (16), (1) is a matter of nonlinear system identification and would be prolix and thus cannot be done here.

6 Controller and observer implementation – experimental results

For mechnical systems the total kinetic and potential energy of the system is *often* a suitable Lyapunov function candidate. For the active joint connection (16) a suitable Lyapunov function is therefore

$$V = \frac{1}{2}c\dot{\theta}^2 + \frac{1}{2}J\theta^2. \tag{17}$$

The control law (10) using (17) was implemented on a DSPACE controller board to maximize the energy dissipation in the active friction joint (Fig. 3). For the estimation of the state $\boldsymbol{x} = [\theta, \dot{\theta}, \varphi]^{\rm T}$ from the signal of the displacement sensor, the introduced algorithm of a stationary extended Kalman filter (15) is used (see Fig. 3).

In Fig. 4 the experimental results of the active joint connection are shown, where the system is excited by an sinusoidal energy input with a frequency of 10 Hz. The friction moment $\tau_{\rm f}$ can be obtained from the measured force with the sensor ③ (Fig. 3) multiplied with the distance of the load cell to the axis. The friction moment $\hat{\tau}_{\rm f}$ estimated from the Kalman filter shows a good agreement with the measured friction moment $\tau_{\rm f}$ as can be seen in Fig. 4(a). Note that a sticking joint dissipates almost no energy, therefore the controller prevents the joint from being stuck to maximize the energy dissipation. The controller renders the active joint sliding which is indicated by the flat regions in the time evolution of $\hat{\mu}(t)$ in Fig. 4(b). The control law is *smooth* which can be seen regarding the time evolution of the control input i. e., the normal force $\hat{u}(t)$ (Fig. 4(c)). The normal force was measured using an appropriate force sensor ④ as can be seen in Fig. 3.

In Fig. 5 the hysteresis loops of a *passive* joint with $u = F_{\rm N}^{\rm min}$ and $u = F_{\rm N}^{\rm max}$ which corresponds to a *constant* normal force of 310 N and 690 N respectively, are shown. The relative displacement θ of the passive joint with $u = F_{\rm N}^{\rm min}$ is rather large compared to the case where $u = F_{\rm N}^{\rm min}$. But the friction moment $\tau_{\rm f}$ is much higher if the normal force is at it's maximum. The energy $W_{\rm d}$ dissipated per cycle (area inside the hysteresis loop) is $W_{\rm d} = 38.9\,{\rm N\,m\,mrad}$ if $u = F_{\rm N}^{\rm min}$ and $W_{\rm d} = 1.6\,{\rm N\,m\,mrad}$ if $u = F_{\rm N}^{\rm max}$. If the joint becomes active by controlling the normal force with (10), (15), the energy dissipated per cycle can significantly increases up to $W_{\rm d} = 68.24\,{\rm N\,m\,mrad}$, which corresponds to an improvement of 75.4%. Roughly speaking the controller works as follows: the normal force in the friction interface is low if the relative velocity $\dot{\theta}$ is low too, therefore sticking can be excluded and rather large displacements θ occur. For high velocities $\dot{\theta}$ in the friction interface the friction moment $\tau_{\rm f}$ increases while the normale

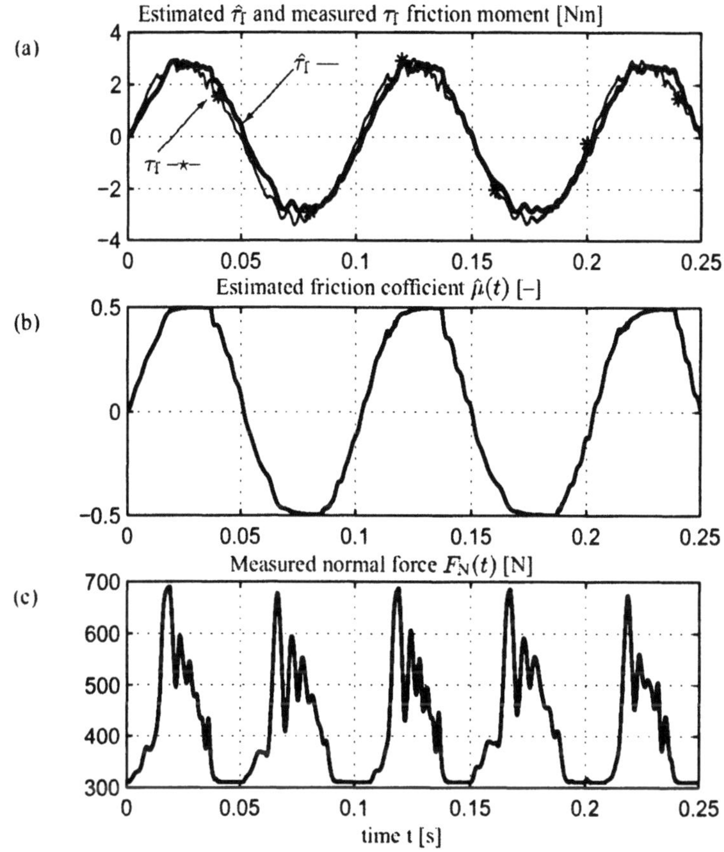

Figure4: Experimental results: time evolution of the (a) measured friction moment τ_f (-*-) and estimated friction moment $\hat{\tau}_f$ (—); (b) estimated state dependend friction coefficient $\hat{\mu}(t)$; (c) measured normal force $F_N(t)$ in the friction interface.

force increases. Note that the stored potential energy in the bristles is analogous to elastic strain energy. To maximize energy dissipation, the controller minimizes the friction moment (e. g the normal force) while energy is returned to the system.

7 Conclusions

A semi-active control design method for friction driven systems has been presented. The friction interface was modelled with a dynamic friction model, the so-called LuGre model. A smooth control law which maximizes the energy dissipation due to friction has been designed. A state estimator, namely an extended Kalman filter has been presented to implement the proposed state feedback. As an application of the semi-active method, a bolted joint connection has been

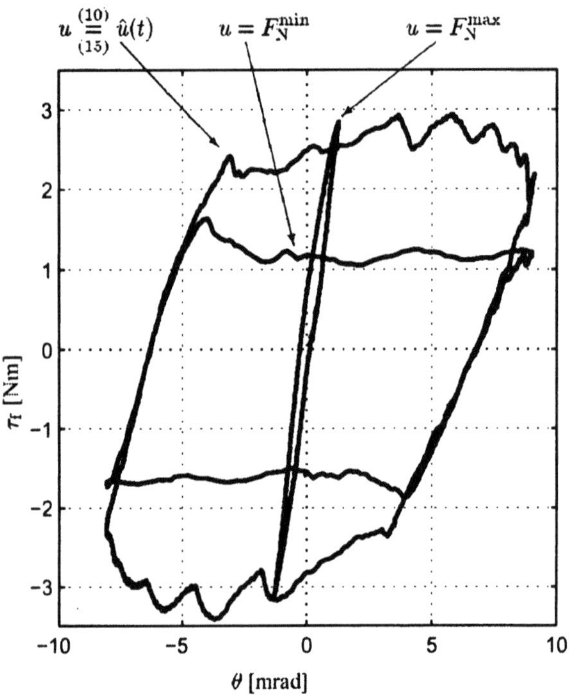

Figure5: Hysteresis loops for the passive and active joint.

changed into an active friction joint by inserting a piezoelectric stack disc as a washer, to control normal force and thus the energy dissipation. The presented method has been applied to the active joint to maximize the energy dissipation instantaneously. The experimental results showed the efficiency of the proposed controller and observer. The activeley controlled joint improves the energy dissipation up to 75.4% compared to a passive joint with a constant normal force.

A MATHEMATICA toolbox for symbolic nonlinear controller and observer design has been developed. With this toolbox it is possible to make the required analysis and calculations for the nonlinear controller and observer design. Furthermore, it is possible to generate software code which can be directly down loaded on a controller board for real-time simulations or controller and observer implementation.

References

Armstrong-Hélouvry, B., Dupont, P., and Canudas de Wit, C. (1994). A survey of models, analysis tools and compensation methods for the control of machines with friction. *Automatica* 30(7):1083–1138.

Canudas de Wit, C., and Olsson, H. (1998). A new model for control of systems with friction. *IEEE Trans. Automat. Control* 40(3):419–425.

Canudas de Wit, C. (1998). Comments on: A new model for control of systems with friction. *IEEE Trans. Automat. Control* 43(8):1189–1190.

Dupont, P., and Stokes, A. (1995). Semi-active control of friction dampers. In *Proceedings of the 34th Conference on Decision and Control*, 3331–3336.

Gaul, L., and Nitsche, R. Friction control for vibration suppression. accepted for publication in Mechanical Systems and Signal Processing.

Gaul, L., and Nitsche, R. (1999). *Computational Methods in Contact Mechanics IV*. Southampton, Boston: WIT Press. 369–378.

Gelb, A. (1974). *Applied Optimal Estimation*. M.I.T. Press.

Jurdjevic, V., and Quinn, J. (1978). Controllability and stability. *Journal of Differential Equations* 28:381–389.

Karnopp, D., Crosby, M., and Harwood, R. (1974). Vibration control using semi-active force generators. *ASME Journal of Engineering for Industry* 96(2):619–626.

Nitsche, R., and Gaul, L. (1999). Vibration control using semi-active friction damping. In *CDROM-Proc. 5th European Control Conference ECC '99*. number CM-4.

Sontag, E. (1998). *Mathematical Control Theory: Deterministic Finite Dimensional Systems*. Number 6 in Texts in Applied Mathematics. Springer-Verlag, 2nd edition.

6. Piezoelectric analysis with FEM and BEM

M. Kögl and L. Gaul

Institute A of Mechanics, University of Stuttgart
Pfaffenwaldring 9, D-70550 Stuttgart
EMail: {M.Koegl, L.Gaul}@mecha.uni-stuttgart.de

Abstract This paper gives an introduction to the numerical solution of dynamic piezoelectric problems using Finite Element and Boundary Element Methods. A comparative approach explains the similarities and differences, as well as advantages and disadvantages of the respective methods. The basic equations of piezoelectricity are summed up, and the Finite Element and Boundary Element Method are formulated for piezoelectric continua. By means of a numerical example, both methods are compared for the case of three-dimensional transient piezoelectricity.

1 Introduction

Since the late 1980s, the importance of piezoelectric materials has steadily increased by a broadening of the range of applications. New fields of research are emerging, as e.g. on smart structures, in which piezoelectric materials are used for sensing and actuation. For this reason, reliable numerical methods are needed to accurately model and solve engineering problems involving piezoelectrics. Besides the Finite Element Method (FEM), Boundary Element Methods (BEM) are a very powerful tool for the numerical solution of field problems of mathematical physics, since they offer some inherent advantages over FEM, like the discretization of the boundary only, as well as an improved accuracy in stress calculations. In this article, both methods are introduced in a comparative manner, and some of their advantages and disadvantages are pointed out. After a general introduction, the solution of dynamic piezoelectric field problems with the two methods is described.

2 Finite element and boundary element methods

This section gives a brief comparison of the Finite Element and Boundary Element Methods. For a more detailed introduction into FEM, the interested reader is referred to the books of Bathe (1996) and Zienkiewicz and Taylor; Zienkiewicz and Taylor (1994; 1991). The BEM is described in detail in Brebbia et al. (1984) and Gaul and Fiedler (1996).

2.1 General features

The most striking difference between FEM and BEM, and one of the important advantages of the latter, concerns the discretization. While in FEM the complete domain has to be discretized, in BEM only the discretization of the boundary is required, as seen in Figure 1. Depending on the complexity of the actual structure and load case under investigation, this simplified discretization can lead to important time savings in the mesh creation and modification process. Other areas in which the BEM possesses certain advantages are problems involving infinite or semi-infinite domains (e.g. acoustics, soil-structure interaction, etc.), as well as stress concentration problems. This is partly due to the use of fundamental solutions, which are *analytical* free space solutions of the governing differential field equations, and can therefore accurately represent far-fields and stresses.

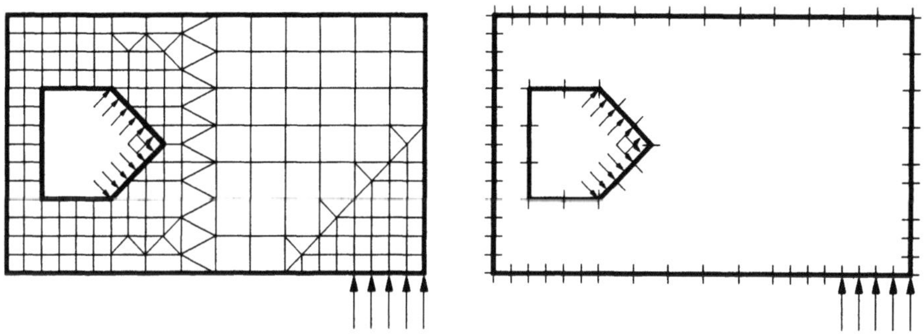

Figure1. Discretization with finite elements (left) and boundary elements (right)

On the other hand, the fact that a fundamental solution is required poses some problems e.g. for inhomogeneous and non-linear analyses, where special methods such as the Dual Reciprocity Method (DRM)Partridge et al. (1992) have to be applied. Since the FEM does not rely on fundamental solutions, it has less difficulties in handling these kinds of problems. Also, in structural mechanics, Finite Element Methods have the advantage of leading to symmetric, positive definite, and sparsely populated matrices, so that special time-saving solvers can be used, whereas the BEM-matrices are usually unsymmetric and fully populated.

2.2 Comparison of FE and BE formulations

This section aims at demonstrating the similarities and differences between the methods by introducing them in a comparative manner, as shown in Figure 2. This is done for the case of linear elasticity without body forces, in order to concentrate on the essentials. The starting point for both formulations is a weighted residual statement, in which the governing differential equation and boundary conditions are weighted with test functions w. Integration by parts leads to the weak statement, which is the basis of the FEM; further integration by parts results in the inverse statement, which forms the basis of the BEM.

The next steps show some significant differences between the methods. The standard FEM as derived here is a pure displacement formulation, in which only the displacements appear as unknowns. These displacements are approximated by functions which have to fulfill the forced boundary conditions. The stresses and tractions are calculated only after the solution of the displacement field. This involves derivation of the shape functions, which results in a loss of accuracy for stress and traction results. On the other hand, the Boundary Element Method is a mixed

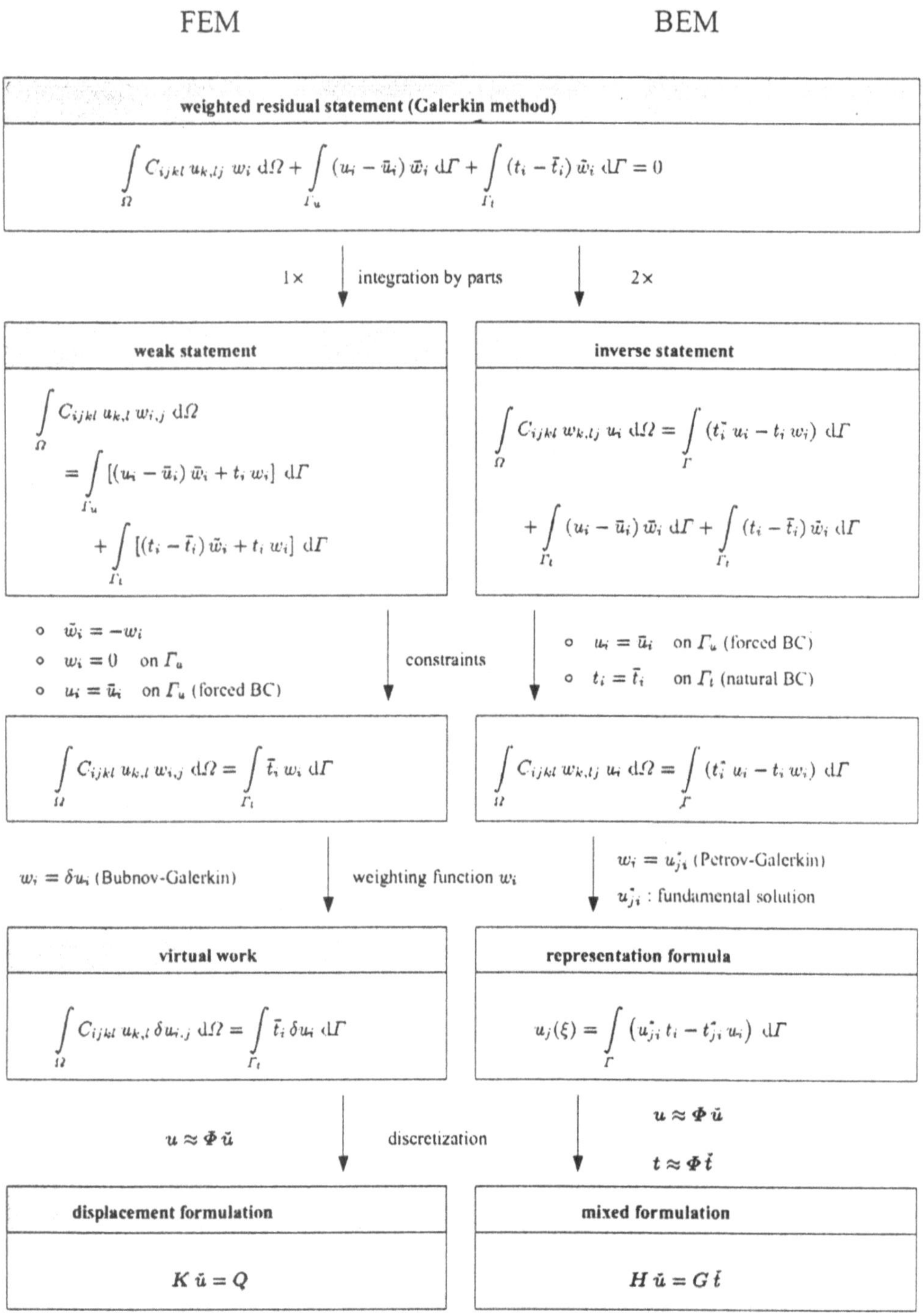

Figure2. Comparison of Finite Element Method and Boundary Element Method

formulation, containing both displacements and tractions as unknowns. Hence, apart from the displacements, the traction field, too, has to be approximated. The approximation functions now have to fulfill the natural boundary conditions. This is why the results for the tractions obtained with BEM are usually more accurate than those obtained with FEM.

Another difference between the methods lies in the choice of the weighting function. The classical FEM is a Bubnov-Galerkin method, using the same set of functions for the approximation of the displacement field as for the weighting functions. On the other hand, the BEM is a Petrov-Galerkin method, where the weighting functions are chosen as the fundamental solutions of the governing differential equations. This leads to an elimination of the domain integral by virtue of the sifting property of the Dirac impulse, which results in a pure boundary formulation for the unknown displacement and traction fields.

3 Basic equations of piezoelectricity

The basic equations of linear elastodynamics are the well-known balance of momentum and moment of momentum respectively

$$\sigma_{ij,i} = \rho \ddot{u}_j \,, \qquad \sigma_{ij} = \sigma_{ji} \,, \tag{1}$$

which describe the dynamics of an elastic body. The infinitesimal strain tensor

$$\varepsilon_{kl} = \frac{1}{2}\left(u_{k,l} + u_{l,k}\right) \,, \tag{2}$$

describes the kinematics of deformation, assuming sufficiently small displacements and displacement gradients. The physical quantities introduced in Eqns (1) and (2) are the Cauchy stress tensor σ_{ij}, elastic displacements u_j, elastic strain tensor ε_{kl}, and mass density ρ. Here and in the following, Einstein's summation convention is applied, the comma denotes partial differentiation with respect to the spatial coordinates, and the superimposed dot denotes partial differentiation with respect to time.

In piezoelectric materials the propagation speed of the electromagnetic waves is several orders of magnitude higher than the propagation speed of the elastic waves. Also, the interaction between the elastic and the magnetic field is usually neglected. Therefore, in piezoelectric calculations, the electric field can be considered as quasi-static. For dielectric materials, which do not contain free electric charges, this is described by the simplified Maxwell's equations

$$\epsilon_{ijk} E_{k,j} = 0 \,, \tag{3}$$

$$D_{i,i} = 0 \,, \tag{4}$$

where E_k is the electric field, D_i is the electric displacement, and ϵ_{ijk} is the permutation tensor. From Eqn (3) it follows that the electrostatic field is irrotational, so that an electric potential φ can be introduced such that

$$E_l = -\varphi_{,l} \,. \tag{5}$$

In piezoelectric materials the elastic and electric fields are coupled, which is described by the linear constitutive equationsNye; Tiersten (1985; 1969)

$$\sigma_{ij} = C_{ijkl}\, \varepsilon_{kl} - e_{lij}\, E_l \,, \tag{6}$$

$$D_i = e_{ikl}\, \varepsilon_{kl} + \epsilon_{il}\, E_l \,, \tag{7}$$

with the elasticity tensor C_{ijkl}, measured at constant electric field, piezoelectric tensor e_{lij}, and permittivity tensor ϵ_{il}, measured at constant strain. The material tensors exhibit the symmetries

$$C_{ijkl} = C_{jikl} = C_{klij} \,, \qquad e_{lij} = e_{lji} \,, \qquad \epsilon_{il} = \epsilon_{li} \,; \tag{8}$$

thus for general anisotropy they contain 21, 18, and 6 independent material constants respectively.

The coupled field equations of dynamic piezoelectricity can now be obtained from Eqns (1),(2), (5)-(7). They are

$$C_{ijkl}\, u_{k,li} + e_{lij}\, \varphi_{,li} = \rho \ddot{u}_j \,, \tag{9}$$

$$e_{ikl}\, u_{k,li} - \epsilon_{il}\, \varphi_{,li} = 0 \,. \tag{10}$$

The flux vectors are given by

$$t_j = \sigma_{ij}\, n_i \,, \qquad q = D_i\, n_i \,, \tag{11}$$

where t_j is the traction vector, q the charge flux density, and n_i denotes a unit normal vector, pointing in the direction in which the flux is evaluated.

Using a condensed notation, where lower case subscripts are ranging from 1 to 3, and upper case subscripts from 1 to 4, it is possible to combine corresponding elastic and electric quantities into generalized piezoelectric vectors and tensors, which simplifies the FEM and BEM formulations. These generalized tensors are defined as follows:

$$\Sigma_{iJ} := \begin{cases} \sigma_{ij} \,, & j = J = 1,2,3 \\ D_i \,; & J = 4 \end{cases} \,; \qquad Z_{Kl} := \begin{cases} \varepsilon_{kl} \,, & k = K = 1,2,3 \\ -E_l \,; & K = 4 \end{cases} \,, \tag{12}$$

$$U_K := \begin{cases} u_k \,, & k = K = 1,2,3 \\ \varphi \,; & K = 4 \end{cases} \,, \qquad T_J := \begin{cases} t_j \,; & j = J = 1,2,3 \\ q \,; & J = 4 \end{cases} \tag{13}$$

$$C_{iJKl} := \begin{cases} C_{ijkl} \,, & j = J = 1,2,3; \; k = K = 1,2,3 \\ e_{lij} \,, & j = J = 1,2,3; \; K = 4 \\ e_{ikl} \,, & J = 4; \qquad k = K = 1,2,3 \\ -\epsilon_{il} \,, & J = 4; \qquad K = 4 \end{cases} \tag{14}$$

In this condensed notation, the motion and electric field of a piezoelectric body can now be described by

$$\mathcal{L}_{JK} U_K(x_i) = \mathcal{B}_{JK} U_K(x_i) \,, \qquad x_i \in \Omega \subset I\!R^3 \tag{15}$$

with boundary conditions

$$U_J = \bar{U}_J \quad \text{on} \quad \Gamma_U \,, \tag{16}$$

$$T_J = \bar{T}_J \quad \text{on} \quad \Gamma_T \,, \tag{17}$$

and initial conditions

$$U_J(t=0) = U_J^0 \,, \tag{18}$$

$$\dot{U}_J(t=0) = \dot{U}_J^0 \,. \tag{19}$$

In Eqn (15), $\mathcal{L}_{JK} := C_{iJKl}\partial_i\partial_l$ is the elliptic operator of static piezoelectricity, and $\mathcal{B}_{JK} := \rho\tilde{\delta}_{JK}\,\partial_t^2$ is a differential operator, which describes the effects of inertia. A modified Kronecker delta has been introduced, accounting for the fact that the electric field is assumed to be quasi-static:

$$\tilde{\delta}_{JK} := \begin{cases} \delta_{jk}\,, & j = J = 1,2,3\,, \quad k = K = 1,2,3 \\ 0\,, & \text{otherwise} \end{cases} \tag{20}$$

4 FEM and BEM for piezoelectric continua

4.1 Finite Element Method

Finite Element Methods for dynamic piezoelectricity are already well established (see e.g. Allik and Hughes; Lerch (1970; 1990)). Therefore, only a brief review is given here, using the condensed notation. The starting point of the FE formulation is the following variational principleAllik and Hughes (1970):

$$\int_{\Omega} (\Sigma_{iJ}\,\delta Z_{Ji} + \rho\,\ddot{u}_i\,\delta u_i)\;\mathrm{d}\Omega = \int_{\Gamma_T} \bar{T}_I\,\delta U_I\;\mathrm{d}\Gamma\,. \tag{21}$$

In order to apply the FEM, the piezoelectric body is subdivided into a finite number N_E of domain elements, the so-called *finite elements*, as shown in Figure 1. These elements are mapped onto reference elements by means of a series of shape functions Φ^q and nodal coordinates $\check{x}_i^q$

$$x_i(\xi,\eta,\zeta) \approx \sum_{q=1}^{N} \Phi^q(\xi,\eta,\zeta)\,\check{x}_i^q\,, \tag{22}$$

where N is the number of nodes of the respective element, $\check{(\cdot)}$ denotes nodal values, and ξ, η, and ζ are the local coordinates in the reference element. The integrals in Eqn 21 can now be expressed by a sum of integrals over the elements.

In most Finite Element analyses, the isoparametric concept is used, which means that the field variables are approximated by the same shape functions Φ^q as used for the mapping of the elements:

$$U_I(\xi,\eta,\zeta) \approx \sum_{q=1}^{N} \Phi^q(\xi,\eta,\zeta)\,\check{U}_I^q\,. \tag{23}$$

The field variables are now replaced by these approximations, and by assembling the contributions of all finite elements one obtains a system of ordinary differential equations in time

$$\boldsymbol{M}\ddot{\check{\boldsymbol{U}}} + \boldsymbol{K}\check{\boldsymbol{U}} = \check{\boldsymbol{Q}}\,. \tag{24}$$

The mass matrix in Eqn 24 has the form

$$\boldsymbol{M} = \begin{bmatrix} \boldsymbol{M}_{uu} & \boldsymbol{0} \\ \boldsymbol{0} & \boldsymbol{0} \end{bmatrix}\,, \tag{25}$$

since the electric field is assumed to be quasi-static and thus does not contain any inertia effects.

4.2 Boundary Element Methods

While the FEM for dynamic piezoelectricity is already well established, a BEM has only recently been developed by the authorsKögl and Gaul; Kögl and Gaul (; 1999). This Boundary Element formulation starts with the piezoelectric reciprocity relation

$$\int_{\Omega} (\mathcal{L}_{JK} U_K U^*_{MJ} - \mathcal{L}_{JK} U^*_{MK} U_J) \, \mathrm{d}\Omega = \int_{\Gamma} (U^*_{MJ} T_J - T^*_{MJ} U_J) \, \mathrm{d}\Gamma \, , \tag{26}$$

where U^*_{MJ} and T^*_{MJ} are the generalized displacement and traction fundamental solutions of the static piezoelectric operator, defined by

$$\mathcal{L}_{JK} U^*_{MK} = -\delta_{JM} \, \delta(x_i, \xi_i) \, , \qquad \text{and} \qquad T^*_{MJ} := C_{iJKl} U^*_{MK,l} n_i \, . \tag{27}$$

By replacing the expressions in the domain integral by Eqns 15 and 27, one obtains a piezoelectric representation formula

$$U_M(\xi) = \int_{\Gamma} (U^*_{MJ} T_J - T^*_{MJ} U_J) \, \mathrm{d}\Gamma - \int_{\Omega} U^*_{MJ} \mathcal{B}_{JK} U_K \, \mathrm{d}\Omega \, . \tag{28}$$

The remaining domain integral, which results from the use of the static fundamental solution and contains the effects of inertia, can be transformed to the boundary using the Dual Reciprocity Method. To this end, the domain term is approximated by a series of tensor functions f^q_{JN} and unknown coefficients α^q_N

$$\mathcal{B}_{JK} U_K(x) \approx \sum_{q=1}^{N} f^q_{JN}(x) \, \alpha^q_N \, , \tag{29}$$

so that, by substituting the approximation (29) into Eqn (28), a new representation formula for the generalized displacement field can be obtained

$$U_K(\xi) = \int_{\Gamma} (U^*_{KJ} T_J - T^*_{KJ} U_J) \, \mathrm{d}\Gamma + \sum_{q=1}^{N} \left(U^q_{KN}(\xi) + \int_{\Gamma} (T^*_{KJ} U^q_{JN} - U^*_{KJ} T^q_{JN}) \, \mathrm{d}\Gamma \right) \alpha^q_N \, , \tag{30}$$

representing $U_K(\xi)$, $\xi \in \Omega$ in terms of field quantities on the boundary only. The particular solutions U^q_{JN} and T^q_{JN}, which appear in Eqn 30, are defined by

$$\mathcal{L}_{JK} U^q_{KN} = f^q_{JN} \, , \qquad \text{and} \qquad T^q_{JN} := C_{iJKl} U^q_{KN,l} n_i \, . \tag{31}$$

From Eqn (30) the interior displacements $U_K(\xi)$ can be calculated at any point $\xi \in \Omega$ if the boundary variables and the coefficients α^q_N are known.

The next step is the discretization of the model into finite elements on the boundary, the so-called *boundary elements*, as shown in Figure 1. Similar to Finite Element analyses, the boundary elements are mapped onto reference elements, and the field variables U_I and T_I are approximated by a series of shape functions and nodal values

$$x_i = \sum_{q=1}^{N} \Phi^q \check{x}_i^q , \qquad U_I = \sum_{q=1}^{N} \Phi^q \check{U}_I^q , \qquad T_I = \sum_{q=1}^{N} \Phi^q \check{T}_I^q , \tag{32}$$

where N is the number of local nodes of the respective boundary element. Now the load point ξ in the representation formula 30 is transferred to the boundary, which leads to the boundary integral equation (BIE). Finally the approximations 32 can be introduced, so that a system of equations

$$\mathcal{M}\ddot{\check{U}} + \boldsymbol{H}\check{\boldsymbol{U}} = \boldsymbol{G}\check{\boldsymbol{T}} \tag{33}$$

can be assembled by summation over the boundary elements. This system is similar to Eqn 24 obtained in Finite Element analysis.

5 Example

The piezoelectric body shown in Fig. 3 is subjected to a Heaviside type loading at the specified face. The material used in the computations is a piezoelectric ceramic, poled in x_3-direction, with transversely isotropic material properties. Its elastic and piezoelectric moduli, and relative permittivities, are given by

$$\begin{aligned} &C_{11} = 107600\,\text{MPa} , \; C_{33} = 100400\,\text{MPa} , \; C_{12} = 63120\,\text{MPa} , \\ &C_{13} = 63850\,\text{MPa} , \; C_{44} = 19620\,\text{MPa} , \\ &e_{31} = -9.6\,\text{N/Vm} , \; e_{33} = 15.1\,\text{N/Vm} , \; e_{15} = 12.0\,\text{N/Vm} , \\ &\epsilon_{11}^{\text{rel}} = 1936 , \; \epsilon_{33}^{\text{rel}} = 2109 , \end{aligned} \tag{34}$$

the mass density is $\rho = 7800$ kg/m^3. Also, the following relations hold between the moduli: $C_{22} = C_{11}, C_{23} = C_{13}, C_{66} = 0.5(C_{11} - C_{12}), e_{32} = e_{31}, e_{24} = e_{15}, \epsilon_{22}^{\text{rel}} = \epsilon_{11}^{\text{rel}}$.

The transient response has been calculated using both the FEM and the BEM formulation presented in Section 4. The discretization of the body is shown in the figure. It consists of 375 finite elements and 350 boundary elements. Two FEM computations have been performed, the first using 8-node brick elements (2304 dof), the second with 20-node brick elements (8304 dof). For the BEM calculation, 4-node boundary elements have been used, along with 99 internal nodes, resulting in a total of 1804 degrees of freedom. The use of internal nodes is not strictly necessary, but enhances the accuracy of the results, since it improves the quality of the interpolation procedure, introduced in the Dual Reciprocity formulation.

For the time integration, a damped Newmark algorithm is chosen, using parameters $\delta = 0.7$ and $\alpha = 0.5$ as proposed by Kögl and Gaul (1999), since it yields more accurate results than Houbolt's algorithm and is more versatile in that it allows an adjustment of its parameters. The time step size was chosen as $\Delta t = 5\,\mu$s. The time histories of the electric potential φ and velocity $\dot{u}_1$ at point $(0, 0, 100)$ are shown in Figure 4. Excellent agreement between the linear FEM and DR-BEM computations can be observed in both cases, with only a slight deviation from the more accurate results of the FEM computation using quadratic element shape functions.

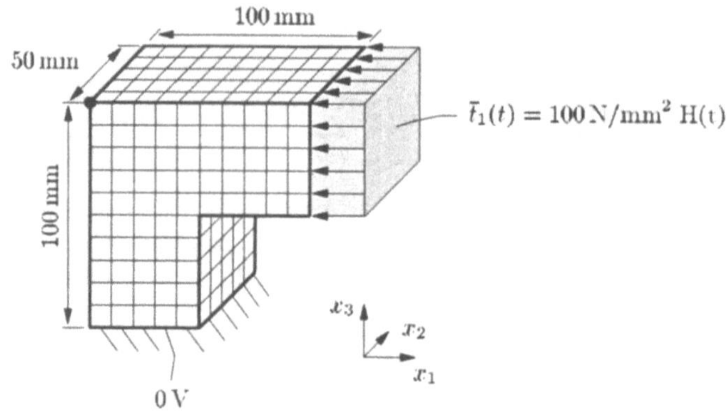

Figure3. Piezoelectric body with discretization and boundary conditions

Figure 5 shows the results obtained for the electric field E_1, and elastic stress σ_{13} at a point with the coordinates $(37.5, 12.5, 62.5)$. Taking the quadratic Finite Element calculations as reference, it can be observed that the Boundary Element computations with linear element shape functions are much more accurate than the corresponding linear Finite Element computations, for reasons which have already been explained in Section 2. This shows that the improved accuracy of the Boundary Element Methods in the calculation of flux quantities is maintained in the present piezoelectric DR-BEM.

6 Conclusions

A comparative study and introduction to the numerical calculation of piezoelectric continua with Finite Elements and Boundary Elements has been given. In a numerical example, excellent agreement could be demonstrated between the Finite Element results and those obtained using a Boundary Element formulation recently introduced by the authors. It could be observed that for the electric field and stress calculations, the results obtained by the new BEM formulation are more accurate than the results from Finite Element computations when using a comparable grid. This BEM formulation therefore presents a powerful numerical tool and an alternative to FEM for the computation of transient piezoelectric phenomena, especially for electric field and stress analysis.

Acknowledgements

Support by the Deutsche Forschungsgemeinschaft (DFG) of the Graduate Collegium 'Modelling and discretization methods for continua and fluids' (GKKS) at the University of Stuttgart is gratefully acknowledged.

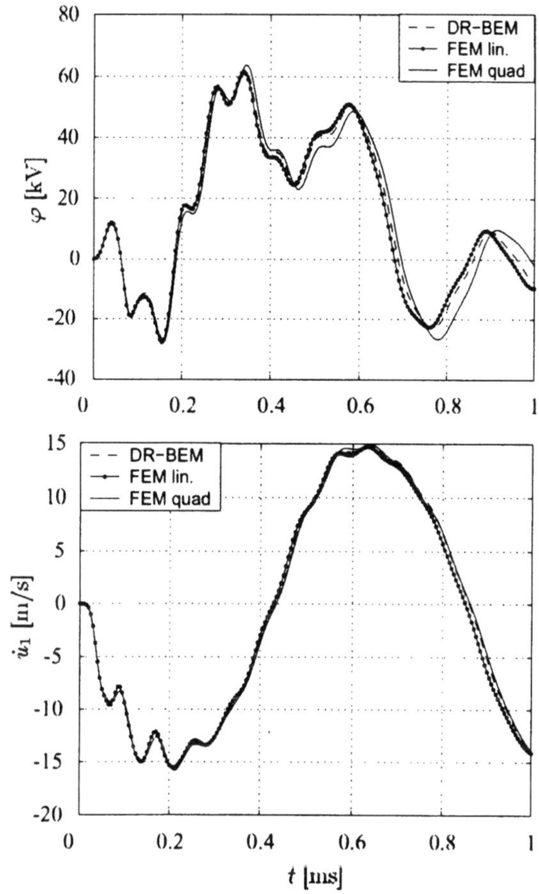

Figure4. Electric potential $\varphi(t)$ and velocity $\dot{u}_1(t)$ at point (0,0,100)

References

Allik, H., and Hughes, T. J. R. (1970). Finite element method for piezoelectric vibration. *Int. J. Num. Meth. Eng.* 2:151–157.

Bathe, K.-J. (1996). *Finite element procedures.* New Jersey: Prentice-Hall.

Brebbia, C., Telles, J., and Wrobel, L. (1984). *Boundary element techniques.* Berlin: Springer-Verlag.

Gaul, L., and Fiedler, C. (1996). *Boundary Element Methods in Statics and Dynamics (in German).* Braunschweig: Verlag Vieweg.

Kögl, M., and Gaul, L. A Boundary Element Method for transient piezoelectric analysis. *Engineering Analysis with Boundary Elements (submitted).*

Kögl, M., and Gaul, L. (1999). Dual reciprocity boundary element method for three-dimensional problems of dynamic piezoelectricity. In *Boundary Elements XXI*, 537–548. Southampton: Computational Mechanics Publications.

Lerch, R. (1990). Simulation of piezoelectric devices by two- and three-dimensional finite elements. *IEEE Transactions on ultrasonics, ferroelectrics, and frequency control* 37(2):233–247.

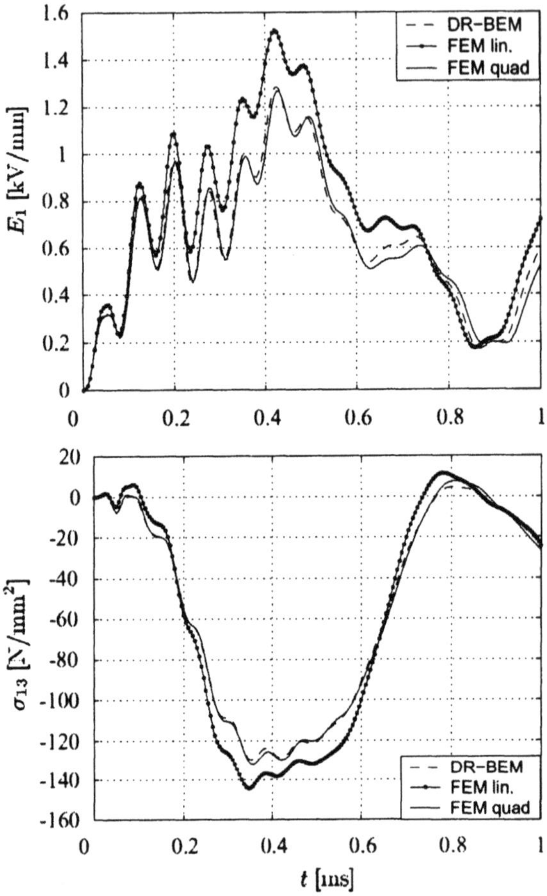

Figure5. Electric field $E_1(t)$ and elastic stress $\sigma_{13}(t)$ at point (37.5,12.5,62.5)

Nye, J. F. (1985). *Physical properties of crystals*. Oxford: Clarendon Press.

Partridge, P. W., Brebbia, C. A., and Wrobel, L. C. (1992). *The dual reciprocity boundary element method*. Southampton: Computational Mechanics Publications.

Tiersten, H. F. (1969). *Linear piezoelectric plate vibrations*. New York: Plenum Press.

Zienkiewicz, O. C., and Taylor, R. L. (1991). *The finite element method*, volume 2. London: McGraw-Hill, fourth edition.

Zienkiewicz, O. C., and Taylor, R. L. (1994). *The finite element method*, volume 1. London: McGraw-Hill, fourth edition.

CHAPTER III

SMART MACHINE SYSTEMS

1. Electrostructured fluids and smart machines.
2. Electrostructured fluid flow quantification
3. ESF clutch driven mechanisms and the ER linear reversing motion demonstrator
4. ER/MR flow mode damper design methodology and railcar lateral suspension application
5. ESF tribology, hydrodynamic lubrication and the flexibly operated lens finisher

W. A. Bullough
Reader and Chairman of SMMART
Smart Machines, Materials and Related Technologies Engineering Network
Department of Mechanical Engineering
The University of Sheffield
Mappin Street, Sheffield S1 3JD
ENGLAND
Tel: 44 114 222 7733
Fax: 44 114 222 7890
e-mail: w.bullough@sheffield.ac.uk

1. Electrostructured Fluids and Smart Machines

W. A. Bullough

The University of Sheffield, Department of Mechanical Engineering
Smart Machines, Materials and Related Technologies Unit, Sheffield, England

Abstract Machines are under the spotlight - and by machines, in the present industrial climate, this means flexibly operated, computer controlled original equipment (unconventional transducer assemblies) capable of delivering significant amounts of force, velocity and displacement in a self cooling arrangement. Also, consideration of methods of implementing the scope and performance, and limitations of the concept and their embellishment by the way of a brief consideration of the first commercial application of ESF - the Lord under-seat shock controller, will be undertaken. Also typical fluid characteristic forms are presented. More detail of the underlying technology is explored and methodologies for the design of the main types of controller are illustrated in subsequent sections.

1.1 Introduction

Recent times have seen a downturn in the activities of the primary machine designer. Attention has turned instead to extracting added value from existing devices by the application of software engineering to them. To support these trends an intensive development of electrical servo motors and control techniques has taken place. The resulting devices and technology have had knock on effects, e.g. in reducing the size of the hydraulics industry by capturing some of its former natural range of application. However, permanent magnet servos and logic cannot accomplish all of the future tasks required of industry and there are signs that once again the pioneering engineering innovator is being encouraged to look for new ways and means.

Some of the prospective enabling technologies arising, covered in the other sections of this book, seek to prove potential with respect to diverse activities such as vibration control via piezo or magnetostrictors, use geometrical adaptive flexibility via shape memory alloys and so on. In this section, machines are under the spotlight - and by machines, in the present industrial climate, this means flexibly operated, computer controlled original equipment (unconventional transductor assemblies) capable of delivering significant amounts of force, velocity and displacement in a self cooling arrangement.

Indications of intent in this direction can be seen in the development of simple hydraulic components. For example, recent announcements have indicated a desire for stepper motor adjusted hydraulic poppet valves possibly for vehicle suspension system computer based control and similar solenoid driven valves acting against springs in e.g. IC engine management systems. The present section considers past work on like control elements, driven by piezo or magnetostrictors, in general, insomuch as their downstream systems are well known and understood, then goes on to consider another current yet all system innovation, ESF, the implication being that none of the other smart materials have the combination of speed of switching and low inertia assembly possibilities required to produce for example a motion range as in Figure 1.1 solely by adjusting the electronics and without any mechanical downtime to change cams and levers etc.

This passive device example is taken from the waveform or signal generator where switching to a series of relatively high capacity voltage sources for a succession of time intervals, produces the output forms flexibly. For small motions where little external resistance is involved an active element such as a relay can be used to select position, all of the power going to overcome the inertia of a light load, e.g. as on the hard disc package. If the disc were drawn to scale with the track width set to the 0.1m size of the target hole on a golf course, then the disc would be say approximately 250m mean effective radius. The precision involved in picking up say 18 bits of disc located information exactly, the right one every time, is equivalent to an 18 hole golf course being played in 18 shots in a round of less than 1 second duration; all the holes are different. What is more, for the same dynamic conditions to apply, the golfer who must play these shots enjoys a situation corresponding to that of being suspended from the undercarriage of a jumbo jet at full speed, flying at circa 800 kph at an altitude of less than 1 cm above each tee in turn. This colourful example serves to illustrate the level of precision which is to be had from the insertion of electronic information into a third age machine and how extreme the operational conditions are.

It is a long way from a compact disc to a golf hole so the intermediate aim is for something in between, say 0.25 m sweep with ~0.1 mm precision - in short the popular range of operations of most mechanical machines or devices of today. Nothing less than an integration of the fastest machines of all (inflexible mechanical single action devices) and electronic information and policing is sought [1] and the devices are to target on areas not easily supplied by more convenient to use, clean and available electromagnetic artefacts. The exploration of these possibilities will bring the observer into contact with catching rather than self driving systems, novel excitation/force transmission devices plus thermally close controlled, inertially loaded machines and a re-consideration of traditional aspects of materials and machine operation.

1.2 Piezo/Magnetostrictor v ESF

A mechanism which involves rapid motion changes plus large displacements is likely to require a force transmitting medium that facilitates cooling and lubrication and affords a high specific thermal capacity density x specific heat - ($\rho \times C_v$) in order to negate the occurrence of excessive temperature excursions. Considering that the machine has to be flexibly operated and non (primarily) electromagnetically actuated, all points towards a hydraulic power transmission which may be controlled adaptively via pre programmed electronic information with a resulting high gain.

Dimensionally analysing this situation shows that some aspect of the controller must have its geometry altered or, for a given fluid velocity, the viscosity of the medium be changed electronically, i.e. if the pressure/shear stress coefficients are to be modulated. Whatever the choice, the stiffness and load spreading properties of the working fluid should not be compromised.

A change of geometry which allows IT governance can be accomplished via servo or solenoid operated valve controlled actuators. The spool construction is complicated, expensive and subject to blockage, Figure 1.2, whilst the solenoid operated valve is inherently non-proportional yet the poppet head offers total shut off in an inexpensive, reliable construction.

In respect of the present task, the poppet is under investigation for matching with a rapid acting inline piezo or magneto strictor driver. Potentially, this would reduce the size and weight of the device and offer proportionality though it has not yet been shown what the frequency response limitations on flow rate of this compact valve would be. To date the outflow poppet only has been investigated (to the point of providing any substantial methodology) in the steady state. Verified CFM work indicates that despite the intense and complex non-isothermal shearing flow therein that the performance may be predicted by the use of relatively simple methods. These show that the valve flow rate range is governed principally by the 0.1% strictor maximum electro-strain [2] and pressure drop by the compliance of the stack viz:

Axial Stack Force $\sim p_i \pi d^2/4 + p_o \pi (D^2 - d^2)/4$ (1)

Flow rate $\sim \pi d \delta \operatorname{Sin} \alpha \quad \sqrt{2(p_i - p_o)/\rho}$ (2)

For sensible flows poppet length considerations impose a large diameter on the poppet with consequent supply capacity problems. With a strictor length of 40d the 'wide open' flow rate of a $2\alpha < 135°$ valve is less than 20% that of a fully withdrawn cone. Whilst a pressure balanced design would eliminate any stack compliance it may over damp the valve frequency response. Practical questions such as what will be the eventual piezo strainment, what is the best form of poppet in that region or can the brittleness of the stack be accommodated without excessive tolerance on eccentricity of cone and inlet port are a matter of continuing investigation. Whatever the outcome, an electron-hydraulic device is in prospect, if only for use as a very compact, lightweight, remotely controlled pilot.

1.3 Electrostructured Fluids (ESF)

A fluid which demonstrates a variable viscosity altered by the application of some kind of electrical signal must involve at least two phases. The base or carrier fluid needs to possess a current limiting or insulating nature and its second phase to be polarisable. The structure set by the distributed second phase influences the overall flow field and determines the relationship between the excitation field, the external flow inducing forces and local hardware geometry. This governance can occur in two main forms depending on whether the second phase is solid or liquid.

If solid, then it will be sized to give a continuum-Newtonian flow characteristic at zero excitation (and thus behave reasonably in a leakage control, heat generation and serviceability sense) and when excited will ideally produce a yield stress type of effect in order to provide the hold position in a machine by virtue of a shear stress or pressure applied to a solid surface. Like the poppet valve the magnitude of the restriction will vary with the severity of the excitation and produce a stable positive slope with respect to rate.

Non Newtonianess is associated with large particles or polymers, in this case ~ 5 µm in a typical 20 - 50% by volume mixture in order to keep the zero excitation viscosity down and the maximum electro-stress τ_e high. In such fluids the zero field viscosity μ tends to be roughly the same as the plastic viscosity - see Section 2. Hence the effect of applying an excitation is to add τ_e and $\mu\dot{\gamma}$ at a given shear rate $\dot{\gamma}$ and excitation E to produce an enhanced total shear stress τ_w or pressure drop resistance capability. It is to be noted that this applies only in a passive or restrictive sense, to impede the flow induced by external forces caused by pump or load pressure or the relative motion of the plates or both etc.

If the fluid particles are ferrous the fluid is a magnetorheological suspension or fluid (MRS/MRF) and needs a magnetic field to be applied transverse to any surrounding valve or shear plate pole pieces. The excitation required to form the structure will involve currents of the order amps to flow through a locally situated coil and the resulting high energy density magnetic circuit will lead to a strong structure and a high electro or yield stress capability. If the particles are non magnetic then an electric field, which may be generated remotely will be required at a high voltage level ~ 3kV + and the particles must afford a means of polarisation by virtue of surface ion movement or similar. The current in this electro-rheological fluid will be < 10 μA per square cm of electrode area and, for a reasonably low base viscosity $\mu <$ 100mPa s, $\tau_e <$ 10 kPa (at present). It is possible to have MR + ER fluids with some degree of synergy arising. - See Fig 3.

It is interesting to note that when spherical particles are densely packed together and subjected to the creeping flow of an oil through them they interfere with one another to the extent that each sphere exhibits a drag force of some 200+ times that due to the Stokes drag on the isolated sphere at the same mean approach velocity. Thus the excitation field generates or fixes the structure and the structure increases the resistance to flow. Since the particles are small they will polarise quickly and structure form as soon as the excitation can be applied < 1m.sec though it may take a little more time for the flow to develop from say a Newtonian velocity profile to the plug flow (zones of non-sheared fluid) incumbent in the flow of the now plastic continuum. When structures are formed, they are reversible insomuch as a rapid removal of the excitation will allow the structure to return to its original loose form, i.e. if the fluid forces remain in place to make it do so. There may be some pseudo-hysteretic behaviour - more anon. These and related phenomenon are considered more in Section 2 and their significance in Sections 3 and 4.

In some devices, notably shock absorbers and vibration isolation dampers there is a need for an electrically controllable viscous or rate characteristic devoid of yield, electro or what may be called in this context, Coulomb frictional stress. This requirement has led to interest in liquid crystal ERF or immiscible liquid-liquid solutions capable of polarisation, Fig. 3. To date these fascinating fluids are very viscous and slow to react to excitation field change. Hence the trend to give an apparent characteristic ex feedback control to fast acting ERF/MRS the result of which may seem like a viscous fluid to the load - see Section 4.

In most fluids there may be three or more phases which may consist of e.g. surfactants to separate the particles which could tend to coagulate in a high centrifugal field say, Figure 1.4 and change properties, extreme pressure additives to improve the tribological performance of the mixture or abrasive compounds to do the opposite - see Section 5, or even to improve the polarisation process through water impregnation of porous particles. Since electrical, thermal and fluid dynamic demands are made on the resulting mixture, the make up is a compromise and the matching of fluid to application requires astute judgement or numerical simulations to be carried out - often both. There are however, limits to compromise and this is where the engineer can make an impact.

1.4 Limits to the Concept

ESF, particularly MRS can enjoy much higher traction stresses than most air-cooled electromagnetic machines. They may or may not be so convenient to control, due mainly to hysteretic behaviour and

ERF is susceptible to impurities, electrophoresis, thermal runaway and over-excitation , see Figure 1.5, like any other (hydraulic) semi-conductor. By and large, temperature control will always be a problem where small amounts of fluid and large amounts of energy are present, unless the fluid can lubricate and be circulated through a reservoir where it can be cooled.

This problem will be endemic in sealed rotational clutch units, which are subjected to high shear rate slip torques, and rapid switching: smart machines always involve large amounts of electricity to be on the move if their change of motion is electrically governed. There is some scope for throughflow cooling (Section 2) when orthogonal drive and cooling flow directions do not severely couple. If high pressure dissipating valve controllers are involved the frequency response of the lateral railcar bogie mounted damper of Section 4 is impaired. If not the under-seat shock absorber is uneconomic. The factor which strongly influences the amount of fluid involved and determines the rate of temperature change, otherwise amounts to τ_e^2 / μ. See later.

Hysteresis can make digital operation obligatory - see Figs. 5 and 6. Sometimes it is confused with the effects of stiffness and/or bulk modulus or compressibility effects.

Where seals and bearings are involved, as they always must be where a fluid is used in linear motion applications where ESF are most likely to score over servo motors etc, and particulate matter is present, then the service lifetime of a product is not easily predicted. As yet these areas have received relatively little attention and due caution must be exercised when rating the load in any particular operation. Similar provisions must be made for possible fluid endurability problems - little is known save from test specific results - serviceability of fluid is mainly application determined. Whilst compaction of ERF has been a problem in high speed clutches, in valve cylinder flow MRF have worked without problems in the first and most tested in service ESF applications, for many thousands of hours.

Digital switching, is required in ERF not only because of the presence of hysteretic type behaviour but as an anti electro-phoretic aid in clutches. Neither voltage or current alone will fix the value of torque. Controversy surrounds the extent of the t_0 clutch speed run up and Rayleigh step pressure rate development situations. It may yet prove to be as high as 10 to 20 m.sec a response after excitation. If so, then the switched motion of Figure 1.1 is hardly likely to apply to these situations. Verification and clarification is required. These are but a few of the extra limitations and idiosyncrasies encountered in ESF/high speed machine design.

Where long stroke lengths of solid material are linked with high acceleration then inertial strain is bound to be a consideration in position control, viz:

Consider a uniform section of straight and uniform material of cross sectional area A, length l and elastic constant E' in inertial tension (stress) σ.

Force to accelerate section at rate a_x $= \sigma A \quad = \rho A l \, a_x$

hence, strain ε $= \sigma / E' = \rho l a_x / E'$

then for say an accelerated belt the elastic distortion Σ is governed by its acceleration, length squared and elastic constant/density property. If say l = 250mm and the positional precision required is 1/2500 an acceleration of say 1000g will identify very few materials flexible enough to say bend round a pulley.

1.5 Fluid Characterisation

Fluids can be characterised in small size samples on Couette type viscometric apparatus. From the data recorded it is possible to model device performance whatever the mode of the controller flow - via computer fluid dynamics (CFD) programme packages. However (Section 2) this work is at present limited to isothermal flow and is still in a state of development. It is important to note that device and fluid characteristic shapes are similar but not the same and, more importantly to check just what specific characteristic data implies.

In the test data provided herein [3] more ERF data is given than MRS. This is simply because the author has more unified ERF test data available, than MRF which come only from one or two commercial sources - see also Section 2. Generally speaking there is no great effect of gap width on either the zero excitation viscosity or yield stress measurements and development length effects are small. These statements ignore the negative wall friction coefficient by virtue of which fluid leaks between wall and particles easier than through a packed bed of circumscribing particles.

Most of the ER fluid data is from tests commissioned by Engineering and Physical Sciences Research Council, Electro-rheological Research Syndicate, Advanced Rail Research Centre and Wolfson Foundation (all of the UK), The British Council and Croucher Foundation (Hong Kong) and the MR suspension data from the public literature of the Thomas Lord Research Centre and QED Inc. (USA).

Typical characteristics of an ERF, in this case Lipol 30/w are given in Figs. 7. Steady state and transient data were taken from a concentric clutch at the same time by applying a step voltage to ERF lying between a constant speed rotating outer case and an inner rotor which was held by a torque meter. The torque transient was used with the torque meter transfer function in post processing - to produce an approximate torque time response t^*_m (Table 1.1). Specimen readings on the same fluid and tests identified electrical properties of R and C [4] (resistance and capacitance) which need non linear correction coefficients to standardise them - dependent on the fluid make up but with RC seemingly faster with increased $\dot{\gamma}$ and temperature. Since identified, different clutch and valve dynamic models are forthcoming.

MRS information is not readily available but like ERF it develops a Bingham plastic type behaviour from polarisation of the soft iron particles - see Figs. 8: reaction between the dipole moments initiates a columnar like structure across the gap width and this resists the flow driven by external forces. The strength of the flow resistance varies with the field magnitude, increasing with $|H|$ but the zero field viscosity is not so shear stable as in ERF. Nevertheless $\tau_e \sim 50/100\text{kPa}$ can be achieved at high fields with the complex modulus in the pre-yield region being significantly field dependent, i.e. $\tau = G'(H).\ \gamma$ below $\varepsilon = 10^{-3}$.

MRS performance is substantially influenced by additives with a much higher $\dot{\gamma} \to 0$ viscosity being involved. Designing with this feature and using the fluid chemistry to advantage is well described in [5, 6]. In particular the excited/zero excitation control ratio, shown to depend on τ_e/μ, and minimum fluid volume $\left(\tau_e^{\,2}/\mu\right)$ are related to the geometrical aspect ratio of a controller though this ideal analysis can be much modulated by the effect of seal friction etc.

Figure 1.8a shows the total steady state shear stress on a rotating Couette viscometer at $\dot{\gamma} = 26\ s^{-1}$ where $\tau_e \rightarrow \tau_0$ at field strengths less than saturation. Saturation commences at 0.2/0.3 Tesla and a stress plateau is reached at complete saturation ~ 1 Tesla. In general the reluctance of the inter pole gap is the largest in the magnetic circuit. Permeability in the linear range of B/H is 5 to 9 times that of a vacuum - this depends proportionally on the concentration percentage of solid - see Figure 1.8b for magnetic properties.

Fluids developed by Lord Corporation have reached commercial application by virtue of the high yield strengths, Figure 1.8c needed for economic acceptance of the product and their undiluted performance in extreme environmental conditions - the like of which would cause breakdown in an ERF. Hence they are usually designed to operate near to saturation flux. $\tau_e \sim 50/100\text{kPa}$ requires typically 150 - 250kA/m and is available across a very wide temperature range with little or no electrical heating effect occurring in the fluid. The coil design can be a limiting factor. Electrical power requirements for equivalent ERF/MRS duty are about the same.

A typical MR damper is shown in Figs. 8d with its open loop characteristic. It is sized for under-seat shock absorbing duty and has a diameter of 35mm, a valve and coil inside the damper, and can dissipate ~ 600W for an input control signal of 4W over a total stroke strength of 50mm. The design current is 1A and the excited fluid consists of about 0.3ml out of a total reservoir of about 50ml - see also Section 4. Such dampers have achieved impressive reliability in use. Small displacement sealess dampers can operate with a water base fluid and seismic dampers with a custard like fluid, designed to keep the particles in suspension over a long period of time. In essence, this entails the development of the final part of ESF methodology where serviceability, the interaction of fluid constituents, dynamic performance required of application and the design and sizing of controller take place alongside active control strategy consideration of the damper for future adaptronic purposes.

References

1. Bullough W A, The Third Wave of Machines, Endeavour (1996), pp 50-55, Elsevier Press.

2. Bullough W A, Chin S B and Wong A P, The Potential of an Electrically Strictive Driven Poppet Valve, Japan CFD Journal, at press, Vol 9, No 1. Apl. (2000).

3. Bullough W A et al, ERF Shear Mode Characteristics: Volume Fraction, Shear Rate and Time Effects. Trans. ASME, Jnl. Dyn. Syst. Meast. & Control, pp 221-225, June (1996).

4. Hosseini Sianaki A et al, Experimental Investigation into Electrical Modelling of ERF Shear Mode. Proc. IEE, Sci. Meast. & Tech., V. 141, No. 6, pp 531-537.

5. Jolly M R, Bender J W and Carlson J D, Properties and Applications of Commercial Magneto-rheological Fluid. Proc. SPIE Ann. Int. Sympo. on Smart Structures and Materials, held San Diego, California (March 1998), pp 1-15.

6. Carlson J D, Catanzariti D M and St.Clair K A, Commercial Magnetorheological Fluid Devices, Proc. 5th Int. Conf. on ERF/MRS and Associated Technology, held Sheffield (July 1995), pp 20-28.

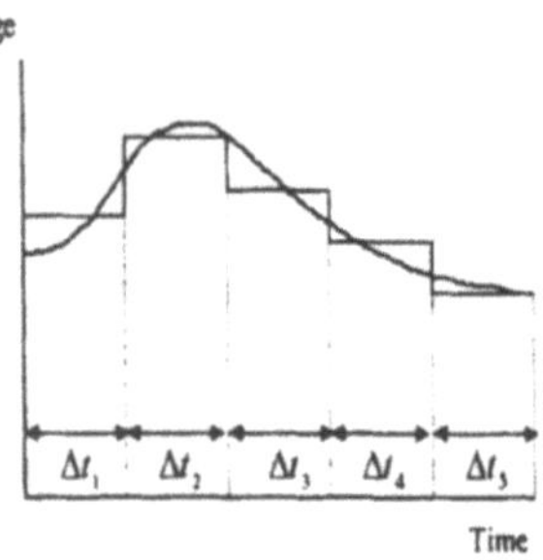

Figure 1.1: Catch concept for force, displacement, velocity

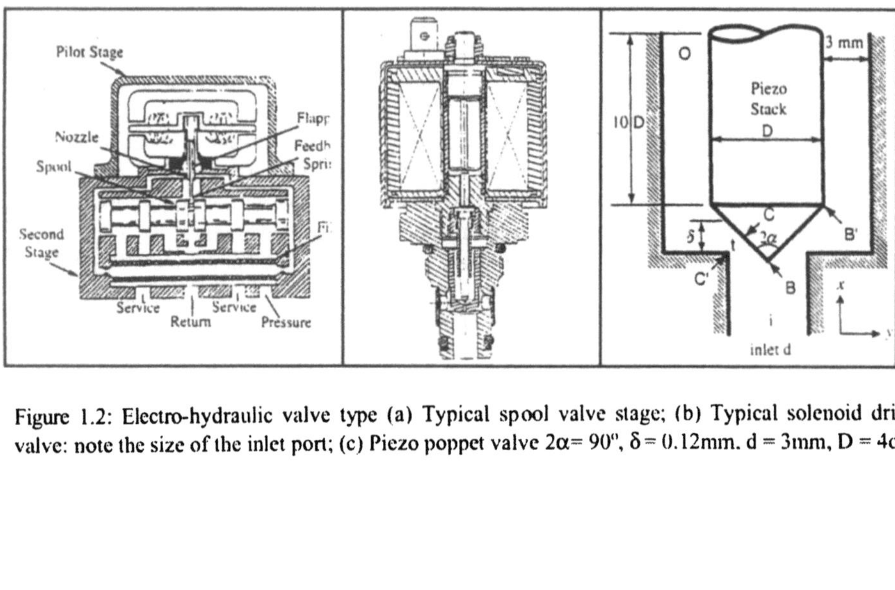

Figure 1.2: Electro-hydraulic valve type (a) Typical spool valve stage; (b) Typical solenoid driven poppet valve: note the size of the inlet port; (c) Piezo poppet valve $2\alpha = 90°$, $\delta = 0.12$mm, d = 3mm, D = 4d

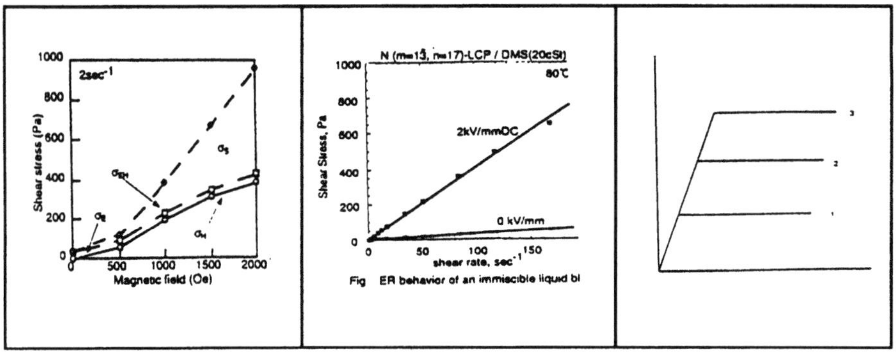

Figure 1.3: New developments in ESF(a) Synergistic effect of EMR fluid in parallel field (particle concentration 40 weight %, temperature is 25°C), when under electric field; E = AC 1.5KV/mm σ_H: magnetic field alone; σ_E: electric field alone; σ_{EH}: simultaneous application σ_S: the synergistic field(b) ER behaviour of an immiscible liquid blend; (c) Shear stress τ versus shear rate $\dot{\gamma}$ for typical liquid crystal ER fluid at various excitation levels 1, 2 and 3.

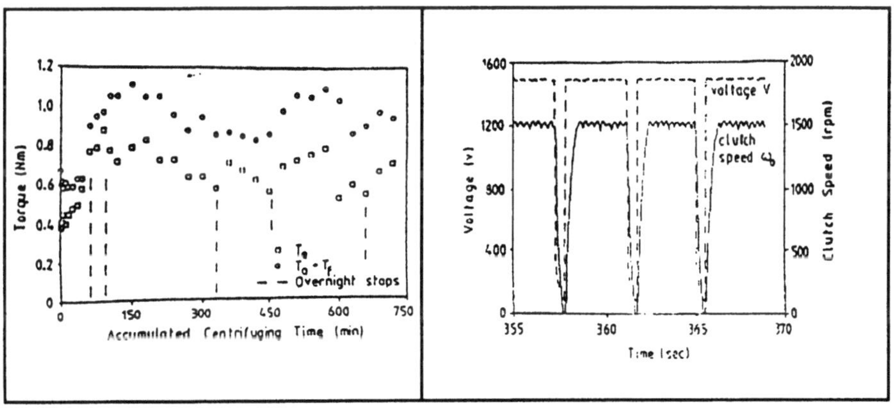

Figure 1.4: Fluid durability/compaction tests: typical form (a) Centrifuging effects in ER clutch: 3000rpm. Te is due to excitation, To is viscous and Tf seal torque; (b) Test cycle for (a) but at 1500rpm $\dot{\gamma}$ at 9500 s^{-1}

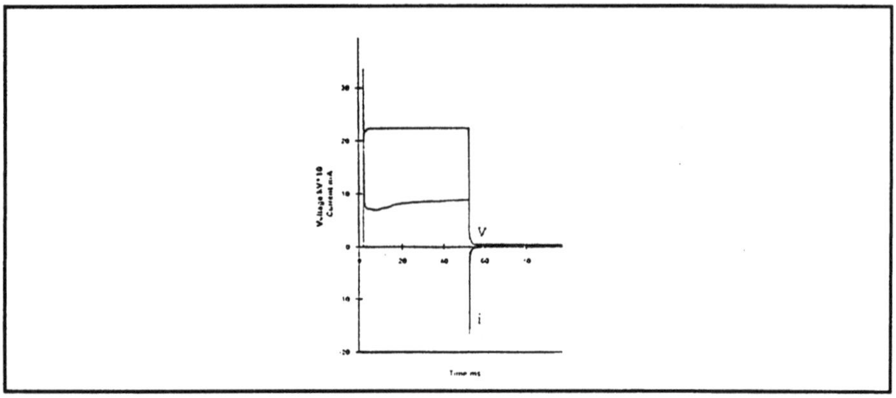

Figure 1.5 High rate switching box output. The extent of the overshoot must be controlled and the off switch must be driven down. For ERF, bi-polar switching can be used to combat electrophoretic tendencies.

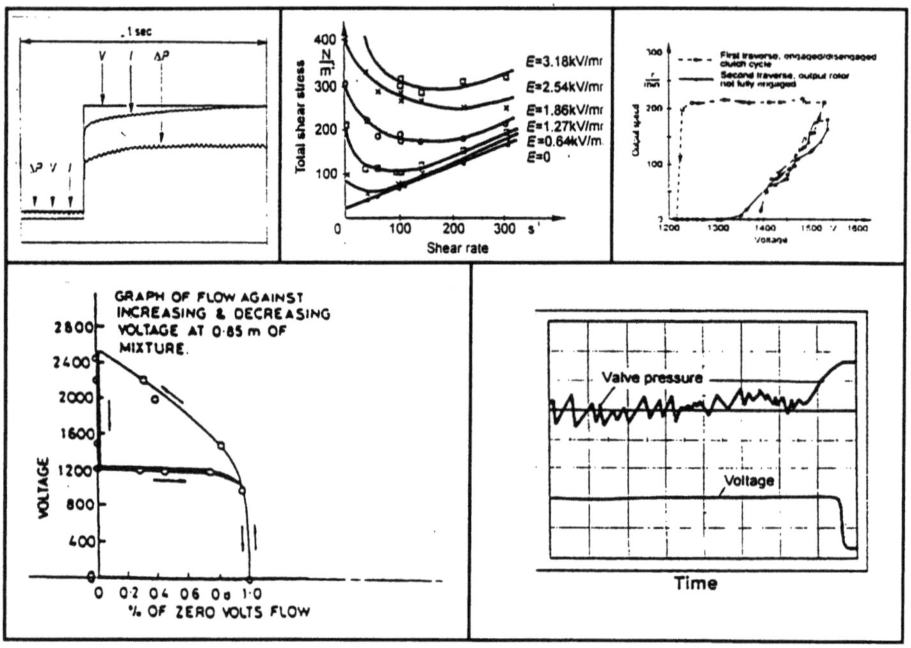

Figure 1.6: Hysteretic tendencies in ERF. (a) Parallel plate valve in constant flow - note (i) current response; (b) Couette viscometer test results; (c) clutch in on-off DC operation; (d) valve fed by header tank-flow stops and starts at different voltages; (e) valve experiencing choking phenomena at nominally constant flow rate - lower trace is ongoing DC voltage and upper valve pressure drop versus time

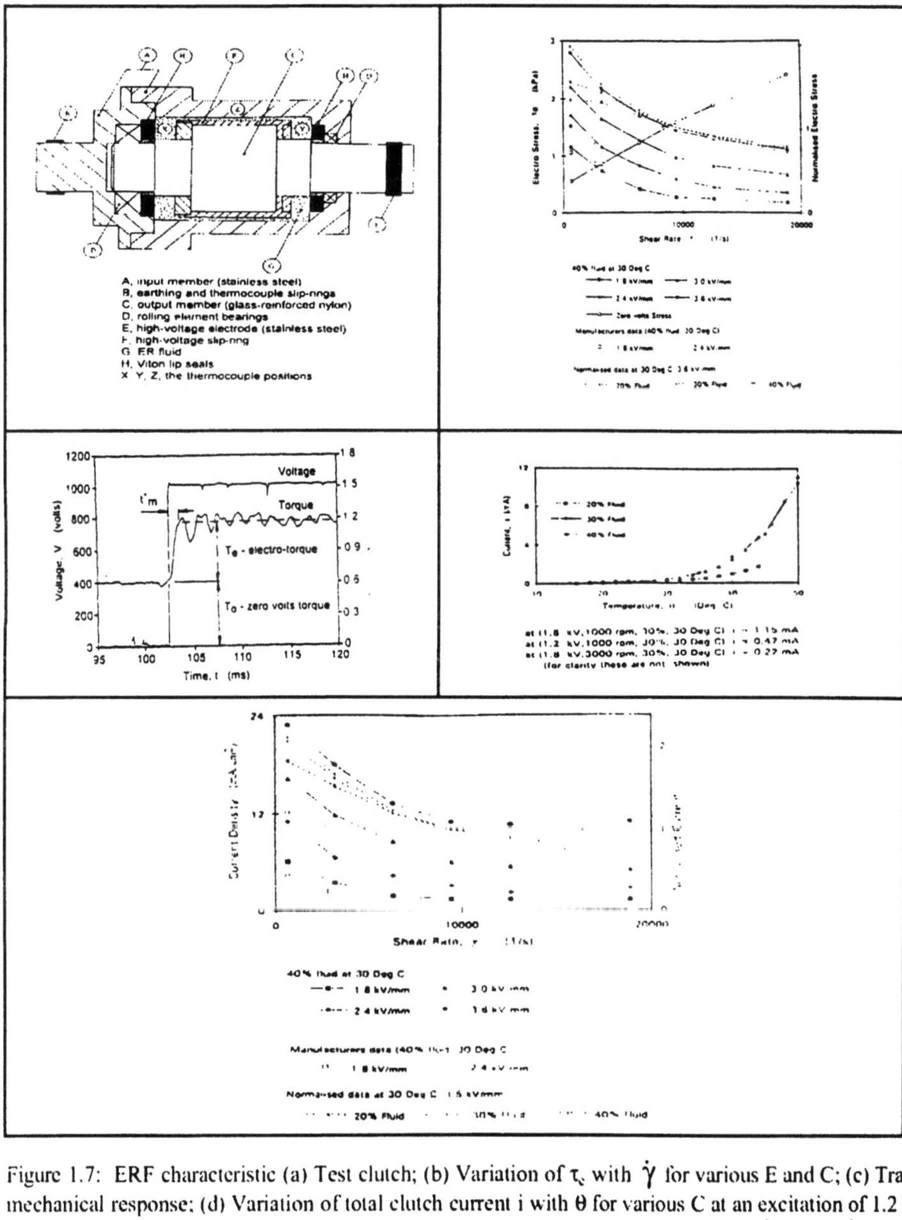

Figure 1.7: ERF characteristic (a) Test clutch; (b) Variation of τ_e with $\dot{\gamma}$ for various E and C; (c) Transformed mechanical response; (d) Variation of total clutch current i with θ for various C at an excitation of 1.2 kV and ω of 1000 rpm; (e) Variation of current density (i/surface area of excited electrode ($\pi d \ell$)) with $\dot{\gamma}$, for various E and C.

Table 1.1 – Values of the electron-hydraulic time delay

Values of the electron-hydraulic time delay (t2 milli-seconds) for step tests carried out to monitor the fluid condition at intervals between intermittent running periods. (a) 1500V, 1000 rpm; (b) 2250V, 1200 rpm

	Accumulated centrifuge time (min)								
	0	8	16	21	[illegible]	[illegible]	102	150	[illegible]
[illegible]	[illegible]	1.15	[illegible]	[illegible]	[illegible]	[illegible]	[illegible]	[illegible]	[illegible]
[illegible]	[illegible]	[illegible]	[illegible]	[illegible]	[illegible]	[illegible]	[illegible]	[illegible]	[illegible]

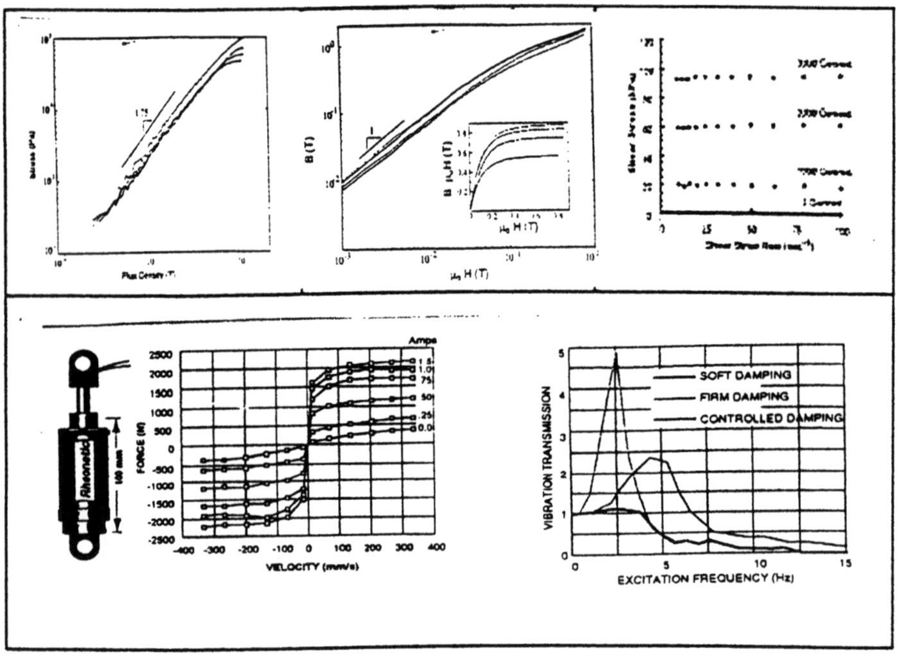

Figure 1.8 MRS characteristics and damper. (a) Fluid shear stress as a function of magnetic flux density at a maximum shear rate of $26s^{-1}$. Ascending order of the plots corresponds increasing iron volume fraction; (b) Flux density within MR fluids as a function of applied field; (c) Yield stress versus shear rate for a typical MR fluid (Lord VersaFlo™ MR-100); (d) Rheonetic™ MR fluid seat damper and measured force versus velocity curves; (e) Vibration transmissibility for a MR fluid controlled suspended seat.

2. Electrostructured Fluid Flow Quantification

W. A. Bullough, D. J. Ellam and R. J. Atkin

The University of Sheffield, Department of Mechanical Engineering

Smart Machines, Materials and Related Technologies Unit, Sheffield, England

Abstract: This chapter first deals with how ESF devices can be sized in a simple but effective way. A more accurate approach is then reviewed in which an ideal ESF (viscoplastic) model is used to gain direct and semi-iterative (CFD method) solutions of the governing equations of motion. A dimensional analysis approach in which an algorithm is used to account for non-ideal behaviour is then explained. The final section explains how the computational fluid dynamics (CFD) technique works and the unique advantages it offers.

2.1 Introduction

Understanding the mechanics of an electro-structured fluid in an engineering context allows scientists and engineers to model the output characteristics of ESF devices from small-scale fluid sample viscometer test results. Development of high tech products can be a very costly process. The ability to test concepts and predict prototype performance and vice versa dramatically aids in the assessment of proposed ESF products. The existence of relevant analytical tools could influence whether a potential ESF device is ever given the go ahead for development. Fluid improvements will also be stimulated, as specified industrial demand can be determined.

2.2 Sizing Techniques

Simple Device Performance Estimation

An initial approach [1,2] is to approximate the total shear stress effect (τ_{eo}) to be a Newtonian contribution (τ_o) due to the shear rate, plus a fixed electro-stress contribution (τ_e) due to the excitation of the ESF. This concept can then be applied to the overall device:

$$\tau_{eo} = \tau_o + \tau_e \quad \Rightarrow \quad T_{eo} = T_o + T_e \quad \& \quad \Delta P_{eo} = \Delta P_o + \Delta P_e$$

This method comes about from looking at the output characteristics of simple ER and MR controllers, all of which employ narrow gap geometry as shown in Figure 2.1 The annular valve operates as a pressure driven Poiseuille type flow and the clutch operates as a boundary driven Couette type flow. Figure 2.2 shows that in the unexcited state the flow curve (τ vs. ?) is that of a Newtonian fluid for which a 'no-field viscosity' μ_o) can be defined. Data for different field strengths can be seen to be translations of the no-field flow curve, the magnitude of which depends on the strength of the excitation (E). The gap size (h) is found to have no significant effect on the fluid properties obtained [3].

The electro-stress (τ_e) is effectively a yield stress which is taken to be constant at a particular excitation, and can be instantaneously manipulated by changing the level of the excitation (E) (refer to section 1). For example, in an ERF one of the following expressions will accurately predict the electro-stress component:

$$\tau_e = \alpha E^2 \quad or \quad \tau_e = \alpha(E - E_0) \quad or \quad \tau_e = \alpha(E - E_0)^2 + \beta(E - E_0) \tag{2.1}$$

α, β and E_0 are properties of the fluid material. The appropriate equation will depend on the particular type of ERF or MRS used.

2.3 Design Formulae

Adopting the above approach allows calculations for a device in which the flow is steady and laminar.

Concentric Clutch (Couette Shear Mode)

A torque (T) is required to maintain a relative speed difference (Ω) between the cylinders, the other parameters are the dimensions of the device (Figure 2.1). Existing clutch data can be used to determine τ_e for a particular excitation:

$$t_e = \frac{T_e}{2pr^2L} \tag{2.1}$$

The form τ_e = f (E) can then be established – for example (2.2). The overall torque for different clutch configurations is given by:

$$T_{eo} = 2\pi Lr^2(\tau_e + \tau_o) \tag{2.2}$$

the Newtonian contribution is given by $\tau_o = \mu_o \dfrac{r\Omega}{h}$ (2.3)

As in Figure 2.2 the resulting characteristic relationship (T_{eo} vs. Ω) is linear.

Annular Valve (Poiseuille Flow Mode)

ΔP_{eo} is the total pressure difference over the valve length (L) that drives a flow rate ($\dot{q}$). For a particular excitation a balance of the forces allows the electro-stress contribution (τ_e) to be determined:

$$2\pi rh\Delta P = 2\tau 2\pi rL \quad \Rightarrow \quad \tau_e = \frac{h}{2}\frac{\Delta P_e}{L} \tag{2.4}$$

The Newtonian contribution is given by $\Delta P_o = \dfrac{12\mu_o L\dot{q}}{bh^3}$ (2.5)

The total effect is a summation of the two pressure contributions - $\Delta P_{eo} = \Delta P_o + \Delta P_e$. Like the clutch the characteristic relationship (ΔP_{eo} vs. $\dot{q}$) is linear (Figure 2.2).

2.4 Applicability

For a given excitation the flow curves (τ vs. $\dot{\gamma}$) determined from flow and shear mode tests do not usually calculate out to be the same. The discrepancy is mainly due to the well known easy to apply Newtonian equations being used to calculated the shear rate ($\dot{\gamma}$); these equations become less valid when non-Newtonian fluid profiles are present. Caution must therefore be taken when faced with any experimentally derived fluid data. Correct viscometer formulation is discussed in section 2.20.

The simplicity of the above approach has its advantages, namely in being able to quickly estimate basic steady state performance of a device controller. For the valve it can clearly be seen that high ratios of L/h are desirable. It will be shown later that this approach can be accurate for estimates of clutch performance, but is less valid for the valve in which non-Newtonian velocity profiles are present.

2.5 Analytical Solutions using the Bingham Plastic Equation

Governing Equations

As determined in the previous section, an excited ESF displays a yield stress value and is therefore a viscoplastic material. For a given excitation the Bingham plastic model can accurately represent the fluid properties of an ESF; this will become evident when solutions are compared to experimental data. The plastic viscosity (μ_p) of an ESF is usually taken to be equal to the no-field (μ_o) value as obtained from unexcited fluid tests. For one-dimensional flow in which τ_e is the yield/electro-stress and $\dot{\gamma}$ is the strain rate the shear stress (τ) can be written as:

$$\tau = \tau_e + \mu_p \dot{\gamma} \qquad \mu_p \approx \mu_o \tag{2.6}$$

The governing equations and the solution of them are explained later. At this point it should merely be noted that the flow is taken to be laminar and isothermal, and that the fluid is assumed to be incompressible. Ideal Bingham plastic flow has been extensively dealt with in the literature for a variety of simple geometries - a good review can be found in [4].

2.6 Annular Valve: 1-d Poiseuille Flow Between Concentric Cylinders

The gap size (h) is considered to be much smaller than the radius of the cylindrical plates. This allows the valve to be modelled as two perfectly flat plates in which one-dimensional flow is caused by the application of a pressure difference (ΔP_{eo}). This solution is very similar to the well known Buckingham solution for one-dimensional Bingham plastic flow in a pipe. Defining x to be the direction of the axial flow and y to be perpendicular to the flow, the shear stress is found to vary within the gap:

$$\frac{\partial \tau}{\partial y} = \frac{\partial p}{\partial x} = \frac{\Delta P_{eo}}{L} \quad \Rightarrow \tau = \frac{\Delta P_{eo}}{L} y + \boldsymbol{constant} \tag{2.7}$$

In the centre of the section the shear stress falls below τ_e, which results in a solid section of fluid flowing along the centre. This characteristic is known as plug flow – Figure 2.3 indicates regions of plug flow. The size of the plug is dependent on the applied pressure gradient $\Delta P_{eo}/L$.

A full analysis shows that the flow rate can be expressed as:

$$\dot{q} = \frac{\pi r \Delta P_{eo} h^3}{6\mu_o L}\left[4\left(\frac{L}{h\Delta P_{eo}}\right)^3 \tau_e^{\,3} - 3\left(\frac{L}{h\Delta P_{eo}}\right)\tau_e - 1\right] \quad (2.8)$$

there is a limiting pressure gradient before which any flow will occur:

$$\frac{\Delta P_{eo}}{L} > \frac{2\tau_e}{h} \quad (2.9)$$

Figure 2.4 shows how this equation compares to the more simplistic approach of section 0 in representing experimental data for an ER valve [5]. The flat plate assumption is found to be quite valid when the gap size (h) is much smaller than the mean radius [6,7].

2.7 Cylindrical Clutch: 1-d Couette Flow Between Concentric Cylinders

The clutch can also be modelled as two perfectly flat plates. The movement of one plate relative to the other causes one-dimensional Couette flow. In this solution the shear rate and therefore the shear stress are constant across the gap; there is therefore no possibility for plug flow to occur. The analysis shows that the torque maintaining a relative speed difference between the cylinders is given as follows:

$$\frac{\partial \tau}{\partial x} = 0 \;\Rightarrow\; \tau = \boldsymbol{constant} = \tau_e + \mu_o \dot{\gamma} \;\Rightarrow\; T = 2\pi L r^2\left(\tau_e + \mu_o \frac{r\Omega}{h}\right) \quad (2.10)$$

The velocity profile is linear (same as that for a Newtonian fluid): the overall torque is equal to the amount required to overcome the electro-stress plus the Newtonian contribution. The solution is therefore no different from that in section 1.

When flow between concentric cylinders is considered a radial distribution in shear stress arrises. This consequently results in a plug being attached to the outer cylinder at low speeds; the plug thickness diminishes as the speed is increased to a point in which it disappears. As shown in Figure 2.5, the presence of the plug results in a non-linear torque-speed relationship at low speeds. A full analysis [8] shows that no plugs are present when:

$$r_y = \sqrt{\frac{T}{2\pi L \tau_e}} > outer\ radius\,(r_2) \quad (2.11)$$

this corresponds to:

$$\Omega > \frac{\tau_e}{\mu_o}\left(\frac{r_2^{\,2}}{2r_1^{\,2}} - \frac{1}{2} - \ln\frac{r_2}{r_1}\right) \quad (2.12)$$

When there are no plugs the torque is given by:

$$\Omega = \frac{T}{4\pi\mu_o L}\left(\frac{1}{r_1^2} - \frac{1}{r_2^2}\right) - \frac{\tau_c}{\mu_o}\ln\frac{r_2}{r_1} \tag{213}$$

The above assumes that $\tau_c \neq f(r)$.

2.8 Radial Distributions In Excitation: 1-d Flow Between Concentric Cylinders

When allowing a radial distribution of the excitation field strength (E) the choice of electric field to electro-stress relationship (see (2.2)) is found to have significant effects on the resulting flow profiles; some allow plugs to be present under certain conditions while others do not. Until the specific relationship is known it is thought best to assume that when the gap size is much smaller than the mean radius, there is a near uniform field strength across the gap [6,7]. This approach does lead to good representation of experimental data and must therefore be quite a valid assumption. Note that for MRS the field strength can be a complex and often has a non-linear distribution and specific solutions may be called for.

2.9 Cylindrical Clutch with Cooling Flow: 2-d Poiseuille and Couette Flow

When the gap size (h) is small an approximate combined solution of section 3.2 and 3.3 can represent a concentric clutch with a pressurised induced axial cooling flow [9]. Depending on the flow parameters plug and parabolic profiles can both occur. For each case the flow rate can be derived in terms of the rotor speed and pressure difference. The analysis shows that for the small flow rates needed to limit temperature rise to 1-2 °C, the output torque is practically unaffected by the Poiseuille cooling flow.

2.10 Limitations

The governing equations for a hydrodynamic step bearing can be analytically derived (combined one-dimensional Poiseuille and Couette flow between flat plates). The solution indicates under what speed and pressure conditions plug flow will occur [10]. A direct analytical solution relating wall speed, flow rate and pressure drop has so far only been achieved when the condition of zero net flow-rate applies. It can be concluded that even for a relatively simple case there is a limit to what can be achieved analytically (see chapter 5 for greater detail).

Approximate solutions are potentially obtainable for various other devices such as a radial plate clutch with a radial cooling flow. A solution of this type will however require the gap size to be very small and/or the inner radius to be quite large and for steady state acceleration terms to be neglected. In an engineering context these assumptions can be quite impractical.

2.11 Computational Fluid Dynamics Steady State Solutions Method

It is evident from the previous section that direct analytical solutions have their limitations. The solution of more complex practical problems requires the use of computational fluid dynamics, for which approximations that may require the gap size to be small do not have to be made. CFD involves numerically solving the full set of governing equations using a semi-iterative method. This is achieved by splitting the domain into a finite number of cells (control volumes) as shown in

Figure 2.6. A nodal point represents a small element of fluid and stores the flow parameters for that particular control volume. During the iterative process neighbouring nodes influence one another, when this 'influencing' reaches a steady state the flow field has converged and a solution has been obtained. One established CFD code [11] has been used [12] to solve Bingham plastic flows for which there are analytical solutions for verification.

2.12Comparison with Analytical Solutions

The CFD solution for flow between flat parallel plates (annular valve) can be compared to the analytical solution of section 2.6. When enough nodal points are chosen, CFD computes the same flow field as the analytical solution to within 0.5%. The same type of accuracy is obtained for the Couette flow (clutch) comparisons. The velocity profiles obtained from CFD for a valve and a concentric clutch are shown in Figure 2.3.

2.13Comparison to Approximate Analytical Solutions

The approximate analytical solution for a concentric clutch with an annular cooling from the previous section has been compared to the CFD solution [13,9]. The flow profiles obtained by each method agree. For the practical flow range and fluid properties tested the torque was verified to be virtually unaffected by a cooling axial flow rate.

2.14Solutions Not Obtainable Any Other Way

As mentioned in the previous section a solution is available for a hydrodynamic step bearing for a specific case for when the net flow rate is zero. The CFD flow field for this model matches the analytical solution to within 0.5%. However, unlike the analytical solution CFD is well capable of solving practical cases for when the flow rate is not zero.

A radial plate ESF clutch with a radial cooling flow is a desirable industrial product. An approximate analytical solution to this problem is currently being finalised. The possible presence of a radial flowing plug that moves with a non-constant velocity as made the derivation more difficult than the solutions already discussed. An exact CFD solution to this problem is easily obtainable [14]. For practical speeds and fluid properties the torque is found to be virtually unaffected by the cooling flow, this is a very desirable characteristic.

2.15 Dimensional Considerations – Real Fluids

Hedström Number

Dimensionless numbers allow experimental data to be represented in a meaningful and graphical form. Hedström [8] used dimensionless parameters to produce a convenient graphical form of representing the Buckingham equation. The parameters are fiction coefficient (Cf), Reynolds number (Re) and Hedström number (He): they can be written in the general form:

$Cf = f(\boldsymbol{Re}, He)$

$$Cf = \frac{\tau}{\rho \bar{u}^2}, \quad \boldsymbol{Re} = \frac{\rho \bar{u} h}{\mu}, \quad He = \frac{\tau_e \rho h^2}{\mu^2} \tag{214}$$

By using an average velocity and taking the viscosity as the no-field value, experimental data can be converted into Cf and Re. Assuming that the ESF is a Bingham plastic He can be obtained from experimental results by using the appropriate Bingham solution – for example (2.8) & (213). If a particular ESF is well characterised by the Bingham model then as expected the technique provides an accurate and useful collation of data as shown in Figure 2.7 for a MR valve [15]. If this is not the case then an algorithm can be used to correct the data.

Algorithm

When this technique was applied to practical ER valve and clutch data [5,16,17] it was observed that He diminished with increasing flow rate. This means that for a real fluid in Couette and Poiseuille flow the 'electro-stress' (τ_e) depends on the 'shear rate' ($\dot{\gamma}$). It is acceptable that for some ESFs τ_e could be falling - the greater the deformation rate the less the particles can contribute to providing resistance to flow. There are other reasons why this behaviour may be observed. For example the ESF is assumed to be fully characterised by the Bingham plastic equation with a plastic viscosity that is always equal to the no-field value; if this is not the case then the above approach can result in τ_e appearing to drop with shear rate.

Determining exactly why this behaviour occurs is unnecessary; the deviation from ideal Bingham plastic behaviour can be accounted for with the use of an algorithm. The variation of He can be approximated by one equation:

$$He = He_{Re=0} + k\boldsymbol{Re} \tag{215}$$

A relationship can be derived to imply the He at Re = 0 in terms of the excitation level (E); Such a relationship can be accurately determined using a static clutch device, or by extrapolation of fluid data to the zero axis. Either the constant k can be taken as an averaged value, or again a more exact relationship can be determined in terms of the excitation level (E).

Applications and Limitations

The technique allows device performance to be predicted regardless of whether the ESF fluid has a constant electro-stress effect or not. The above technique has been effectively applied to valves and clutches. The approach is however limited to devices in which the geometry and flow field can be simply characterised. Accounting for combined Couette and Poiseuille flow will be complicated, if not impossible.

2.16 Continuum Mechanics & Rheology

Electro-Structured Fluids

To accurately model the flow of ESFs using continuum mechanics it is important to understand their rheology. Although these fluids are suspensions, in the non-excited state the fluids display Newtonian behaviour over a wide range of flow conditions and geometry. The particles within the dispersant do not introduce severe non-Newtonian fluid characteristics such as the Weissenberg effect – this is thought to be because the particles are micron as opposed to macro-sized.

It is also accepted that above their yield stress as is evident from the matching of experimental data (Figure 2.4), that ESFs do not display any noticeable visco-elastic properties. Considering the microstructure one would expect that a shift in particle position would result in a near instantaneous attraction to a new neighbour and would therefore not be attracted back to its previous neighbour.

Governing Equations

When ESFs are viewed on a microscopic scale the micron-sized particles are visible; there is little prospect in the immediate future for deriving applicable fluid behaviour laws for difficult geometries from microstructural considerations. However on the macroscopic scale the particles are no longer visible, this allows the ESF to be considered as a homogenous continuum. The behaviour can be described in terms of continuum properties such as velocity, pressure, density, temperature and the space and time derivatives – the ratio of gap size (h) to particle size is 50 +.

By considering the forces on, and the flow in and out of a small fluid control volume, the governing equations for isothermal flow can be stated and written in Cartesian tensor notation as:

$$\overbrace{\frac{\partial \rho}{\partial t}}^{\textbf{\textit{rate of gain of mass flux}}} + \overbrace{\frac{\partial}{\partial x_i}(\rho u_i)}^{\textbf{\textit{mass flux}}} = 0 \tag{2.16}$$

$$\overbrace{\frac{\partial}{\partial t}(\rho u_j)}^{\textbf{\textit{rate of change of momentum flux}}} + \overbrace{\frac{\partial}{\partial x_i}(\rho u_j u_i)}^{\textbf{\textit{momentum flux}}} = \overbrace{-\frac{\partial p}{\partial x_j}}^{\textbf{\textit{stress term due to static pressure}}} + \overbrace{\frac{\partial \tau_{ij}}{\partial x_i}}^{\textbf{\textit{stress term due to deformation stresses}}} + \overbrace{\rho g_j + F_j}^{\textbf{\textit{body forces}}} \tag{2.17}$$

The tensor notation allows the equations to be expressed in a neat and visible form. This set of differential equations is the building block upon which any useful numerical tool would be built.

Non-Newtonian Viscosity

If the above differential equations are to be solved a rheological expression is needed that will accurately represent the deformation stress tensor (τ_{ij}). In many fluid flows the viscous shear stress (τ) can be expressed as a function of the local shear rate ($\dot{\gamma}$).

$$\tau = f(\dot{\gamma}) \tag{2.18}$$

$$\tau = \mu_{nn}\dot{\gamma} \quad \text{in which} \quad \mu_{nn} = f(\dot{\gamma}) \tag{2.19}$$

The function μ_{nn} is named the 'non-Newtonian viscosity' and is also referred to by various other names but is often just called the 'viscosity'. When the viscosity is simply a constant value the fluid is classed as Newtonian. No matter how complex (2.18) is a value of μ_{nn} can always be calculated for a given shear rate. When the Newtonian law is used to predict τ_{ij} the differential equations (2.16) & (2.17) are referred to as the Navier-Stokes equations.

Obtaining the function μ_{nn} for non-Newtonian fluids is one of the main tasks in the science of rheology. To know such a relation is to know the basic governing fluid property. Equations that fit this function usually arise from curve-fitting exercises.

2.17Viscoplastic Constitutive Equations

As determined earlier the viscoplastic constitutive model that approximately describes the behaviour of an ESF at any given excitation is the Bingham plastic constitutive equation. For one-dimensional shear flow it can be expressed as:

$$\tau = \tau_e + \mu_o \dot{\gamma} \qquad (2.20)$$

The Bingham plastic equation is a two-parameter model in which τ_e is the yield stress and μ_o is the plastic viscosity. Successful predictions of experimental data [5] support the use of this model. Other models that can be used are listed below; they are all variations of the Bingham model.

Casson

$$\sqrt{\tau} = \sqrt{\tau_e} + \sqrt{\mu_o}\sqrt{\dot{\gamma}} \qquad \text{2 parameter} \qquad (2.21)$$

Herschel-Bulkley

$$\tau = \tau_e + m|\dot{\gamma}|^{n-1}\dot{\gamma} \qquad \text{3 parameter} \qquad (2.22)$$

Shulman

$$\tau = \left[\tau_e^{\,1/M} + (\mu_o \dot{\gamma})^{1/N}\right]^M \qquad \text{4 parameter} \qquad (2.23)$$

Using one of these slightly more elaborate models may further complicate the mathematics. One particular constitutive equation may suit a given ESF fluid better than one of the others. Considering the assumptions that need to be made in order to obtain practical solutions it is proposed that in most cases the use of any other equation but the Bingham model in an attempt to gain more accurate analytical solutions is tenuous. A good review of viscoplastic constitutive equations can be found in [4].

One-Dimensional Analytical Solutions

To derive an analytical solution the governing equations are further simplified by assuming the fluid to be incompressible and the flow to be laminar. Further simplifications are usually made by considering the flow to be one-dimensional and time-independent. In simple geometry steady state acceleration terms are often zero; when they are not it is usually assumed that they are negligible which places a limit on speed parameters. Using these above assumptions the equations reduce to:

$$\frac{du}{dx}=0 \quad , \quad \frac{\partial p}{\partial x}=\frac{\partial t_{xy}}{\partial y} \tag{2.24}$$

Equation (6.9) can then be integrated to obtain τ_{xy}, which is also given by the Bingham plastic constitutive equation:

$$\tau_{xy}=\frac{\partial p}{\partial x}y+C \quad , \quad \tau_{xy}=\tau_e+\mu_o\frac{\partial u}{\partial y}$$

The constants and plug regions are obtained by applying the appropriate boundary conditions. The velocity profiles can then be obtained and integrated to determine the flow rate.

Stress Tensor (2-d and 3-d Solutions)

Solving other than one-dimensional flow situations requires the constitutive equation to be expressed in a more generalised form. Here is it possible that the resultant shear stress is greater than τ_e while the individual components are below. The magnitudes of the 'shear stress' and 'deformation rate' tensors are therefore used [4].

The 'stress tensor' can be written as:

$$t_{ij}=\mu_{nn}e_{ij} \ , \tag{2.25}$$

$$\text{where } e_{ij}=\left(\frac{\partial u_i}{\partial x_j}+\frac{\partial u_j}{\partial x_i}\right) \tag{2.26}$$

Here e_{ij} is the 'local deformation rate tensor and μ_{nn} is the 'non-Newtonian viscosity'. The magnitudes of the tensors τ_{ij} and e_{ij} are defined as:

$$\tau=\sqrt{\tfrac{1}{2}\tau_{ij}\tau_{ij}} \tag{2.27}$$

$$\dot{\gamma}=\sqrt{\tfrac{1}{2}e_{ij}e_{ij}} \tag{2.28}$$

For when the shear stress is above τ_e, the Bingham model can be used to give the non-Newtonian viscosity:

$$\mu_{nn}=\mu_o+\frac{\tau_e}{\dot{\gamma}} \quad \tau>\tau_e \tag{2.29}$$

When τ_e has not being reached the local deformation rate tensor is taken to be zero:

$$e_{ij}=0 \quad \tau<\tau_e \tag{2.30}$$

2.18 Carreau & Bi-Viscosity Constitutive Equation

When the Bingham model is used μ_{nn} approaches infinity as the shear rate approaches zero; this is impossible and therefore indicates that there must be an upper limit. An alternative approach to thinking in terms of a yield stress limit below which no flow occurs is to have a shear rate limit below which very highly viscous flow occurs – as shown in Figure 2.8. This model is called the bi-viscous model and can be written as follows:

$$\mu_{nn} = \mu_{\dot{\gamma}\to\infty} + \frac{\tau_o}{\dot{\gamma}} \qquad \tau > \tau_e$$

$$\mu_{nn} = \mu_{\dot{\gamma}\to 0} \qquad \tau < \tau_e \tag{2.31}$$

$$\boldsymbol{R} = \mu_{\dot{\gamma}\to 0} / \mu_{\dot{\gamma}\to\infty}$$

Flow problems that contain plugs can be dealt with by determining the velocity of the plug and then applying the equations of fluid motion in the non-plug regions. This approach is valid until complex geometry problems are under consideration for which violations of the constitutive equation (due to none constant velocity plugs) can occur [18]. A way round this problem is to use the bi-viscosity model. Practical flow solutions are obtained by setting R = 10^4 for ER plate squeeze film dampers [19].

A four-parameter model such as the Carreau [20] equation can model the full viscosity flow curve as shown in Figure 2.9.

$$\frac{\mu - \mu_{\dot{\gamma}\to\infty}}{\mu_{\dot{\gamma}\to 0} - \mu_{\dot{\gamma}\to\infty}} = \frac{1}{\left(1 + (K\dot{\gamma})^2\right)^{m}} \tag{2.32}$$

Such a model is usually too complex to be used in direct analytical solutions, but can be utilised in CFD solutions. As shown in Figure 2.9 the Bingham model predicts most of the flow curve quite well; imposing a cut of viscosity results in the bi-viscous model.

2.19 Time Effects

All ESFs have time-dependent properties - a finite time is required for the particles to realign as they are sheared. This time is however so small that for engineering calculations it can be taken to be zero. Fluids that do show a drop in viscosity with time at constant shear rate are said to be thixotropic. Such characteristics can indeed be measured [8], and can potentially be made useful in CFD. Thixotropic materials are however rarely encountered.

When ESF devices are allowed to lock (fluid at zero relative velocity) hysteresis effects are encountered [3]. A complete loss in deformation rate (shear rate) is alleged to result in the electro-stress having to be reduced until it is less than the internal stress of the device; when this point is reached the fluid structure rapidly disintegrates and the device output jumps back to the expected output for that particular field strength and shear rate. This hysteresis phenomena does not seem to constitute a true fluid time property.

2.20 Viscometers

Viscometers are used to obtain fluid data. Two predominant methods are used for ESFs, the annular valve and cylindrical viscometer (concentric clutch).

Cylindrical Viscometer (Couette Flow)

The shear rate and shear stress can be accurately determined using the equations below [8,21]:

$$\dot{\gamma} = \frac{r_1 \Omega}{h}, \ \tau = \frac{T}{2\pi r_m^2 L} \tag{2.33}$$

r_1 must be the radius of the rotating cylinder (the other cylinder is stationary) and r_m the mean radius. At very low rotational speeds a plug may be attached to the outer cylinder; for an ideal Bingham plastic the point of transition between plug and no-plug flow is given by equation (2.11). Below this transition point, non-linearity is observed and the shear rate equation (2.33) does not really apply.

Annular Valve (Poiseuille Flow)

$$\tau_w = \frac{h}{2} \frac{dp}{dL}, \ \dot{\gamma}_w = \frac{2\dot{q}}{bh^2} \left(2 + \frac{d(\ln \dot{q})}{d(\ln \tau_w)} \right) \tag{2.34}$$

These two equations give the conditions at the wall [21]. The second term in second equation is relevant for viscoplastic flow, the shear rate is therefore calculated using numerical integration which makes this method less accurate than the Couette flow method.

Bingham Plastic

An alternative approach is to assume that the ESF is an ideal Bingham plastic fluid. The plastic viscosity can be assumed equal to the no-field viscosity (μ_o); the analytical solutions for a concentric clutch and annular can then be used to determine the electro-stress τ_e.

2.21 Unsteady Flow Solutions

Response Time

When the excitation is changed a finite time t_m* is required for the fluid properties (τ_e) to change. This time is found to be less than one millisecond [3] and offers the potential for very rapid response mechanisms to be built. The fluid inertia thereafter imposes a limit to how fast a device could actually respond – time to reach steady state (t_o).

The separation of the true response time (t_o) from unsteady recordings of a step type excitation input is a complex and tedious affair [22]. By assuming an instantaneous change in rheological properties (t_m*=0) CFD can be used to directly determine the t_o response by solving the governing equations in the time domain. This is potentially an alternative method to time-consuming and expensive experimentation.

For one-dimensional flow it is possible to obtain analytical solutions to time-dependent flows with reservation. This allows investigation and prediction of the fluid inertia effect. Due to the complexity of the analytical solution less dominant terms are neglected and inexact relations are thereby used resulting in only approximate solutions. Such t_o solutions are available for the annular valve [23] and the concentric [24] and radial [25] clutch.

2.22 Annular Valve

The response time (t_o) is taken to be the time while operating at a constant pressure difference, to change from an initial steady state (unexcited) flow rate to 99% of the final (new excited) steady state value due to a step input in excitation. The analytical solution [23] gives the expected time response which is not dependent on the electro-stress (τ_e) or the applied pressure (ΔP_{eo}) as:

$$t_o \approx \frac{3}{32}\frac{\rho h^2}{\mu_o} \qquad (2.35)$$

The CFD solution which makes no approximation other than an instantaneous change in fluid properties agrees with this result, giving for practical fluid properties the expected order of magnitude for t_o to be ~1 ms. The changing velocity profiles and flow rate response are shown in Figures 2.10 and 2.11.

2.23 Concentric Clutch

If the time for τ_e to change is taken to be zero then a clutch operating at a 'constant relative speed' for which no plugs are present, will results in an instantaneous change in output torque due to a step input in excitation. This is because the velocity profile is linear and remains the same after excitation: the new value of output torque results only from the change in τ_e – a CFD solution to this problem shows this to be true.

If it is necessary to operate in the low speed plug region the response time t_o to a step input in excitation for a constant speed will no longer be instantaneous a time is now required for the velocity profile to change. Such a time would be expected to be similar to the valve (less than 1 ms). The plug region which as explained in section 2.7 only occurs at very low speeds; if it became necessary to know the t_o response when operating in this low speed plug region CFD could be used to determine it.

It is often required to know the time (t_o) for the stationary cylinder (output) to run up from zero velocity to the same speed as the input cylinder. The analytical solution [24] tends to indicate that the fastest possible time is given by:

$$t_o \approx 4\frac{\rho h^2}{\mu_o} \qquad (2.36)$$

A preliminary CFD solution to this problem which uses a free surface boundary condition to represent an output shaft with no inertia shows that ρ, h and μ_o have the expected effect on t_o, but also suggests that τ_e and the angular speed (Ω) have an effect on the solution. For a typical speed and fluid parameters the changing velocity profiles produced from CFD are shown in Figure 2.12.

Further work is required to fully understand the discrepancies but ultimately verification will be required. This result (run up to speed) should be considered controversial until experiments are considered.

2.24 Hydrodynamic Bearing

It is desirable to know how rapidly the pressure can be changed in a hydrodynamic ESF step bearing due to a step change in excitation. The time it takes while operating at a constant speed to change from a steady state (unexcited) lifting force to 99% of the final (new excited) lifting force was found experimental [26] to be of the order of 10 ms. It would be expected to be similar to the valve (order of 1 ms). The CFD solution to this problem for which there is no analytical solution suggests that the response time of such a device is indeed about 1 ms. Experimental work is currently under consideration to re-measure t_o on a new set of apparatus thereby clarifying the position.

2.25 The Next Generation of CFD Solutions

How CFD Solves Non-Newtonian Flow Situations

Equation (2.28) allows the shear rate to be determined at every cell location. Using the constitutive model (e.g. Bingham plastic equation) the non-Newtonian viscosity μ_{nn} ($=\tau/\dot{\gamma}$) can be determined at every cell location. Each individual cell can then be treated as a Newtonian fluid for which the Navier-Stokes equations apply. The velocity values at each cell face influence the shear rate of the adjoining cells. For a domain in which enough correctly shaped cells are chosen the problem can be solved using a semi-iterative method until the whole domain converges.

Constitutive Model

All of the CFD solutions referred to in this chapter were obtained using the bi-viscous model for which the ratio R was chosen so that it is practically indistinguishable from the Bingham model. Directly using the Bingham equation can result in the viscosity at certain cell locations going to infinity making the computing process more difficult.

Within the CFD method complex constitutive equations (Carreau model) that accurately represent experimental viscometer data can be readily used. The plastic viscosity parameter in the Bingham type models does not have to be assumed to be independent of field strength or indeed shear rate. A constitutive model that fits the relationship (μ_{nn} vs. $\dot{\gamma}$) allows it to be easily inputted into the CFD code. All the CFD method requires is the relationship between μ_{nn} and $\dot{\gamma}$.

Using Small Scale Cylindrical Viscometer Data

The annular valve type viscometer allows higher shear rate data to be obtained than the cylindrical viscometer. A valve operates with high shear rates at the wall but within the valve the shear rate drops to zero at the centre line therefore allowing the electro-stress effect to be utilised. A suitably designed cylindrical viscometer will allow more accurate determination of the relationship (μ_{nn} vs. $\dot{\gamma}$) than an annular valve viscometer – there is no need to numerically integrate the data to obtain the shear rate in the former.

For devices that do operate with shear rates above those measured in the cylindrical viscometer the overall flow curve Figure 2.9 can be assumed for which high shear rate data can be predicted. The final viscosity value $\mu_{\dot{\gamma}\to\infty}$ can be taken to be equal to the no-field value μ_o.

The advantages of this technique are as follows:

A drop in the electro-stress contribution with $\dot{\gamma}$ will be automatically accounted for.

There is no need to measure a yield stress or plastic viscosity value.

There is no need to assume that Newtonian behaviour occurs at the onset of yielding.

There is no need to use the zero field viscosity as the plastic viscosity.

Small samples of fluid can be used.

Provided that the assumptions of laminar flow and time-independent fluid properties are valid this approach will provide the most accurate solutions.

Complete Model

Computational fluid dynamics offers the potential to model machines that could involve elaborate geometry; non-steady and non-isothermal flow; and in which the fluid properties vary with applied field and temperature as well as shear rate.

Commercial CFD code [11] is universally accepted to accurately model non-isothermal flow and can quite easily accommodate complex geometry by virtue of in package meshing facilities and is regularly used to obtain real engineering solutions. The parts of the code that do this can be switched on when modelling ESFs. The required inputs will be the thermal conductivity of the walls and their known temperatures, as well as the specific heat coefficient of the fluid.

The viscosity can be set to be a function of shear rate only. The ultimate aim is to develop an entire map of the viscosity function using the proposed small sample viscometer method:

$$\mu_{m} = f(\dot{\gamma}, T, E) \qquad (2.37)$$

This function for a particular ESF could then be inputted into the CFD code. In setting up a model the field strength (E) at each cell location could be inputted that could even be set to vary in the time domain. The code will need adjustments to enable such models, but the vast majority of the time consuming programming has already been done.

References

1. Bullough, W.A. and Peel, D.J. "Electro-rheological oil hydraulics", Proc. Japan Soc. Hydraulics and Pneumatics, Vol. 17, 1986, pp 30-36.

2. Bullough, W.A. "Adaptronics and Smart Structures, Chapter 6", Editor: Janocha, H., Springer-Verlag Berlin Heidelberg New York, 1999, pp 161-179.

3. Bullough, W.A., et al. "The ER catch: design, performance, considerations and operation" Proc. IMechE., Vol. 207, 1993, pp 253-266.

4. Bird, B., Dai G.C. and Yarusso, B.J. "The rheology and flow of viscoplastic materials", Rev. Chem. Eng. 1, 1983, pp 1-70.

5. Peel, D.J. and Bullough W.A. "Prediction of electro-rheological valve performance in steady flow", Proc. Instn. Mech. Engrs., Vol. 208, 1994, pp 253-266.

6. Atkin, R.J., Shi, X. and Bullough W.A. "Effect of non-uniform field distribution on steady flow of an electro-rheological fluid", Jnl. of Non-Newtonian Fluid Mechanics, Vol. 86, 1991, pp 119-132.

7. Atkin, R.J., Shi, X. and Bullough W.A. "Solutions of the constitutive equations for the flow of an electro-rheological fluid in radial configurations". Jnl. of Rheology, Vol. 35, 1991, pp 1441-1461.

8. Wilkinson, W.L. "Non-Newtonian fluids", 1960, Oxford: Pergamon Press.

9. Bullough, W.A., et al. "Two dimensional flow of an ESF: experiment, CFD, Bingham plastic analysis", At press, Int. Jnl. Mod. Phys. B.

10. Leek, T.H., et al., "Solutions for one dimensional combined Couette and Poiseuille flow of an electro-rheological fluid modelled as a Bingham plastic", Private communication.

11. Fluent Incorporated: Fluent 4.4, hand book

12. Kinsella, J. "CFD modelling of an electro-rheological fluid: feasibility study", MEng thesis, Department of Mechanical Engineering. Sheffield University (1996).

13. Urang, S. "Electro/magneto-rheological clutch and valve computational fluid dynamics", BEng thesis, Department of Mechanical Engineering. Sheffield University (1999).

14. Ellam, D.J. "Electro-structured radial clutch with induced cooling flow: Bingham plastic analysis and CFD", MEng thesis, Department of Mechanical Engineering. Sheffield University (1999).

15. Gorodkin, S., Lukianovich A., and Kordonski W. "Magneto-rheological throttle valve in passive damping systems", Proc. Smart materials and structures: 4th European and 2nd MIMR conference, held Harrogate, June 1998. Institute of Physics Publishing Ltd, 1998, pp 261-266.

16. Peel, D.J., Stanway R., and Bullough, W.A. "A generalised presentation of valve and clutch data for an ER fluid, and practical performance prediction methodology", Int. Jnl. of Phys. B, Vol. 10, 1996, pp 3103-3114.

17. Peel, D.J. and Bullough, W.A. "A technique for the normalisation of electro-rheological fluid performance data in cylindrical/shear and pressure/flow modes of steady operation", Proc. of SPIE: Smart Materials, Structures and MEMS, Bangalore, India, 1996, pp. 511-523.

18. Lipscomb. G.G. and Denn. M.M. "Flow of Bingham fluids in complex geometries", Jnl. Non-Newtonian Fluid Mechanics, Vol. 17, 1984, pp 337-346.

19. Sproston, J.L., et al. "A numerical simulation of electro-rheological fluids in oscillatory compressive squeeze-flow", Jnl. Phys. D: Appl. Phys., Vol. 27, 1994, pp 338-343.

20. Carreau, P.J. "Rheological equations from molecular network theories", Trans. Soc. Rheol., Vol. 16, 1972, pp 99-127.

21. Barnes, H.A., Hutton, J.F. and Walters K. "An introduction to rheology", first ed, New York: Elsevier Science Publishers (1989).

22. Whittle, M., et al. "Electo-rheological relaxation times derived from pressure response experiments in the flow mode", Jnl. Non-Newtonian Fluid Mechanics, Vol. 57, 1995, pp 1-25.

23. Whittle, M. "Dynamics of an electro-rheological valve", Proc. 5th Int. Conf. on ERF/MRS and associated technology, held Yamagata. World Scientific Publishing, 1995, pp 100-115.

24. Whittle, M., Atkin, R.J. and Bullough W.A. "Fluid dynamic limitations on the performance of an electo-rheological clutch", Jnl. Non-Newtonian Fluid Mechanics, Vol. 57, 1995, pp 61-81.

25. Whittle, M. "Dynamics of a radial plate clutch", Proc. 6th Int. Conf. on ERF/MRS and their applications, held Yamagata. World Scientific Publishing, 1999, pp 100-115.

26. Leek, T.H., et.al, "The time response of an electro-rheological fluid in a hydrodynamic film", Int. Jnl. Mod. Phys. B., Vol. 10, Nos. 23 & 24, 1996, pp 3343-3355.

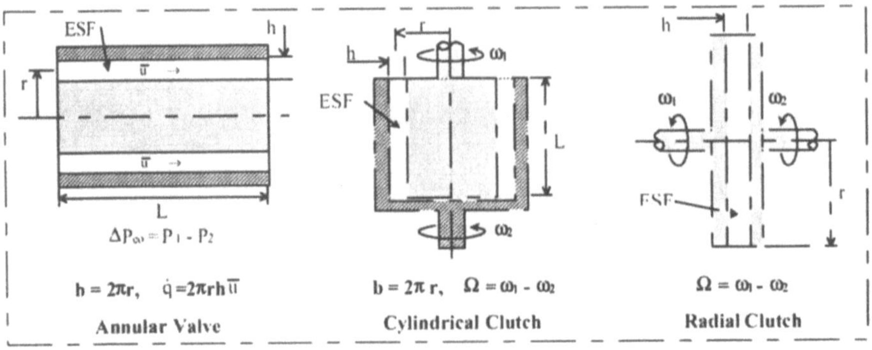

Figure2.1 Geometry and notation for simple ESF devices

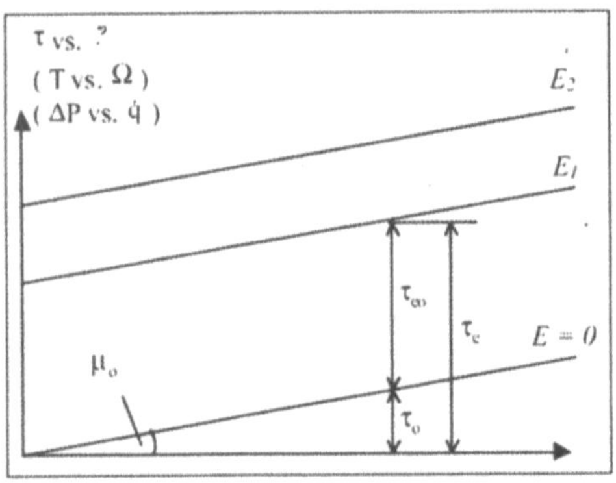

Figure 2.2 Idealised flow curves for an ESF annular valve and clutch

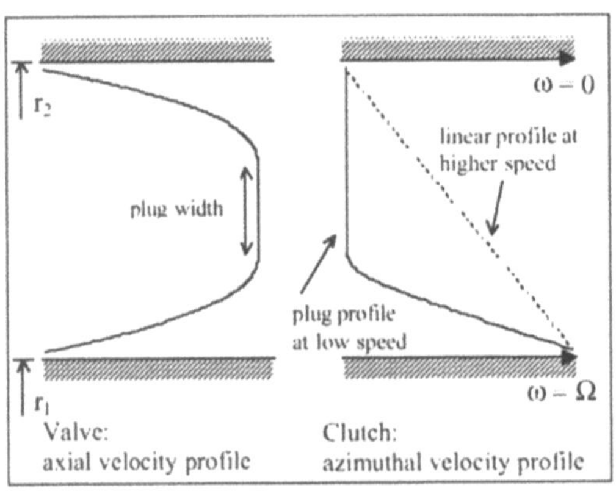

Figure 2.3 Plug velocity profiles produced by CFD for an annular valve and concentric clutch

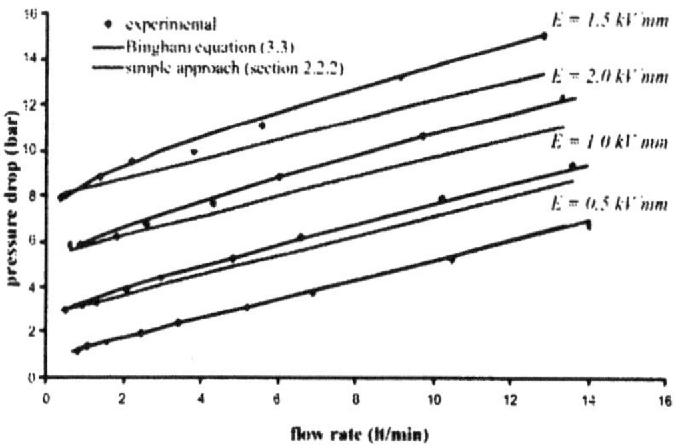

Figure 2.4 Bingham plastic solution versus ideal model and experimental data from an ER valve

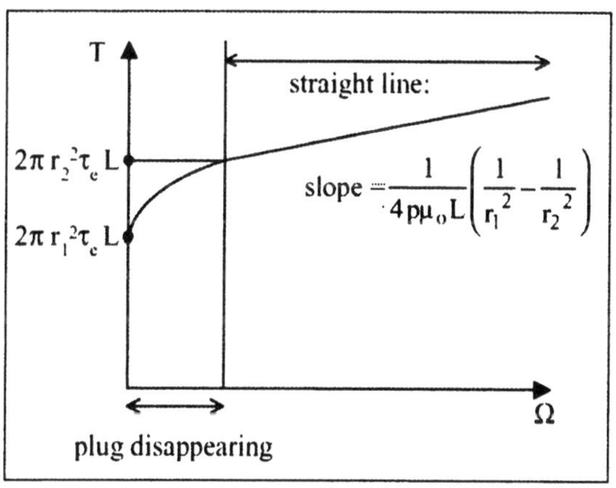

Figure 2.5 Non-linear T_{eo} vs. Ω Couette clutch relationship due to plug presence at low speeds

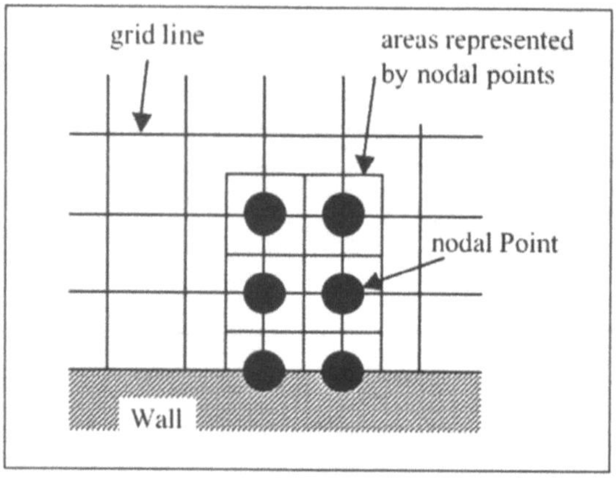

Figure 2.6 Typical grid as used in the CFD analysis to obtain semi-iterative solutions of the governing equations of motion

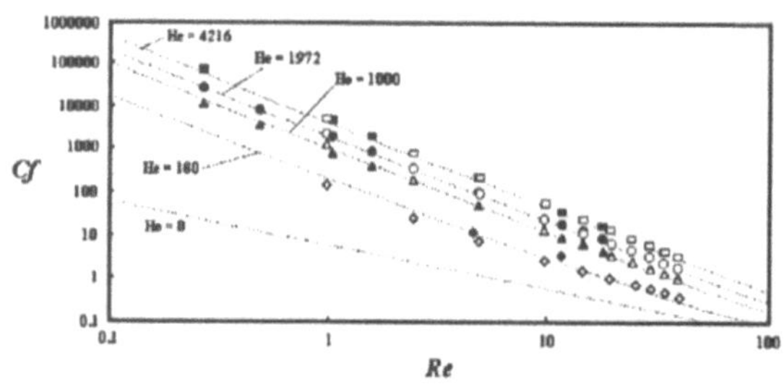

Figure 2.7 MR valve data plotted using the dimensionless parameters: Re, Cf & He

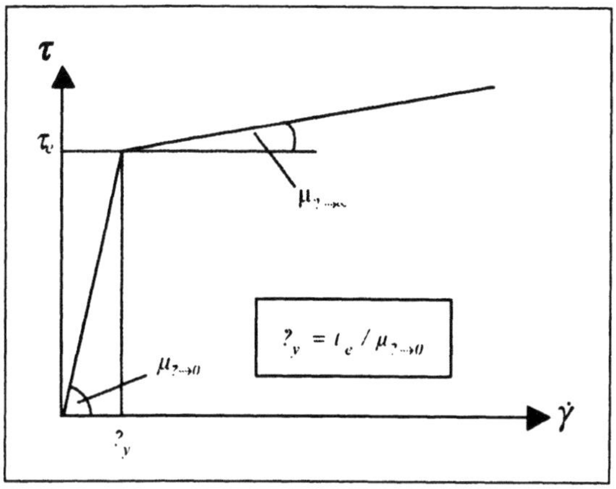

Figure 2.8 Flow curve for the bi-viscosity model

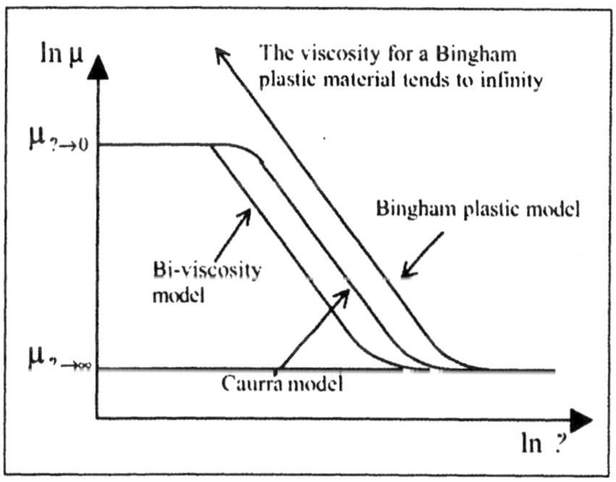

Figure 2.9 Full flow curve for high and low shear rates

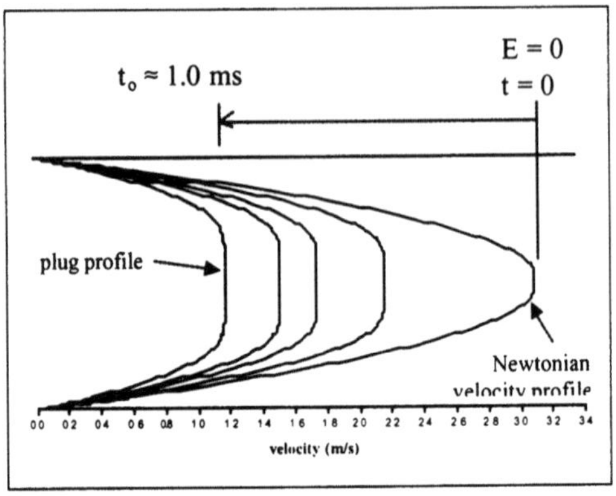

Figure 2.10 CFD annular valve prediction: change in velocity profile due to a step input of excitation at constant pressure difference

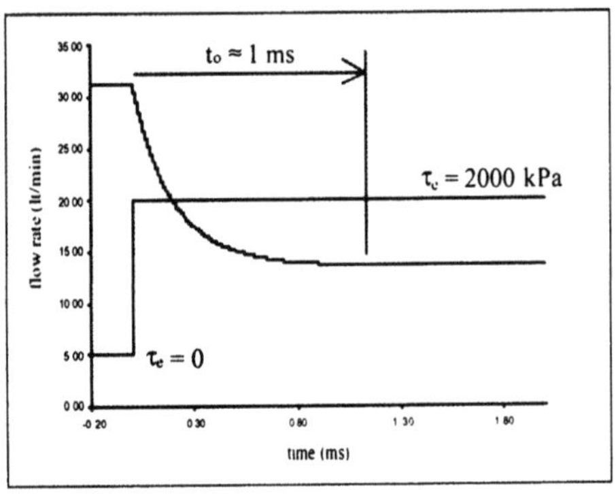

Figure 2.11 CFD annular valve prediction: flow rate response due to a step input of excitation at constant pressure difference across valve

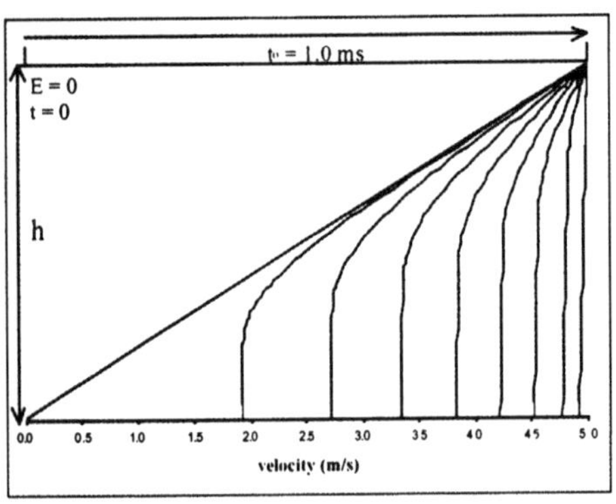

Figure 2.12 CFD concentric clutch prediction: minimum time required for both cylinders to run up to the same speed (velocity profiles)

3. ESF Clutch Driven Mechanisms and the ER Linear Reversing Motion Demonstrator

W. A. Bullough, A. R. Johnson, J. Makin and R. C. Tozer

The University of Sheffield, Department of Mechanical Engineering
Smart Machines, Materials and Related Technologies Unit, Sheffield, England

3.1 Introduction

The ESF clutch is formed by enclosing fluid between driver and driven plates and exciting it magnetically or electrostatically, or both, thereby enabling a torque or force transmission depending on whether the applied motion is rotary or linear respectively. Generally the more rewarding ESF clutch applications drive an inertial type of intermittent load, otherwise highly developed, mechanical and electrically closed or opened clutches are preferable on a cost and convenience basis. If a shaped motion is called for servo or stepper motors would be natural first choices. It follows that if the ESF clutch is to be employed it must have something special to offer and this derives from a combination of rapid switching capability and low output side inertia, see Figure 3.1.

It will be noted that the ESF clutch does not need to close in the sense that its plates must touch, thus allowing a concentric design aspect, and that the majority of the clutch inertia can be due to the fluid mass: on 'no load' and with negligible solid parts in the output shafting the response time will be limited by the fluid inertia (which up to a point depends on the properties of the fluid) and the rate of electron-hydraulic (t_m*) torque engagement in which the polarising rate is implicit - see Section 1. Also, when excitation is not applied relatively high parasitic shear stress associated losses are present in the slipping clutch with consequent heat generation problems arising. Since the electrical energisation of an ERF also has a heating effect, then the cooling of an ESF clutch is always likely to be a prime consideration - whether natural or forced. The presence of coil/field windings in the proximity of an MRS clutch may impede and indeed magnify the heat transfer provision.

3.2 The ERF Demonstrator Clutch Example

One example in particular [1], the linear motion reversing mechanism, demonstrates the potential flexible operational advantages of an ESF clutch driven machine and also the practical disadvantages of, in this case, the present generation of ERF. Additionally, a study of the design methodology involved in arriving at even a prototype device will give a good, if extreme example of the interdisciplinary, multivariable, non-linear nature of the problem faced by the design engineer. This runs all the way through the design from, for example, the fluid characteristics which include perhaps hysteretic tendencies, to position control which is made digital in order to accommodate them. Likewise the exponential current drawn/semi-conductor tendency of the ER fluid (thermal runaway) is inverted to allow temperature control by cutting down the voltage input once periods of unsteady operation that necessitate a high yield stress are over and varying the switching time. Cooling is assisted by having the input clutch which always rotates at full speed, as the outer. The choice of ER rather than MR medium is in the present case related to the requirement for multi-head yarn winding machines which could enjoy a common input drive shaft and ± excitation bus-bars and possibly because of the t_0 run up time limitation which may be more severe for MRS - the much higher yield stress capacity of these fluids being otherwise unutilisable on the very low output inertia system.

For the purposes of the numerical design of a clutch since $t_m^* <<$ the run up time (1 m.sec compared to ~ 10 m.sec) and power supply RC $<< t_m^*$ (typically 100 μ.sec compared to 1 m.sec) the yield or electro stress can be assumed to apply immediately the voltage is switched on and vice versa. High capacity H T electrical sources and drive down (earth shorting assistance) are required respectively to achieve the required level of performance.

Choosing an inter electrode spacing (h) of 0.5 mm in order to keep the supply voltage (E) to below 4kV and to limit the indicated theoretical value of $t_0 \sim 4\ \rho h^2/\mu$ whilst producing only a sensible amount of heat from $\int V \times i.\ dt$ and viscous τ_0 losses leads to a large clutch surface area and effective radius (r) product. A compact and relatively stiff lightweight design is achieved by the adoption of a dual channel concentric design. Given that the r/h ratio is high enough to eliminate radial effects in velocity and electric fields, the shear rates and the field strengths are ~ uniform.

Experience [2] has shown that for the present design whether the fluid properties are taken from low shear rate Couette or high shear rate Poiseuille flow tests makes little difference to the outcome, the change of τ_e and μ over the shear rate range 0 ~ 13,000 s^{-1} being insignificant in terms of errors in comparison to say those caused by the effects of temperature. The acceleration of the belt is ~ 100g and the centrifugal field ~ 30g, both have some effect on the fluid performance which is invariably derived from steady state tests. Also, the dimensions of the clutch (gap) are unlikely to remain constant given these severe mechanical conditions of operation.

The design involves limiting technology in many of its aspects and has not yet resulted in an optimum machine. The iterative nature of the ESF flexible machine design field is the reason why the yarn wind prototype has often been referred to as the 'linear motion reversing demonstrator' or even the 'universal ERF rheometer', on account of the unsteady state and the thermal, electric and overall fluid requirements. The approach to such problems will usually remain iterative, solving individual sectors of the problem (e.g. electrical switching) and relaxing them against the overall scheme of things. This is the method used in the following text.

3.3 Dynamic Performance Considerations

A measure of hydraulic balance is bestowed by the twin concentric channel clutch design which leads to a sensible radius and non overlong inverted lightweight cup output clutch connected to a pulley via a short bearing/shaft arrangement. The pulley is arbitrarily given a diameter equal in circumference to the sweep length of the motion, with the rotational speed of 1200 rpm nominal, giving the traverse a velocity of 5 meters per second, at its periphery. This is the starting point, the analysis that follows being used as a continual design tool for similar arrangements once it has been verified by experiment.

In more detail, the continuous belt links one pulley to its twin - Figure 2. Motion is supplied, for the time being, by two contra rotating variable speed motors which drive relatively high inertia constant speed input clutches. As each clutch is energised alternatively the clutch with the yield stress component causes the previous motion to reverse, thus the yarn feed speed/position changes;

sometimes it has a steady rate and sometimes not (during the turn round process) the position of the point of reversal being adjusted by the synchronised timing of the on/off voltage waves.

At the outset it is assumed that the heat generated by excitation and shear will not prove excessive in comparison with the motion assisted cooling capacity of the outside of the outer clutch. A practical design point in the fluid engineering is that the cap fitted on top of the outer (driving) clutch, serves to reduce the radius to that of the inside channel (or less) should any fluid, under any circumstance, be induced to centrifuge out of the assembly. Also, when and if an anti-contamination or hermatic seal is fitted between the input and the output clutch shaft, the seal would be of a small diameter and friction type.

Switching is directed by an adjustable timing system fed from a high capacity DC source with a bi-polar ± HT output which is fed to the clutch by carbon brushes. In this way it is hoped that the constant reversal of the velocity and of the current directions will act so as to eliminate or reduce any electrophoresis or particle separation effects in the fluid. The drive down assistance is required in order to drain the charge from the fluid rapidly.

The dynamic analysis assumes a constant yield stress- $\tau_e \neq f\ (\dot{\gamma})$ and is given only in outline. Since MRS and ERF have similar constitutive forms the approach is common to both media. Results of simulations are given in Figs. 3.3.

During a run of the system the performance is taken from a speed/position measurement tachogenerator located on a third, belt tensioning pulley, with an infrared thermometer reading on the outside of the blackened clutches. All data is logged (including current and voltage) on a fast data logging system which is controlled and outputs via a PC. Verification is in the form of Figure 3.4.

Given confidence from the verification of the analysis [3], this is then used to check the degree of power optimisation of the system, Figs. 3.5. Overall the position now becomes one of dynamics programme accuracy plus thermal imbalance. ER fluid gives a dis-proportionate increase in current at higher temperatures. MRS would not be so sensitive to temperature in an electrical supply sense. It should be noted that some fluids respond poorly in yield strength magnitude and time at off design temperature conditions. Since the viscosity of the fluid and its conductance are both temperature dependent the simultaneous dynamic, thermal, electric and fluid property interactive simulations are ponderous though only quasi steady in some respects. Hence optimisation modelling tends to be at constant (but varied) temperature. Any hysteretic [4] tendency in the fluid is countered by the digital mode of operation.

3.4 Thermal Balance

Over one cycle of steady operation of the yarn traverse the energy rates crossing the control boundary will balance; $\dot{Q}-\left(\dot{W}_e + \dot{W}_x\right) = 0$, where the subscripts e and x apply to electrical and mechanical energy rates $\dot{W}$ respectively. Strictly speaking this equation applies to the system defined in Figure 3.2 and is for both clutches.

From symmetry, since electrical power is always (at this stage) applied to one or other clutch, neglecting the overshoots associated with the charging and discharge currents and taking the heat transfer process to be uniform with time:

$$2\times \dot{Q} \text{ per clutch} \times t_{cy} = (i \times V)\, t_{cy} + 2 \int_{cycle} \Omega T_{\Lambda} \,.\, dt \tag{3.1}$$

wherein t_{cy} is the period of one complete cycle and i and V are the steady current and step voltages applied to excite τ_e respectively.

Given the continuing assumption that the Bingham plastic constitutive equation applies instantly the fluid is excited and Newtonian when not, that $\dot{\gamma} = (\Omega \pm \omega)\, r / h$ and that the triangular portions of the Fig 3 power curve are identical then:

$$\dot{W}_x \text{ per clutch} = \Omega \left[2\pi\tau_e \left(r_o^2\, l_{eo} + r_i^2\, l_{ei} \right) t_{trt} + 2\pi\mu \left(l_{vo}\, r_o^3 / h_o + l_{vi} \,.\, r_i^3 / h_i \right) 2\Omega \left(t_{cy} - t_{trt} \right) \right] \tag{3.2}$$

where t_{trt} is the time for the clutch to reverse from the +ve to -ve maximum velocities.

It will be noted that a high τ_e value will increase the height of the trapezoid on the power curve and reduce its period. For the parameters associated with the present problem the relationship is inversely proportional within a percent or so. Hence testing to validate the thermal equilibrium procedure need not take place at the optimal turn round time. Perhaps more importantly it means that $\tau(\theta)$ and the duration of the turn round time are not factors in the iterations of the thermal model. Note also the dependence of power on the inertia term – Figure 3.5.

The heat transfer model per-se was checked by running the outer (drive) rotors at speed whilst holding the belt stationary with unexcited ER fluid of known viscosity. At the equilibrium temperature the shaft work equals the heat transfer rate. The blackened outer rotor has its temperature taken and plotted against a calculated value based on the viscous shaft work input. A full verification of the thermal package could then be carried out, Figure 3.6.

3.5 Summary

Essentially a maximum ERF electro-stress ~ 7kPa is needed in order to turn round the postulated minimum inertia output clutches in the minimum possible time: the magnitude of τ_e does not greatly affect the amount of heat generated. Limitations on the existing main voltage supply and switches determine that the gap shall be 0.5mm. For want of more authoritative guidance the viscosity has to be not a long way below ~ 100 mPa. s 1 cm, for t_o (run up time performance) reasons. Cooling problems dominate any optimisation process.

Geometry, speed and inertia are the important parameters to be set by designers. Simulations (Figs. 5) show that little is gained by changing the pulley diameter and the speed of rotation. This part constitutes but one element of the output rotor inertia and a lower angular speed reduces the cooling capacity of the outer surface of the output rotor, roughly in proportion to any reduction achieved in the amount of mechanical work done. A smaller radius inverted output rotor cup indicates a more convenient drop in inertia though its length must be increased to maintain the turn round time. However, the cup soon becomes overlong given the severe mechanical stresses that the cantilevered assembly has to tolerate. The heating balance is not effected as the outer rotor length must increase in line with the cup length.

The greatest gain available is in cooling capacity, by a factor of 6 × achieved by increasing the outer diameter of the input clutch and finning its circumference: this also gives a welcome added non accelerated inertia should the hesitation seen at the end of the belt acceleration prove to be due to speed regulation on the prototype test arrangement rather than a t_0 effect - Figure 4b.

Finally, the electrical load can be reduced after a pair of clutch rotors lock. This seems also to provide a means of trimming fluid temperature, by the timing of the reduction. A conservative voltage reduction (5.6 to 2 kV) of that required to accelerate the output clutch to speed is proposed for the uniform speed regions.

Given that the accelerating parts will be reduced to near their minimum practical mass (without using exotic materials) a turn round time of 14 ms and an equilibrium temperature of ~ 35°C should be achievable, Figure 7. This temperature level plus the reduced average field magnitude and amount of discharge current short circuited at switch off, the voltage reduction now takes place in two equal steps, may go some way towards improving the serviceability of the fluid in an anti-breakdown/electrophoretic sense. With this, the procedure should provoke a high level of repeatable output - See Figure 8.

The hoped for final stage of development outlined above will necessitate a part rebuild which has not yet been implemented. Though there may be some way to go (sealing of the fluid and control strategy development) before the machine reaches commercial reality, as say a yarn package winder, the methodology adopted has given the concept its best chance of reaching viability and the reliability standards necessary for industrial success.

References and Bibliography

The contents of this chapter have drawn extensively from

1. Bullough W A, Johnson A R, Tozer R and Makin J. Methodology, performance and problems in ER clutch based positioning mechanisms. Plenary lecture at 6th Int. Conf. on ER Fluids. MR Suspensions and their applications, held Yamagata, published in Proc. pp 623 - 630 (1997). World Scientific Pub. Co.

2. Makin J, Fast acting controllers, rheologically based devices, interactions between fluid, ancillary materials, auxiliary devices and operating requirements, draft PhD thesis, University of Sheffield, Department of Mechanical Engineering (2000).

3. Johnson A R, Bullough W A and Makin J, Dynamic simulation and performance of an ER clutch based reciprocating mechanism, Smart Mater. Struct 8 (1999), 591 - 600.

4. Tozer R C, Orrell C T and Bullough W A, On-off excitation switch for ER devices, Int. Jnl. Mod. Phys. B, V8, No 20 and 21, (1994), pp 3005 - 3013.

5. Wang H Y, Heat Transfer for Engineers, Longman (1977).

6. Rogers G F C and Mayhew Y R, Thermodynamic and Transport Properties of Fluids, Blackwell, Oxford (1982).

Appendix 1 - Dynamic Model

Considering clutch A to be energised and slipping (see Table 3.2 for nomenclature) the electro (e) torque T on the output rotor by the fluid is:

$$T_e = \tau_e (\pi d_o . l_{eo}) \frac{d_o}{2} + \tau_e (\pi d_i . l_{ei}) \frac{d_i}{2} \tag{3.3}$$

And the viscous (o) torque

$$T_{oA} = \pi\mu . \left(\frac{l_{vo} d_o^3}{4h_o} + \frac{l_{vi} d_i^3}{4h_i} \right) (\Omega - \omega) \tag{3.4}$$

Since $\dot{\gamma}_{oA} = \dfrac{(\Omega - \omega)(d_o/2)}{h_o}$ and $\dot{\gamma}_{iA} = \dfrac{(\Omega - \omega)(d_i/2)}{h_i}$ (3.5)

and $I\alpha = T_e + T_{oA} - F_A r$ from Newton 2nd law where F_A is the

belt/pulley force, and α the angular acceleration.(3.6)

For the unenergised rotor B the fluid torque

$$T_{oB} = \pi\mu \left(\frac{l_{vo} d_o^3}{4h_o} + \frac{l_{vi} d_i^3}{4h_i} \right) (\Omega + \omega) \tag{3.7}$$

$T_{oB} = F_B r - I\alpha$ and since (3.8)

$$F_A - F_B = mr\alpha \tag{3.9}$$

$(2I + mr^2)\alpha = T_e + T_{oA} - T_{oB}$ which, since $\alpha = d\omega/dt$(10)

$$\left[(2I + mr^2)d\omega\right] / \left[(\pi\tau_e/2)\left(d_o^2 l_{eo} + d_i^2 l_{ei}\right) - (\pi\mu\omega/2)\left((l_{vo} d_o^3/h_o) + (l_{vi} d_i^3/h_i)\right)\right] = dt$$
(11)

i.e. the basic differential equation defining the motion of the clutch A when it is energised and slipping.

A similar equation can be described to define the clutch B torque and slipping time. The electron-hydraulic time delay is taken to be zero and $\tau_e \neq f(\dot{\gamma})$ in which case (3.11) can be integrated to give the motion of belt and rotors

Appendix 2 - Heat Transfer Model

Heat transferred from the conducting outer cylindrical clutch which rotates about a vertical axis proves to be the dominant mode of cooling: the central drive shaft is small and the clutch base plus top end are secondary, especially so since the top is covered by an (insulating) nylon cap fluid spill arrester. The prime assumption is that the heat transfer simulation verified for a horizontal shaft single channel clutch [2] could be modified to suit a vertical axis twin channel model, radiation effects being negligible in both.

The principal difficulty in calculating the rate of transfer lies in the determination of the temperature of the ERF and air properties which in turn determine the mode of convection (free or forced). It was accomplished thus:

Stage1: An initial estimate of the temperature of the fluid in the outer gap θ_o and an estimate of the temperature difference between the inner and outer fluid gaps $\Delta\theta_{io}$ was made. These estimated temperatures were then used to evaluate the ER fluid viscosity and the electrical conductivity in each of the inner and outer fluid gaps, from which an estimate of the rate of energy input within each gap is determined. In the steady state, this rate of work input equals the rate of heat conduction to the clutch surface. The thermal conductivity of the outer drive rotor was evaluated using equations derived from [5] from which an estimate of the outer rotor surface temperature was determined. The mean of this estimated surface temperature and the ambient air temperature was used to determine the air properties from [6] and thus the rate of heat transferred to the surroundings at that estimated surface temperature was calculated.

Stage 2: The fluid viscosities at the inner and outer fluid gaps at the estimated temperatures together with the air properties calculated in stage 1 were then used to calculate values of the ER fluid temperature in the outer electrode gap, θ_o and the temperature difference between the gaps θ_I - θ_o. See following equations.

Stage 3: The estimated and calculated temperatures were then compared and new estimates were made for input to stage 1. The computational process was repeated until sufficiently close estimated and calculated values were obtained. (<1°C difference for most work).

An initial underestimation of the fluid temperature led to a high calculated value of fluid viscosity in the heat calculation and thus an overestimate of the calculated fluid temperature. The simple procedure of using the mean of the initial estimate of θ_o and the computed value of θ_o as the new estimate in the iteration procedure quickly converged or indicated that thermal equilibrium would not be achieved.

The relationship between the clutch geometry, heat transfer within the clutch and the heat dissipation from the clutch outer surface are derived in [3] from empirical relationships for heat transfer and the equations for the non dimensional numbers of [5].

For the purpose of the thermal analysis per se the heat conduction rate through the aluminium light-weight output rotor $\dot{Q}_i$, assumed to be a hollow circular cylinder of radii r_i and r_o, length ℓ_{vo} and thermal conductivity k_n, was related thus:

$$\dot{Q}_i = (\theta_i - \theta_o) 2\pi \ell_{vo} k_n / \ln(r_o / r_i) \qquad (3.12)$$

The cylindrical wall of the constant speed steel drive rotor was similarly treated as was its base. Using L_o as the length of the outer rotor of radius R_o, A_b as the area of the base of d_b thickness plus k_s the thermal conductivity of steel and θ_w as the outer wall temperature:

$$\dot{Q}_{io} = (\theta_o - \theta_w)\left(\frac{2.\pi.L_o.k_s}{\ln(R_o/r_o)} + \frac{A_b.k_s}{d_b}\right) \qquad (3.13)$$

where $\dot{Q}_{io}$ is the heat conduction rate through the outer drive rotor.

The heat transfer to the surroundings $\dot{Q}_c$ from the cylindrical outer case of area A

$$\dot{Q}_c = h_c A(\theta_w - \theta_a) \qquad (3.14)$$

where

$$h_c = \overline{Nu}_w k_a / 2R_o \qquad (3.15)$$

The thermal conductivity of air k_a is linearly interpreted from tables [6] by taking the mean of the ambient air temperature θ_a and the wall temperature θ_w.

The mean Nusselt number $\overline{Nu}_w = 0.18.\left[\left(0.5.Re_w^{\ 2} + Gr_w\right).Pr\right]^{0.315}$ (3.16)

where Reynolds number $Re_w = \Omega(2R_o)^2/\nu_a$ (3.17)

and Grashof number $Gr_w = \beta g\,(2R_o)^3\,(\theta_w - \theta_a)/\nu_a^{\ 2}$ (3.18)

for the present clutch size and operating parameters combined free and forced convection is assumed, i.e. $Re_w \leq 5 \times 10^4$, ν_a is kinematic viscosity of the surrounding air.

For the outer rotor base, modelled as a disc of radius R_o with a hole of radius r_{bi} in the centre, the defining Reynolds number used was $\Omega(2R_o)^2/\nu_a$ which indicated laminar flow for a Prandtl number between 0.7 and 5.0. In this chosen case:

$$\overline{Nu}_{bo} = (0.277 + 0.105\,Pr)Re_{bo}^{\ 0.5} \quad \text{and} \quad \overline{Nu}_{bi} = (0.277 + 0.105\,Pr)Re_{bi}^{\ 0.5} \qquad (3.19)$$

giving $h_{bo} = \overline{Nu}_{bo}.k_a/2.R_o$ and

$$h_{bi} = \overline{Nu}_{bi}.k_a/2.r_{bi} \qquad (3.20)$$

hence

$$\dot{Q}_b = (\theta_b - \theta_a)\left[0.5.k_a.\pi.\left(\overline{Nu}_{bo}.R_o - \overline{Nu}_{bi}.r_{bi}\right)\right] \qquad (3.21)$$

The summation of all elements of the heat transfer rate problem related the total heat dissipated $(\dot{Q})$ and $\theta_o - \theta_a$ and formed the basis of the iteration procedure.

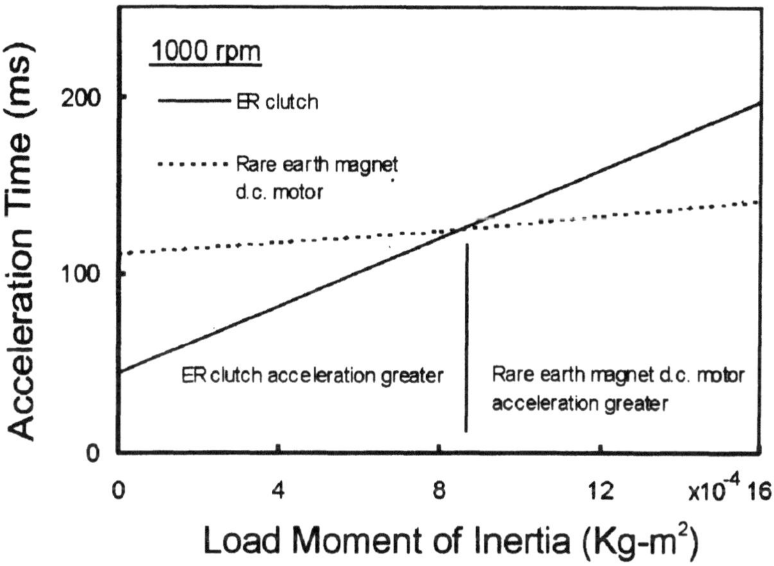

Figure 3.1 Comparison of the acceleration times of an ER clutch against a typical 0.5 kW 1.6 Nm rare earth magnet dc. motor for a range of load moment of inertia. Common output speed 1000 rpm, fluid Lipol 30/w @ 3.6kV/mm, h = 0.5mm. The inertia of the ER clutch output rotor = 4×10^{-4} kgm^2; The inertia of the armature of the motor = 4×10^{-3} kgm^2.

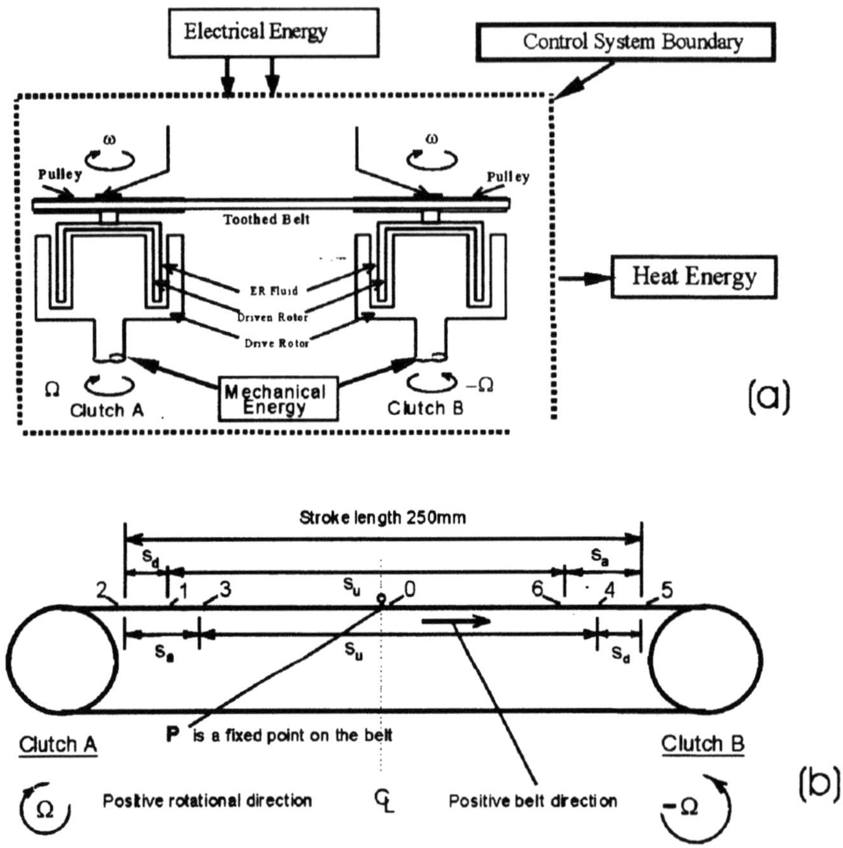

Figure 3.2 ER yarn wind system/linear motion demonstrator. (a) The control system boundary ; (b) The motion cycle

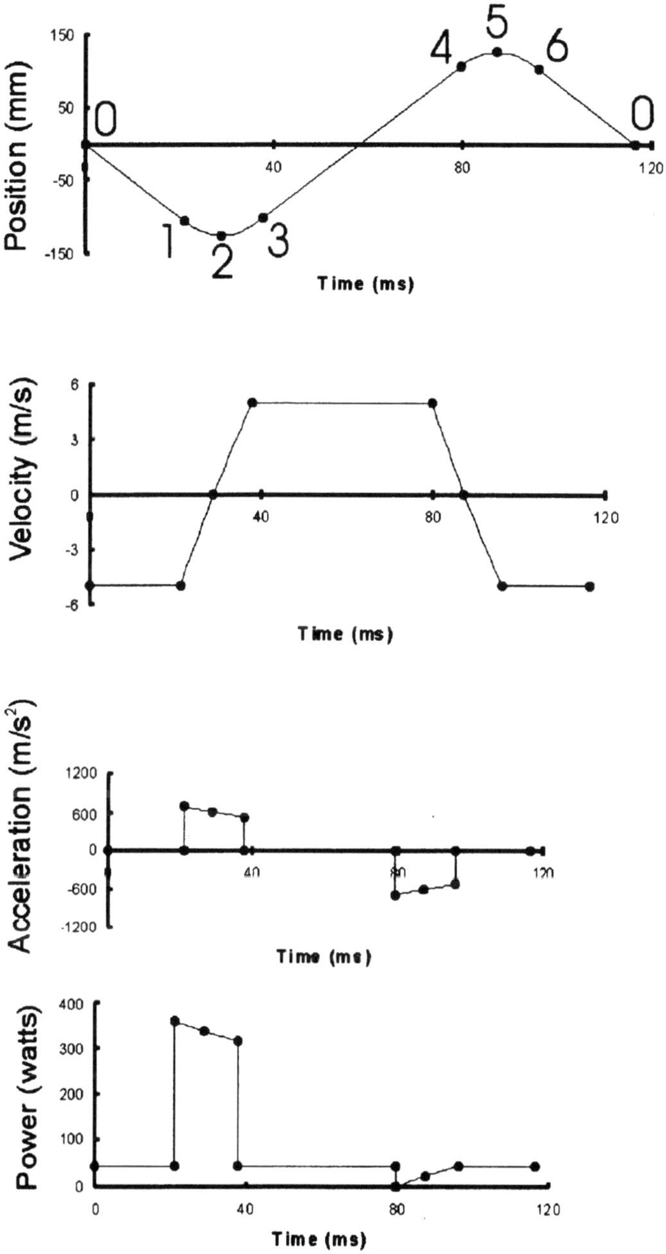

Figure 3.3 Typical results from dynamic analysis performed on short sweep prototype: Assumed values of τ_c and μ. Data as Table 3.2.

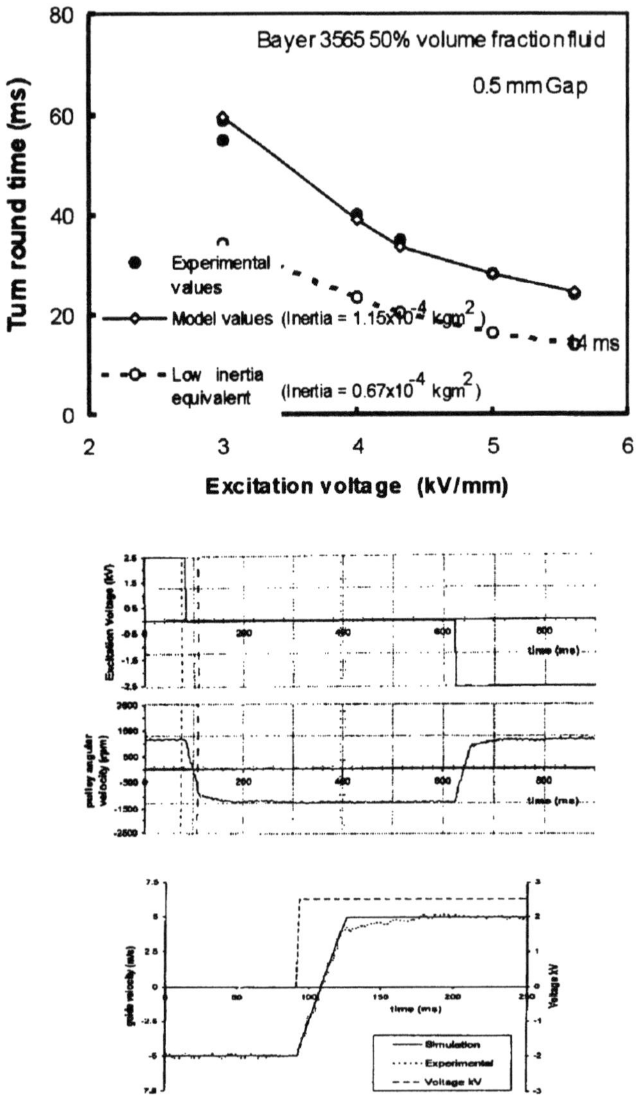

Figure 3.4 Verification of Dynamic Analysis. (a) - Experimentally derived turn round times with the linear traverse test rig together with the model derived times and 'low inertia equivalent' turn round times. Temperatures between 49°C and 60°C. Speed 1200 rpm/5ms (b) Speed changes following application of a step change in the excitation (Bayer Rheobay® VP AI 3565 50% fluid at 50°C, geometric values given in Table 3.2). (i) Experimental speed changes displayed on a digital oscilloscope (for clarity scales have been super-imposed on the traces); (ii) Comparison of experimental and theoretical speed changes.

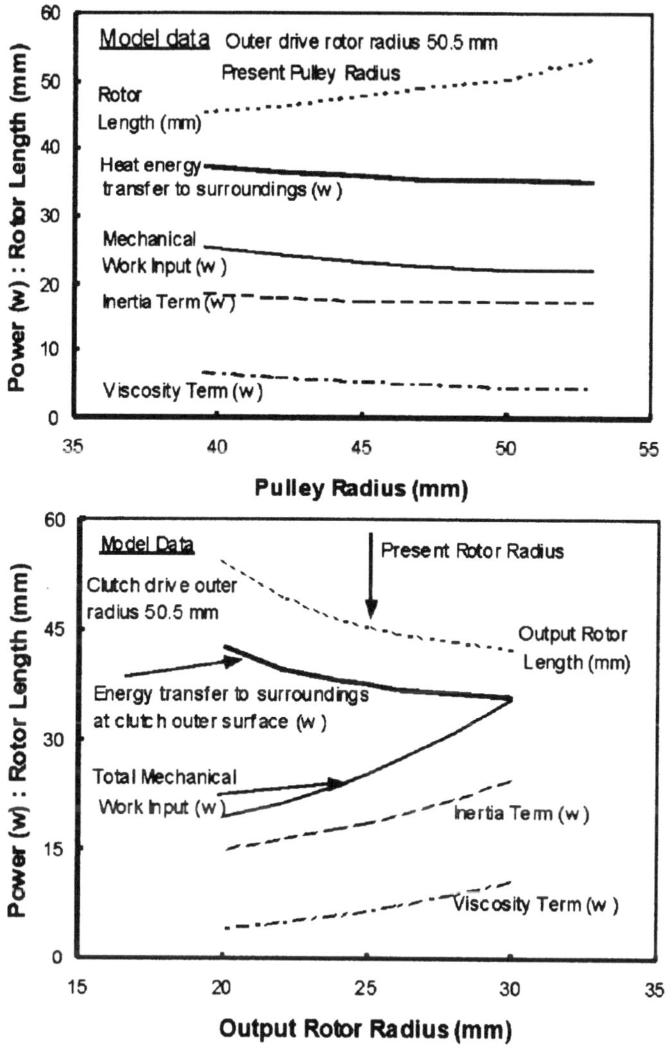

Figure 3.5 Power/Heat Transfer Optimisation Simulation (a) The effect of a change in the pulley radius with the rotational speed reduced to maintain the belt speed at ± 5 m/s, with the rotor lengths increased to maintain the turn round times at less than 14.25 ms with the increase in pulley diameter and thus moment of inertia. The fluid properties used in the modelling were those for the Bayer 3565 50% fluid at 65°C and 5.6 kV/nm. Clutch outer drive rotor radius 50.5 nn. Nominal output inertia ~ 0.67×10^{-4} kgm^2, clutch radius R~ 25.5 mm. (b) The effects of a change in the clutch output rotor radius with the rotor lengths changed as required to maintain the turn round time (at belt velocity nominal ± 5 m/s) at less than 14.25 ms. The fluid properties used in the modelling were those for the Bayer 3565 50% fluid at 65°C at 5.6kV/m excitation with the moment of inertia equal to 0.67×10^{-4} kgm^2 at the present rotor radius of r ~ 25.5 mm.

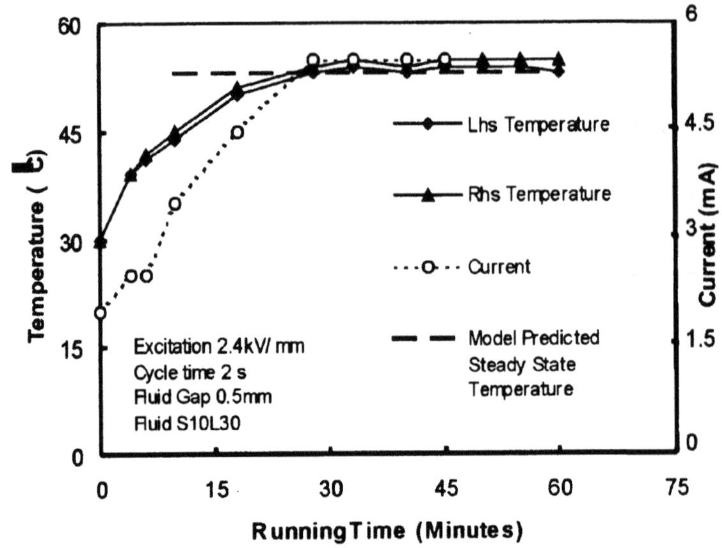

Figure 3.6 Experimental and theoretical steady state running temperature of the left and right clutches with time during the running period. Also shown is the current drawn. S10L30 ER fluid. 0.5 mm gap rotors. 2.4 kV/mm excitation. Output moment of inertia of 1.62×10^{-4} kgm^2. Rotational Speed 1200 rpm.

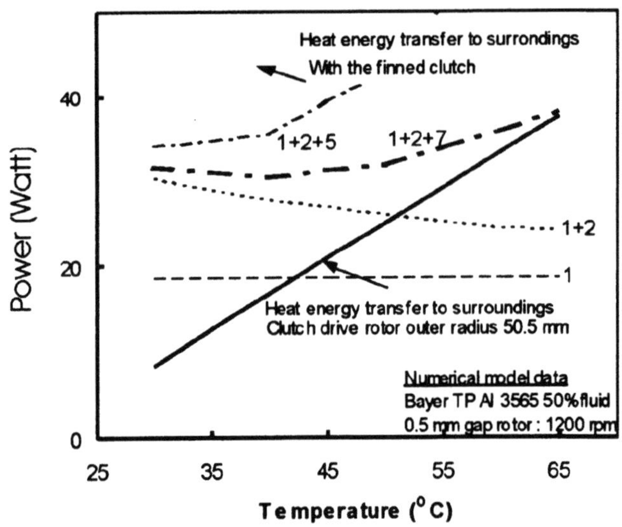

Figure 3.7 The effect of introducing fins to the outer surface of the clutch drive rotor on the heat transfer to the surroundings. Ambient temperature 20°C. Total output moment of inertia 0.67×10^{-4} kgm^2. Bayer TP AI 3565 50% Fluid. Rotor gap 0.5 mm. Motor speed 1200 rpm.. (1) Mechanical work input due to inertia (low inertia equivalent); (2) Mechanical work input due to viscosity term; (5) Electrical energy input @ 5.6kV/mm; (7) Electrical energy input @ 5.6kV/mm with excitation volts reduced to 2kV.

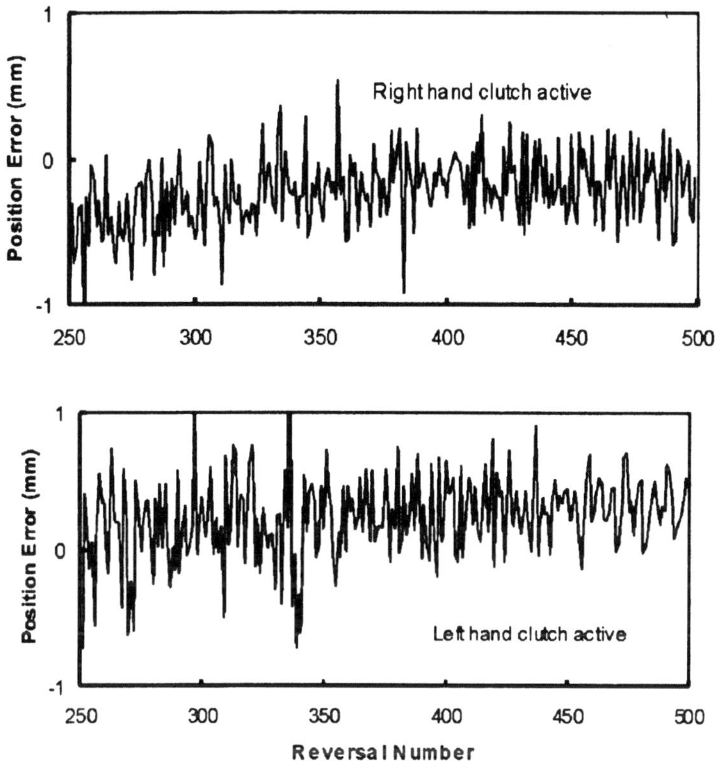

Figure 3.8 The difference between the calculated distance to turn round based on the average of the previous 5 turn round distances and the actual distance to turn round. For the section of the cycles at 1200 rpm and 4 kV/mm – compensated for temperature difference. Fluid Bayer 3565 × 50%. Fluid temperature rising 30 to 40°C h = 0.5mm

Table 3.1 The mode of work input to the ER fluid within each section of the motion cycle.

Belt	Motion	Clutch A			Clutch B		
position	Event/Status	$\|\Delta\omega\|$ **	Mechanical Work	Electrical work	$\|\Delta\omega\|$	Mechanical Work	Electrical Work
0 to 1	A off, B **on**	2Ω	Yes	No	0	No	Yes
1 to 3	A **on**, B off	$2\Omega \rightarrow 0$	Yes*	Yes	$0 \rightarrow 2\Omega$	Yes	No
3 to 4	A **on**, B off	0	No	Yes	2Ω	Yes	No
4 to 6	A off, B **on**	$0 \rightarrow 2\Omega$	Yes	No	$2\Omega \rightarrow 0$	Yes*	Yes
6 to 0	A off, B **on**	2Ω	Yes	No	0	No	Yes

* Includes the work input to the ER fluid by the shear of the electrically activated ER fluid.

** $|\Delta\omega|$ The relative rotational velocity between the drive and driven rotors.

Table 3.2 Typical values of geometric and fluid properties

- effective radius of pulley $r = 39.55$ mm

- constant speed belt velocity $u = 5$ m/s

- input rotor angular velocity $\Omega = 1207$ rpm

- moment of inertia of one output rotor and pulley assembly $I = 5.66 \times 10^{-5}$ kg m^2

- mass of belt $m = 16$ grams

- electro-rheological shear stress of fluid $\tau_{eo} = 10$ kPa,

- electron-hydraulic time delay $t^*_m = 0$ ms

- fluid viscosity $\mu = 100$ m Pa.s

inner surface of output rotor	**outer surface of output rotor**
- diameter $d_i = 47.0$ mm	- diameter $d_o = 51.0$ mm
- inter electrode gap $h_i = 0.5$ mm	- inter electrode gap $h_o = 0.5$ mm
- electro-rheologically active length $l_{ei} = 33.0$ mm	- electro-rheologically active length $l_{eo} = 33$ mm
- viscously active length $l_{vi} = 33.0$ mm	- viscously active length $l_o = 40$ mm

4. ER/MR Flow Mode Damper Design Methodology and Railcar Lateral Suspension Application

W.A. Bullough, D.J. Peel, N.D. Sims and R. Stanway

The University of Sheffield, Department of Mechanical Engineering
Smart Machines, Materials and Related Technologies Unit, Sheffield, England

Abstract This section presents a case study and theoretical example to illustrate how control may be applied to achieve a specific vibration isolation performance, with a linearised, viscous type, overall damper characteristic as an alternative to that of the steadily excited Bingham plastic fluid. The study is followed through from the point of view of an ER fluid, but for an MR fluid the consideration is almost exactly the same.

4.1 Introduction

Many tasks can been forseen for the electrorheological and magnetorheological based vibration damper, over wide ranges of force/displacement amplitude and disturbance frequency. The groundwork needed before such a device can proceed to its maximum potential in pre-service, in-situ testing, is common to all cases of valve and accessories design: a railcar suspension is chosen as a test vehicle to represent most of the considerations that will be encountered, viz:

1. A method of fluid characterisation, using small samples of fluid, to be applied over practical valve controller operating ranges.

2. Damper modelling and design techniques which relate damper shape, size and dynamic performance to fluid characteristics.

3. Verification of the fluid characterisation and damper modelling techniques, on a realistic scale, by comparison with experimental test data.

4. Determination of a control strategy.

Further, the case study and theoretical example will illustrate how control may be applied to achieve a specific vibration isolation performance, with a linearised, viscous type, overall damper characteristic as an alternative to that of the steadily excited Bingham plastic fluid. The study will be followed through from the point of view of an ER fluid, but for an MR fluid the consideration is almost exactly the same. The ER fluid is chosen because its availability has furnished ample data to demonstrate the methodology: an MR fluid would equally have been suited to the task.

From this procedure, the interaction of damper size, displacement stroke, control voltage level, fluid characteristic, and damper load can be appreciated. Although dynamic simulation of a semi-active suspension using continuous feedback/adaption is plausible - digital switching of excitation during a test cycle demonstrates this - nevertheless, verification of this requires a specific test vehicle.

4.2 Practical Damper Configuration

Damper piston, cylinder and control valve dimensions will be determined by the envelope requirements of the operating conditions, the fluid characteristics, the maximum cylinder pressure, and the available space for the complete damper, together with practical constructional and operational conditions. To maximise controllability the effects of friction and fluid inertia plus compressibility must be minimised.

A test damper configuration is shown in Figure 4.1. The control valve is located externally for prototype development purposes only: proof of the arrangement adopted implies verification of other more compact configurations. The double acting piston is combined with a through piston rod, eliminating asymmetry and minimising mass transfer into and, out of the high pressure fluid region, as the piston direction of travel reverses. Friction effects (and also inertia) are reduced by the use of short, large diameter fluid ducts, and particularly by the elimination of sliding seals between the piston and cylinder. At the piston rod, the relative motion of piston and cylinder is accommodated by flexible bellows, together with static 'O' ring seals. The piston circumference forms sliding bearings, and the small leakage flow associated with these, and with fluid compressibility and thermal expansion effects, is exchanged with a small fluid reservoir maintained at a pressure level a little above atmospheric. This reservoir regulates pressure in the region of the seals, situated in the coupled outer pair of the two pairs of volumes into which the cylinder is divided - this contributes to extension of the seal life. The inner pair of cylinder volumes, seen in Figure 4.1, act in turn as the high pressure cylinder, driving the flow through the control valve, according to the direction of piston motion - pressure in the other inner volume is maintained at the reservoir pressure, the reservoir communicating with the two inner volumes via a pair of non-return valves to return such leakage flow. The inner pair of cylinder volumes with the ER control valves and connecting pipes forms the high pressure circuit which will be modelled. A typical high-pressure cycle for the test damper is shown in Figure 4.2, A and B subscripts identifying the pair of inner cylinder volumes.

4.3 Fluid Parameters

In predicting the performance of an ER damper, the first task is to assemble the fluid characteristic data [1], [2]: this may not be comprehensive, and rarely is available in the time response domain. However, it has been shown with three fluids that non-dimensional plotting of the steady state controlling parameter of Hedström Number against Reynolds Number provides an effective technique for translating fluid characteristic test data from a field controlled Couette viscometer shear mode motion, to predict Poiseuille valve/flow mode performance. Only a few millilitres of fluid are required for such Couette type tests, which may be easily extended to include the time response domain.

In the time response domain, it is assumed that all ES fluids will respond fully in 1 msec or less when excited sufficiently rapidly, even under the influence of rapidly changing shear or flow conditions. It is difficult to dissociate the fluid response from the test rig dynamic effects in both shear and flow modes of operation. ER fluid electrical characteristics are important for the design of excitation/control equipment, in both steady and time response domains, and these can be determined simultaneously with fluid flow characterisation. Conduction in an ER fluid is temperature, excitation level, and shear/flow rate dependent, but tends to be similar for either DC, or AC operation up to 50 Hz. In the absence of design data, it is normal to assume (in the optimum temperature range) that electrical and mechanical response time delays/time constants, for say a step of excitation, are similar and might be neglected for cycle time calculations of under 30 Hz or so.

For a range of excitation, the energy associated with excitation/conductance overshoot is of generally small significance. However, for excitation and control system calculations, the high capacity supply and discharge/earth facility are major considerations - especially with respect to the ER fluid.

4.4 Damper Modelling

The large number of variables to be taken into account in the design and performance testing of a damper, suited to any particular task, presents a difficulty in itself. This difficulty is reduced by the non-dimensional plotting of fluid performance data. Given the adequacy of such data, including bulk and shear modulus performance as necessary, a modelling analysis which includes inertial, compressibility, and frictional effects can generally be carried out with the further simplification of a quasi-steady approach. The approach taken is to reduce parasitic friction as far as possible, to maximise controllability, while the design is considered in isolation from the compliance of other components; this is minimised in performance testing with a stiff and high inertia test rig drive.

The damper model comprises the three elements shown in Figure 4.3. The combination of high stiffness of the damper components, low frictional pressure loss, and a short acoustic length (less than 1 msec) of the fluid circuit which dispenses with the need to seriously consider wave action, means that the entire system of fluid can be lumped into the mass m, on which the piston/damper force acts directly. The ER control valve receives the fluid displaced by the piston, regulating the cylinder pressure/pressure drop: this is represented by the viscous dashpot with resistance function f(...). The cylinder pressure leads to compression of the fluid by an amount which depends on the fluid bulk modulus: the fluid stiffness is represented by the spring constant k. Finally, two displacements are taken into account, x defining the piston displacement, and x_l defining the fluid displacement at the control valve. For harmonic motion at frequency ω, the basic equations for the system can be written:

$$F = F_o \sin \omega t$$

$$x = X \sin (\omega t + \phi)$$

$$x_l = X_l \sin (\omega t + \phi_l)$$

(where ϕ, ϕ_l are phase angles)

$$m\dot{x} + f(\dot{x}, \dot{x}_l) = F$$

$$-f(\dot{x}, \dot{x}_l) + kx_l = 0 \qquad (4.1)$$

The solution is via the numerical procedure given in Figure 4.4.

4.5 Damper Operating Envelope

A simplified characteristic for a conventional, fixed hydraulic damper for a vehicle suspension as illustrated in Figure 4.5. For a given suspended mass and suspension stiffness, the second order, linear, single degree of freedom equation of motion leads to a fixed damping rate c - the constant of proportion between the damping force applied and velocity of the suspended mass. The damping rate may be symmetrical or may have different (but fixed) values in compression or rebound of the suspension.

Also shown in Figure 4.5 is the performance envelope for a controllable damper based on ER/MR fluid technology. Here a minimum damping rate - a 'fail-safe' condition - occurs under conditions of zero excitation. Under field control and maximum excitation, the damper is capable of developing a maximum level of force across the entire velocity range. The boundaries of zero excitation, maximum excitation, and maximum positive and negative velocity define a symmetrical envelope for the force/velocity/displacement map required of the damper, within and across which the damper will operate under excitation control as demanded.

4.6 Damper Design Optimisation

The damper operating envelope is defined in terms of the minimum damping (fail-safe) coefficient c, maximum piston force F_{max} and velocity $\dot{x}_{max}$, and maximum cylinder pressure, that is approximately maximum ER control valve pressure drop Δp_y. As piston velocity $\dot{x} \rightarrow 0$, a piston 'yield force' F_y which is an upper limit of the electro-force component F_e is defined by $F_{max} = F_y + c\,\dot{x}_{max}$ (2). The control valve electrode dimensions are length ℓ and width b and inter-electrode gap h. The high pressure (half) cylinder dimensions are length L, diameter D, and cross-sectional area A.

The fluid properties, taken at a nominal operating temperature, are yield stress, τ_y as the upper limit of electro-stress τ_e for the maximum control field E, no-field viscosity coefficient μ, (i.e. an approximately Newtonian characteristic), density ρ, and bulk modulus K. Then neglecting connecting ducts and wall elasticity, it can be shown that the fluid effective mass and stiffness are respectively approximated as:

$$m = 2\rho AL\left[1 + \left(\frac{\ell}{2L}\right)\left(\frac{A}{bh}\right)\right] \tag{4.2}$$

$$k = (AK/L)\left[1 + \left(\frac{\ell}{2L}\right)/\left(\frac{A}{bh}\right)\right] \tag{4.3}$$

The maximum values of the Hedström Number and Reynolds Number via which the fluid is characterised are given by

$$He_{max} = \frac{\tau_y \rho h^2}{\mu^2}$$
$$Re_{max} = \frac{\rho A \dot{x}_{max}}{\mu b} \tag{4.4}$$

Where ℓ' and t' are respectively development length and development time for a steady flow at the valve inlet, and δ_y is a piston displacement increment associated with development of cylinder pressure $\Delta p_y = F_y / A$, rapid response of the damper to control signals will require that ℓ', t' and

δ_y be small. It can be shown that ℓ' and ℓ increase with increasing Re and h, while δ_y is proportional to Δp_y. Thus, to optimise the valve time response to control signals:

A and b should be maximised;

h and Δp_y should be minimised.

Practically the range of h is limited to 0.5<h< 1.0mm, from the points of view of particle size and control valve power requirement – the same range applies for the MR control valve. The remaining valve dimensions follow as

$$\text{with } \tau_y = \frac{\Delta p_y h}{2\ell} = \text{fn}(E^2) \text{ in general} \tag{4.5}$$

where E and J are respectively the maximum constant control field strength and the associated current density between the valve electrodes, the electrical control power requirement is

$$P_I = EJ\ell bh \tag{4.6}$$

Since from equations (5)

$$\ell bh = \frac{3\mu F_y^2}{c\tau_y^2} \tag{4.7}$$

the maximum control power requirement is given in terms of the performance envelope as

$$P_I = \frac{3\mu F_y^2 EJ}{c\tau_y^2} \tag{4.8}$$

This maximum control power is more or less independent of the piston velocity, and since in general F_y, τ_y and J can all be expressed approximately as quadratic functions of field strength E, then to a first approximation, at constant temperature,

$$P_I \approx \text{fn}(E^3) \tag{4.9}$$

so that it is advantageous to the control power requirement, as well as to the rate of response, to minimise the cylinder pressure. Where ω and X are the circular frequency and the displacement amplitude of the piston motion, with limiting value $\omega X = \dot{X}_{max}$, at maximum excitation E with (an approximately) constant force F_{max}, the damper hydraulic power dissipation will be given approximately as

$$P_E \approx \frac{2F_{max}\omega X}{\pi}$$

with maximum

$$P_E \approx \frac{2F_{max}\dot{x}_{max}}{\pi} \tag{4.10}$$

from which a corresponding equivalent viscous damping coefficient can be expressed as

$$c_{eq} \approx \frac{4F_{max}}{\pi\omega X} \tag{4.11}$$

which has its highest values for $\omega X << \dot{x}_{max}$ at low values of the piston displacement frequency and amplitude. Since generally $F_{max} \approx F_y$ for a wide range of the control ratio c_{eq}/c, then to a first approximation, at constant temperature,

$$P_E \approx \text{fn}(E^2\omega X)$$

$$\text{and } c_{eq} \approx \text{fn}\left(\frac{E^2}{\omega X}\right) \tag{4.12}$$

The ratio of the electrical control power to the hydraulic power dissipation – which is a measure of 'inefficiency' – is then

$$\frac{P_I}{P_E} \approx \frac{3\pi\mu F_y{}^2 EJ}{2c\omega XF_{max}\tau_y{}^2}$$

$$\text{so that } \frac{P_I}{P_E} \approx \text{fn}\left(\frac{E}{\omega X}\right) \tag{4.13}$$

In the absence of field excitation, the minimum damping coefficient is

$$c = \frac{12\mu\ell A^2}{bh^3} \tag{4.14}$$

so that the damping coefficient control ratio is

$$\frac{c_{eq}}{c} \approx \frac{bh^3F_{max}}{3\pi\mu\ell A^2\omega X} \approx \text{fn}\left(\frac{E^2}{\omega X}\right) \tag{4.15}$$

Thus although there is some sacrifice of efficiency in operating at higher control field strengths, more significant gains result in the range of the control ratio, both effects being greatest for small piston displacement frequency and amplitude.

4.7 Experimental verification of the model

In the test damper configuration described above, the stationary parallel plate control valve is mounted externally to the cylinder for ease of modification: the model does not require distinction between external or internal valve locations. Test data where shown is for control valve dimensions $\ell x b x h = 70 x 102 x 0.5 mm$. The nett piston area is $1355mm^2$, providing for nominal maximum piston force 6.5kN and velocity 0.5 m/sec. Temperature stability in the testing is assisted by ER fluid circulation from a temperature controlled reservoir; the ER control valve also has a water cooling jacket supplied from the fluid reservoir temperature control system. Tests carried out are for either constant or variable control voltage, at selected constant piston displacement frequency and amplitude, and selected approximately constant temperature.

4.8 Experimental test results

The test damper and an appropriate procedure provide verification of the damper modelling technique on a realistic, industrial scale.

Figure 4.6 shows a test result for piston pressure vs., speed at constant control field strength, compared with a model predicted characteristic and a test result for zero excitation. Fluid inertia effects are not very significant, but the plots show a prominent loop around zero velocity, which is associated with compression of the fluid at reversal of the piston direction of motion; referring to equations (1) in section 4 above, at this position $\dot{x}_1 \approx 0, \dot{x} \neq 0$. The pressure pulse seen at the point where the control valve flow resumes after reversal of the piston direction $(\dot{x}_1 \neq 0)$ is at present not fully explained, and indicates the possibility of a pseudo-hysteresis or structuring behaviour in the fluid, with the development of a 'static' type yield stress at near-zero shear conditions. Figure 4.7 shows the corresponding plot of piston pressure vs. displacement, with near constant maximum piston force defining the damping through the cycle. The approximately Newtonian fluid test characteristic at zero excitation is also seen in these figures.

In Figure 4.8, a damper test cycle measurement is used as damper model data to determine the characteristic plot of Hedström Number against Reynolds Number for the damper test fluid at constant field excitation. Superimposed on this is a steady flow characteristic predicted for the control valve flow from small scale clutch test data for a similar fluid from a different batch of fluid manufacture. The test result shows the prominent effects of fluid compression and valve flow initiation discussed above, while in addition the damper test fluid attains a given yield stress at a lower field strength than required by the small scale clutch test batch of fluid. Nevertheless, the common underlying fluid characteristic can be seen over almost the entire piston stroke.

The piston pressure response to a step change of control field strength during piston motion is shown in Figure 4.9. Here a pressure rise of ≈ 10 bar occurs in a time of ≈ 7 msec – the static stiffness F_y/δ_y, and the prevailing piston speed, will require a pressure rise time of ≈ 5 msec, so that the pressure response is seen to depend largely on the stiffness constant k in equation (3). The good approximation to mass and stiffness of equations (3) is found to be verified by amplitude and phase response of the valve pressure drop signal over the piston displacement frequency range in further tests, using a hydraulic oil of known Newtonian viscosity characteristic.

Finally in Figures 4.10 and 4.11 the damping coefficient control ratio c_{eq}/c and the control/hydraulic power ratio P_I/P_E are shown, plotted against control voltage, for a range of piston displacement frequencies, for a single displacement amplitude and (nominally) constant temperature. The plots show the types of relationship suggested in equations (4.13) and (4.15). However, for the control ratio c_{eq}/c, at other than small values of the control field (voltage) strength, the control ratio is little more than doubled for a doubling of the control field, at both high and low displacement frequencies. For the power ratio P_I/P_E, at low displacement frequency the power ratio is roughly proportional to the field strength, but at high frequency the power ratio rises much more rapidly than either the field strength or the control ratio. Contributing factors include the form of the Hedström Number vs Reynolds Number characteristic (of Figure 4.8) and non-linearity of the zero-excitation characteristic (of Figure 4.6) for the fluid, the more rapid rise of current density J than of yield stress τ_y with increase of the field strength E, but chiefly the fact that in the higher displacement frequency tests there is an increase of fluid temperature, which has a marked influence on the current density in particular, showing that ideally an ER fluid is selected for the temperature range within which it will normally operate.

Case study for a vehicle suspension

A simplified suspension model is represented by the single-degree-of-freedom system of Figure 4.12, for which the undamped natural frequency of free oscillations of mass m_s on a spring of stiffness k is approximately $f_0 = (1/2\pi)\sqrt{(k/m_s)}$.

With viscous damping, coefficient c, in parallel with the spring, if the support is subject to steady harmonic oscillation with displacement amplitude X_i at frequency f, then the mass will be subject to forced oscillation at frequency f, with a displacement amplitude X_o which depends on X_i, f and c, in accordance with the transmissibility characteristic shown in Figure 4.13. The transmissibility is $X_o / X_i = 1$ at the cross-over frequency $f=f_0\sqrt{2}$.

In Figure 4.13 the transmissibility is calculated for mass $m_s \approx 10^4$kg and stiffness $k \approx 2x10^5$ N/m, giving an undamped natural frequency $f_0 \approx 0.7$ Hz and a crossover frequency $f_0\sqrt{2} \approx 1.0$Hz. This may be typical of the lateral suspension of a light rail vehicle, and a fixed, compromise damping coefficient of $c_{mean} \approx 30$ kN.s/m will provide rapid damping of free vibrations. The operating range for such a suspension damper is determined by considerations such as comfort and loading gauge requirements, so that for lateral oscillations peak displacement amplitudes may be $\approx \pm$ 50mm, and with the low natural frequency rms values of displacement, velocity, acceleration and damper power dissipation may be typically $\pm$ 12.5mm, $\pm$ 0.05m/sec, 2%g and 50 watts respectively, with rms damping force $\pm$ 1.5 kN corresponding.

The transmissibility characteristic of Figure 4.13 shows that very significant reductions in the transmitted displacement amplitude are attained if (i) the damping coefficient is increased at low frequency to $c \approx 4c_{mean}$, and (ii) the damping coefficient is decreased at high frequency to $c \approx c_{mean}/4$. This approach is particularly suited to the ER fluid damper, with which changes of this order of magnitude are readily obtainable by means of the field controlled yield stress, with high levels of damping coefficient attainable particularly at low displacement frequencies – as is illustrated by Figs. 10 and 11.

Simple control strategy

The simplest approach to controllable damping is that of switching the damper control field strength between two (or more) constant levels, to generate respectively a high or low level of damping, accordingly as disturbances lie below or above the suspension transmissibility characteristic crossover frequency. Referring to the force/velocity operating envelope defined in Figure 4.5, such a switchable damper would operate either on the upper boundary of maximum field excitation, or on the lower boundary of the no-field or 'fail-safe' condition.

A switching criterion for such a damper might be based on information from a velocity sensor; or it might be based on acceleration and displacement, which must be controlled from considerations of comfort and loading gauge requirement respectively – an example of this case is illustrated in Figs. 14 and 15. An input displacement amplitude disturbance comprises (i) a step of 40mm; (ii) an amplitude of ±10mm (offset) at frequency 3.5 Hz; (iii) an amplitude of ±25mm (offset) at frequency 0.55Hz; and (iv) simultaneous amplitudes of ±10mm at frequency 3.5 Hz and ±25mm at frequency 0.55Hz. The figures show this input disturbance sequence and compare the amplitudes of acceleration and displacement transmitted for the two cases of: firstly, a fixed, compromise damping coefficient which corresponds to a typical damping factor $\zeta = c_{mean}/4\pi m_s f_o \approx 0.4$; and secondly, a viscous (equivalent) damping coefficient switchable between four levels, respectively c_{mean} x4, x2, x2/5 and x1/5 from the highest to the lowest level.

The principle switching criterion applied is that of input acceleration $\ddot{x}_i$: if $\ddot{x}_i > 0.5\text{m/sec}^2$ for > 20 msec, then $c = c_{mean}$ x2/5 is set; if $\ddot{x}_i < 0.5\text{m}/\text{sec}^2$ for > 20 msec, then $c = c_{mean}$ x 2 is set; the time delay (larger than the control valve response time in Figure 4.9) reduces unwanted switching associated with brief input transients. A secondary criterion based on the relative displacement of input and output $(x_o - x_i)$, with a time delay related to the suspension natural frequency, is used together with the acceleration criterion for switching to the maximum or minimum levels of the damping coefficient.

The calculations show that, for the switchable damper in comparison with the fixed damper:

i.) the acceleration response to the individual frequency components is reduced;

ii.) the acceleration response to the simultaneously applied frequency components is reduced;

iii.) the displacement response to the individual frequency components is reduced; but

iv.) the displacement response to the simultaneously applied frequency components might increase – however, the higher frequency component is effectively eliminated, so that this represents a trade-off to eliminate unwanted transmission of large high frequency accelerations.

The example given is not optimised. The reduction in the acceleration response to the higher frequency input in particular is appreciable, while no significant transients are seen at switching of the damper rate. Increases of acceleration response at the switching points occur before switching and are associated with the damping level during the delay period – nevertheless the acceleration response for the switched damper remains within the overall limits of the response for the fixed damper.

One feature of a simple switched, constant field strength approach to damping control, the significance of which depends on the application, may be a 'stiction' effect associated with the fluid yield characteristic. This will be relieved to a degree by any piston leakage flow – this can be enhanced by means of a viscous flow channel in parallel to the ER control valve, which will thus define an upper limit to the damping coefficient. It is also possible in fact to develop a control strategy capable of linearising the force/velocity relationship within the envelope defined, adjusting the force to that desired by means of field control. Thus, in the limit as the number of switched levels of damping is increased, the damper characteristic might become effectively that of a continuously variable viscous damping coefficient, with possible benefits to efficiency of operation (cf. equations (13) and (15), and Figs. 10 and 11).

4.9 Advanced Control Strategy

It is well-known that a proportional feedback control strategy is capable of maintaining a linear system's performance under the influence of a disturbance signal. However, it is less well-known that the same control strategy can be used to linearise an inherently non-linear system. The latter case is applicable to the ER damper described in this article, where the output (piston force) is a non-linear function of the inputs (applied electric field and piston velocity) The theory that describes this linearisation process is illustrated using graphical techniques in [3].

At this stage, it is useful to develop the control strategy by using a block diagram approach. Noting that the 'controllable' input to the damper is the electric field strength, E, the damper can be expressed as shown in Figure 4.16(a). Applying a proportional feedback control strategy is now straightforward, and results in the system shown in Figure 4.16(b). The 'set-point' or controlled input to this system is now a desired piston force (expressed as a voltage). To render this set-point more physically meaningful, it can be made proportional to the piston velocity, which still acts as a 'disturbance' input. This resulting system is shown in Figure 4.16(c), and it is clear from this figure that providing the feedback strategy linearises the damper, a force will be generated which is proportional to the piston velocity. The constant of proportionality can be adjusted via the gain D, and so the result is linear, variable damping with an adjustable 'viscous damping' coefficient.

The theory developed above has been researched extensively in recent years. In a preliminary study to assess the feasibility of the concept, a modelling approach was taken, and it was shown how linear variable damping could be achieved [4]. Following on from this study, a comprehensive model was developed and validated [5], which paved the way for detailed simulations of the controlled response [6]. Finally, the control strategy was implemented on the experimental facility. Typical experimental results are shown in Figure 4.17. It can be seen that linear, variable damping is achievable, within the control limits defined by the maximum and zero electric field strength conditions set for the device.

4.10 Summary

The major impact of the verified design methodology in the above sections is that a low piston pressure/large piston area combination is needed to give a high frequency response. Then, sensibly eliminating acceleration terms and frictional effects in the design produces as fast a system for fluid and damper as it is possible to obtain. The result is a capability for the shaping of the damper characteristic by electronic means alone.

References

The content of this chapter has drawn extensively from a review paper:

1. Peel D.J.. Bullough W.A. and Stanway R., "ER/MR Long Stroke Damper: Performance Testing, Modelling and Control Strategy Simulation". Proc. 6th International Conference on Electrorheological Fluids, Magnetorheological Suspensions and their Applications. held in Yonezawa (1997), pp 744-751, World Scientific Publishing Co.

2. Several other relevant papers are included in the source of [1] and in the Proceedings of the corresponding 7th International Conference, held in Honolulu (1999), World Scientific Publishing Co.

3. West J C 1960, "Analytical Techniques for non-linear control systems", The English Universities Press Ltd, London.

4. Sims N D, Stanway R and Beck S B M, 1997, "Proportional feedback control of an electro-rheological vibration damper" *Journal of Intelligent Material Systems and Structures*, **8**, Technomic Publishing Co Inc, Lancaster PA USA. p. 426-433.

5. Sims N D, Peel D J, Stanway R, Johnson A R and Bullough W A, 2000, "The Electrorheological Long Stroke Damper: A New Modelling Technique With Experimental Validation" *Journal of Sound and Vibration*, **229**, p. 207-227.

6. Sims N D, Stanway R, Peel D J, Bullough W A and Johnson A R, 1999, "Controllable viscous damping: An experimental study of an electrorheological long-stroke damper under proportional feedback control" *Smart Materials and Structures*, **8**, Institute of Physics Publishing Ltd, Bristol England. p. 601-615.

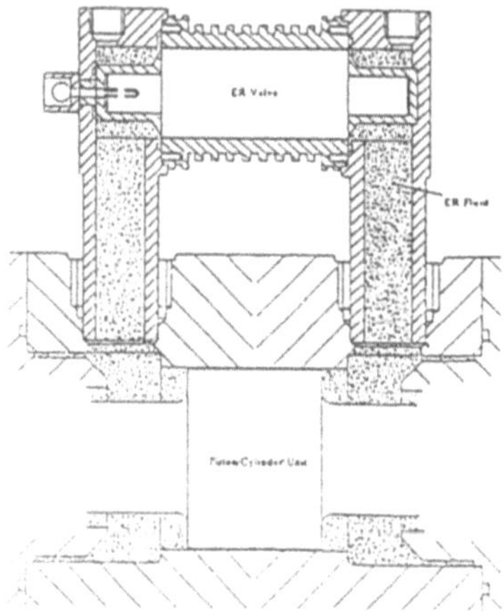

Figure 4.1 Section through ER damper, piston-and-cylinder unit and control valve

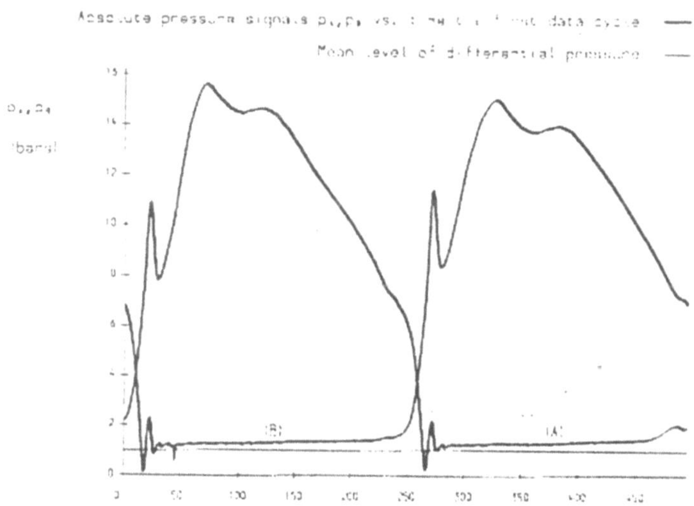

Figure 4.2 Cycle variation of pressure at each piston, constant control field strength

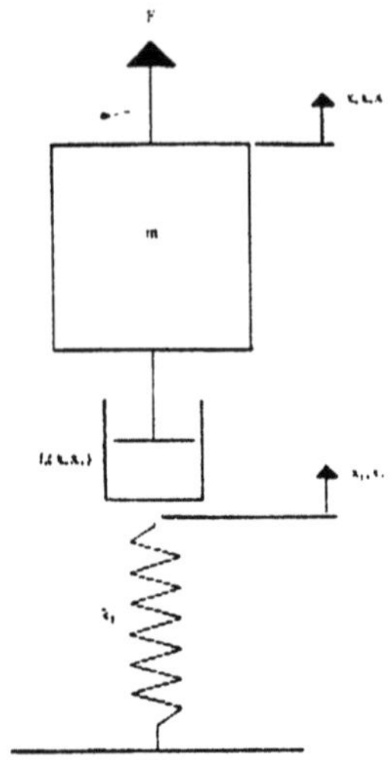

Figure 4.3 Schematic of damper model using electro-structured fluid

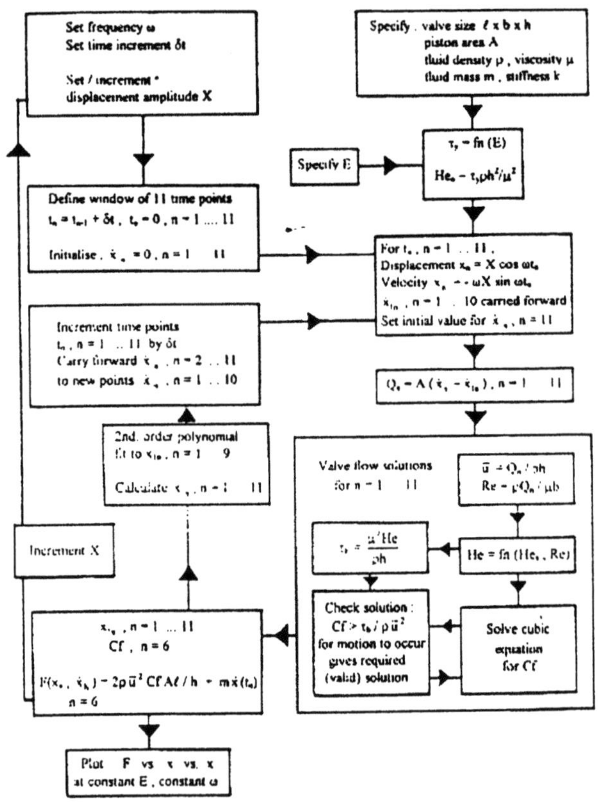

Figure 4.4 Flow chart for computation of ER fluid damper model data shown in Figs. 6 and 7

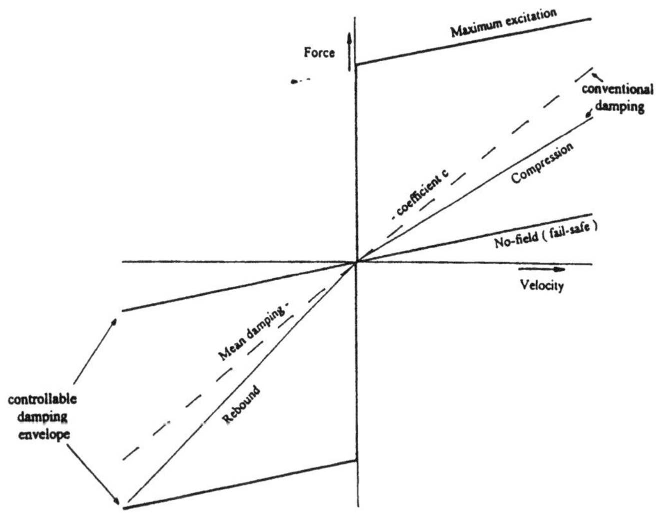

Figure 4.5 Force/velocity characteristics and envelope for a suspension damper: conventional and controllable

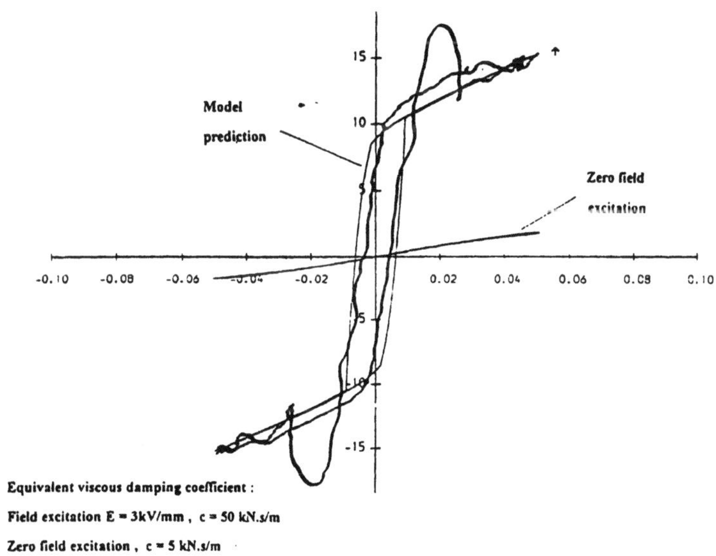

Figure 4.6 Damper rig test data Differential pressure vs. piston velocity; Constant field excitation E = 3kV/mm; Piston frequency 1Hz and displacement ±8mm; Comparison with model prediction and zero field excitation

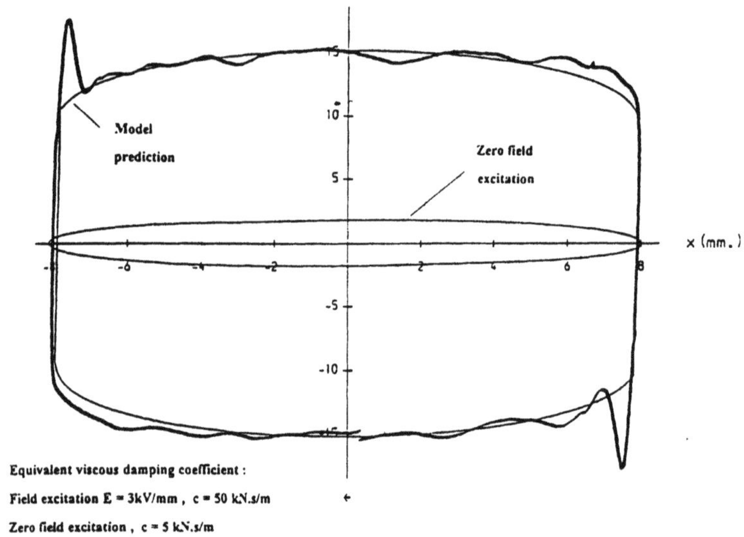

Figure 4.7 Damper rig test data Differential pressure vs piston displacement Constant field excitation E = 3kV/mm Piston frequency 1Hz and displacement ±8mm Comparison with model prediction and zero field excitation

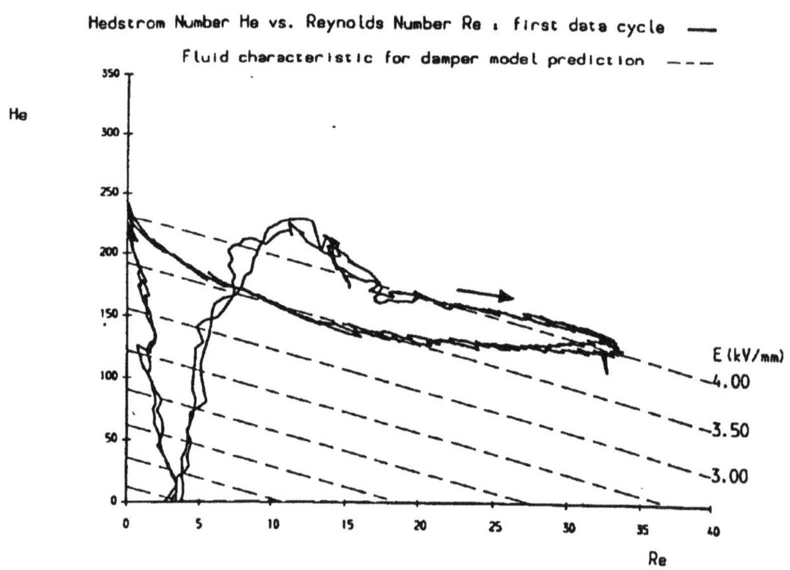

Figure 4.8 ER damper test, ER fluid characteristic Hedstrom Number He versus Reynolds Number Re

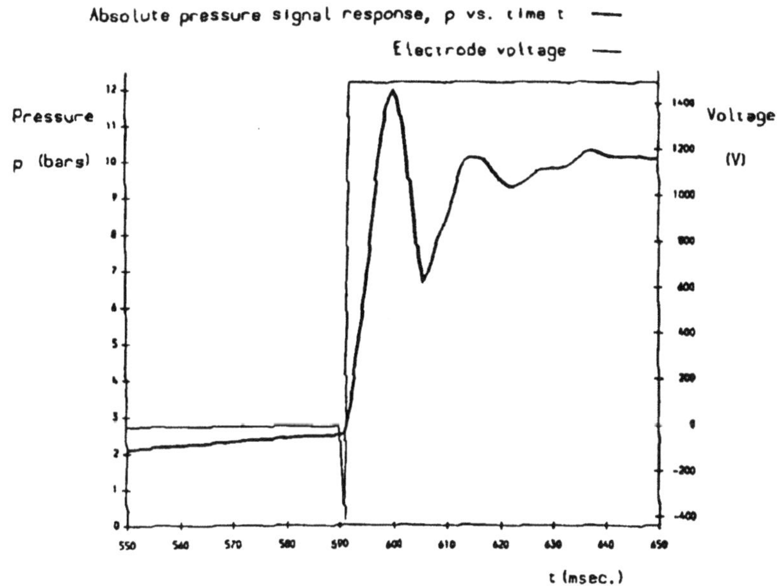

Figure 4.9 ER Damper test, pressure response at valve inlet to voltage step, versus time. Displacement frequency 1Hz and amplitude ±8mm. Bayer 3565 x 50% fluid at 40°C.

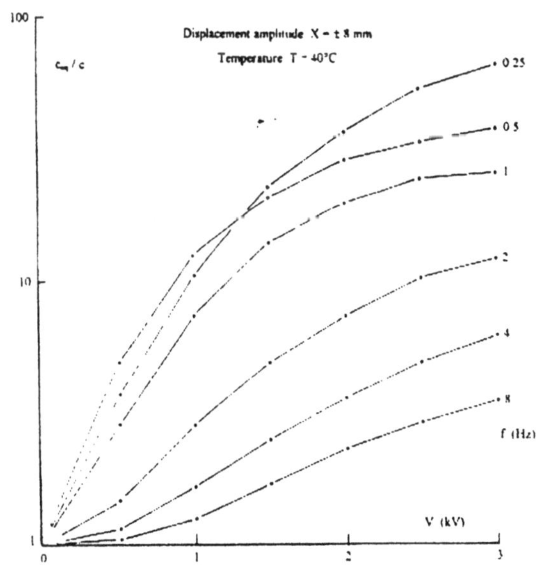

Figure 4.10 Experimental measurements, field-excitation/no-field damping ratio c_{eq}/c. vs. damper control voltage V, at constant values of the displacement frequency f

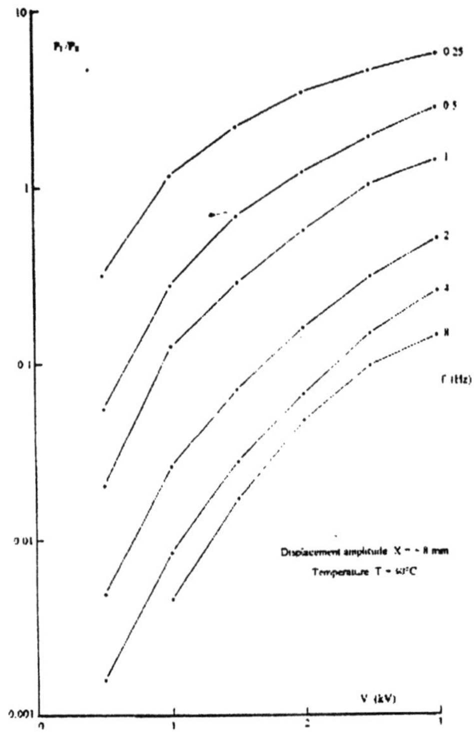

Figure 4.11 Experimental measurements, electrical control/hydraulic dissipation power ratio P_I/P_E, vs. damper control voltage V, at constant values of the displacement frequency f

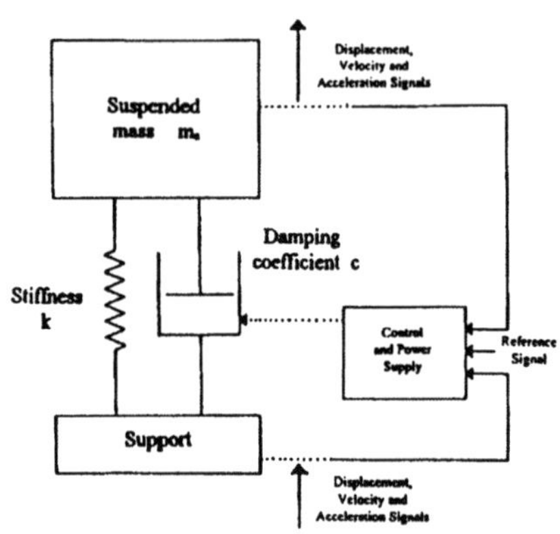

Figure 4.12 Simplified model of a suspension with passive damping, and schematic showing adaptation to controlled passive damping

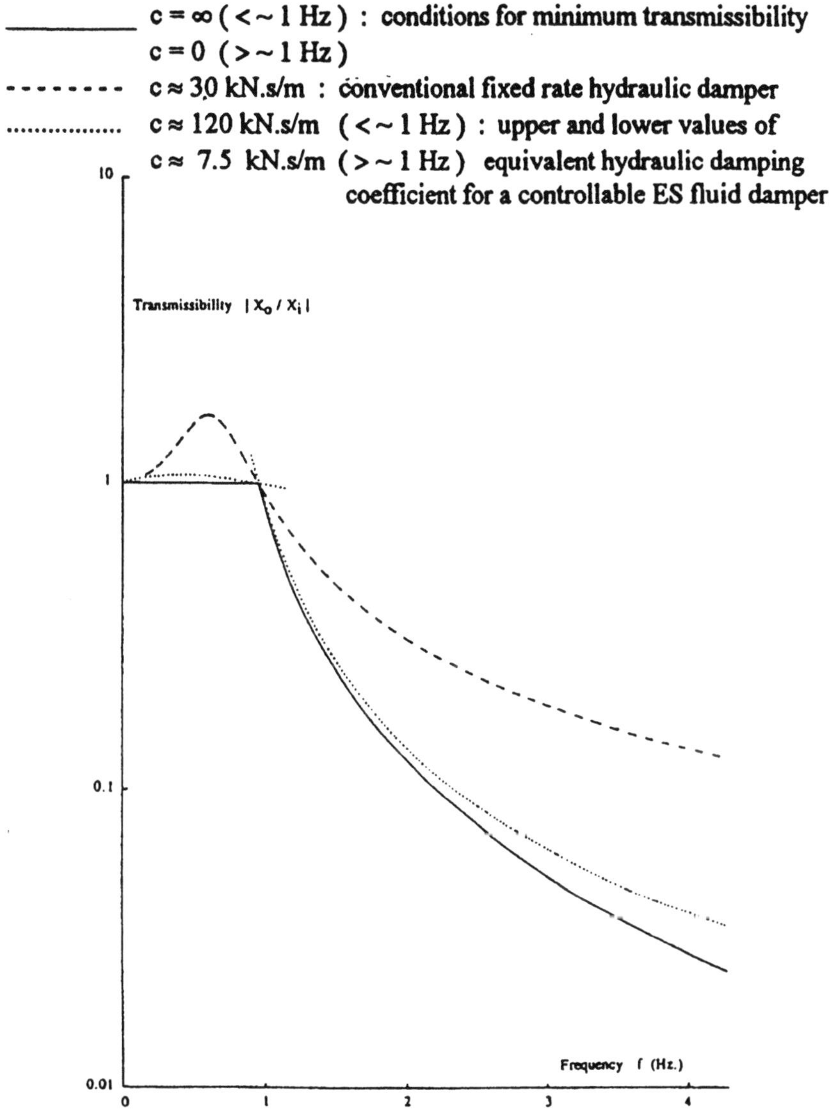

Figure 4.13 Comparison of transmissibility for a suspension, with a fixed rate conventional hydraulic damper, and with a controllable (switched) ES fluid damper, showing the minimum attained transmissibility level for a fixed spring rate

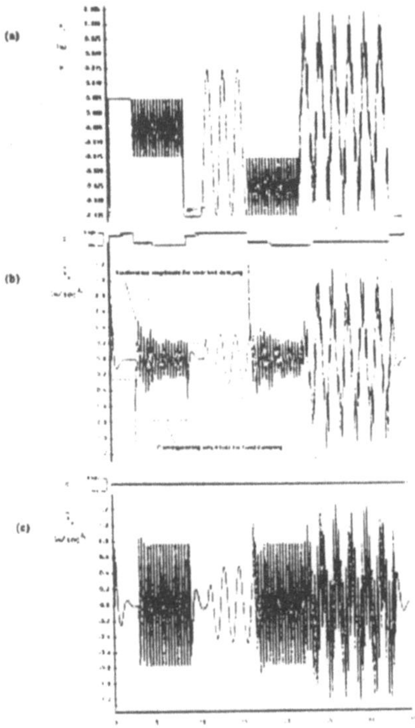

Figure 4.14 Comparison of acceleration transmissibilities, for suspensions with switched and fixed damping, showing (a) input displacement amplitude vs. time, with (b) output acceleration amplitude for switched damping, and (c) output acceleration amplitude for fixed damping

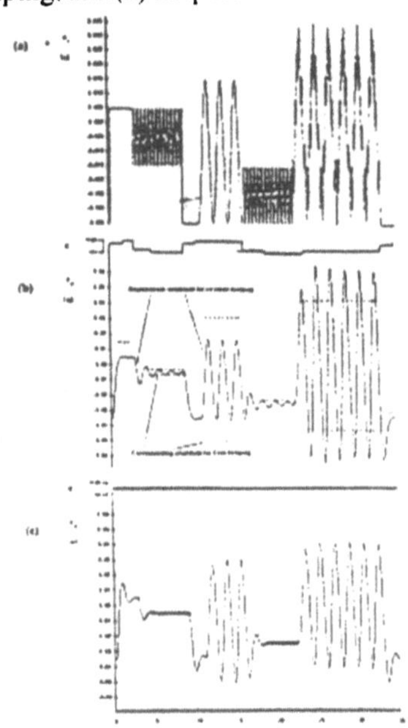

Figure 4. 15 Comparison of displacement transmissibilities, for suspensions with switched and fixed damping, showing (a) input displacement amplitude vs. time, with (b) output displacement amplitude for switched damping, and (c) output displacement amplitude for fixed damping

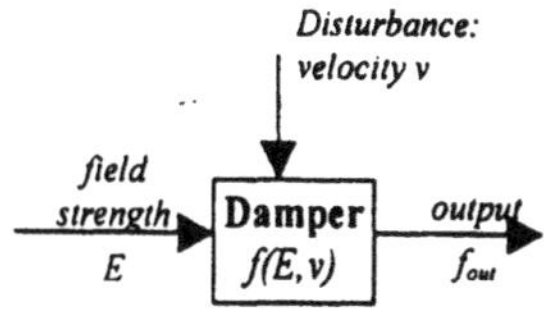

(a): ER damper expressed in block diagram form

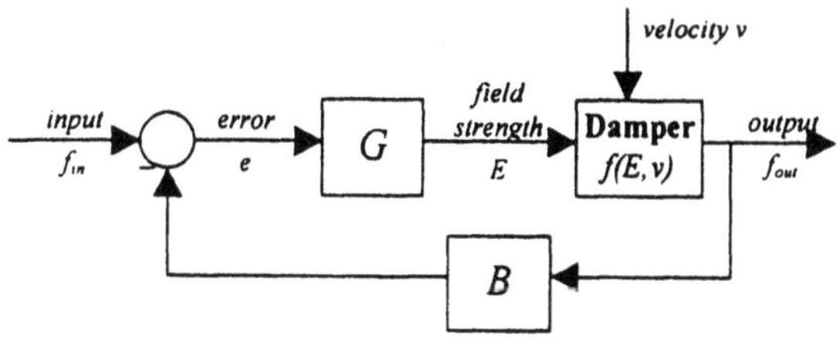

(b): Applying proportional feedback control to the ER damper.

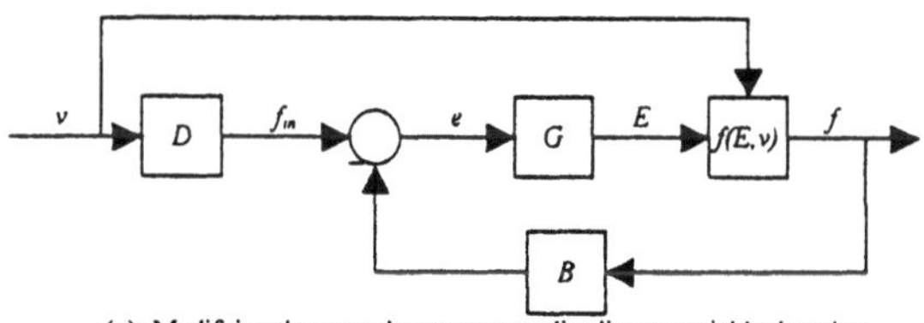

(c): Modifying the control strategy to realise linear, variable damping.

Figure 4.16 Typical control strategy: linear solution

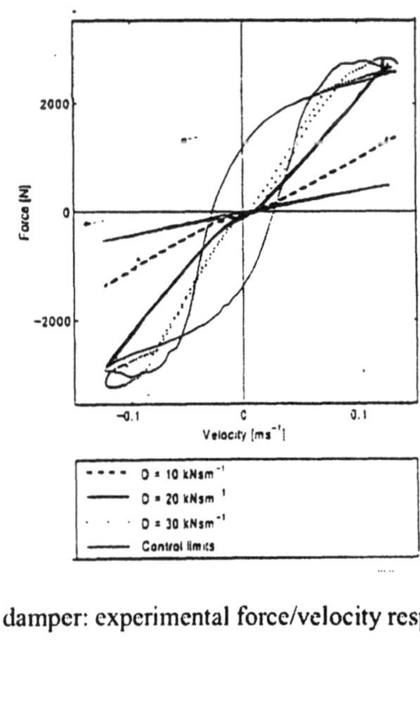

Figure 4.17 Linearised damper: experimental force/velocity response at 5Hz, 4mm amplitude, 40°C

5. ESF Tribology, Hydrodynamic Lubrication and the Flexibly Operated Lens Finisher

W.A. Bullough and P.L. Wong

The University of Sheffield, Department of Mechanical Engineering
Smart Machines, Materials and Related Technologies Unit, Sheffield, England

Abstract: The aim of this section is to present what information is available and relevant to the tribological performance of moving surfaces in or in near contact with ESF. This is done by first reviewing the sparse amount of theoretical work in the literature and then concentrating on the experimental results achieved from practical tests. Though the outcome is not in the fonn of a precise friction and wear control design methodology it probably constitutes the best guide to the selection of suitable or possible operating conditions at the present time. The above is done in loose connection with forced tribological situations/abrasive particle machining aspects and an outline description of the first ESF based flexible machine in this sphere of endeavour - the QED litho-lens polisher

5.1 Introduction

The tribology of electrostructured fluids in conjunction with close proximity bounding surfaces in relative motion has proceeded only to the most basic stage of characterisation. Any potential user who scans the public literature is faced with but a few non-systematically produced data, gathered at various centres and on different apparatus, to point the way towards effective wear rate/load/speed and friction estimation. This position is likely to improve substantially now that MRS's are being sold by several manufacturers. Indeed, it is possible that the design juncture has already been reached in terms of information already accrued by the commercial/industrial producers of the fluids, dampers, lens-polishing machines, prostheses/body and other prototype devices that are in the post appraisal pipeline.

With implants, these limitations in mind the subject of ESF tribology divides itself into forced wear, in which the flexible machining of a surface in terms of the computer controlled and purposeful removal of material is involved, and the prevention of wear and reduction of friction via lubrication respectively. The lens polishing process [1] is shrouded in industrial secrecy but is generally associated with the final production of surfaces by rotating, e.g. a lens in contact with relatively hard particles carried in a polarisable controllable rheology process fluid.

Every new initiative in commercial development needs some kind of financial impetus to start it off. In the present example it is believed to be strategic insomuch as USA industry and defence agencies were faced with a dwindling source of human lens polishers for military and microcircuit-litho photo lens production.

5.2 The QED Finishing Machine

Material removal in the resulting machine is caused by 1-3 μm size MR suspended particles interacting with a chemically reacting surface and takes place at the molecular level. The equivalent to the mechanical pad pressure in a conventional hand polishing is provided by the magnetically controlled hydrodynamic pressure generated by the fluid in the converging flow channel, Figure 5.1. The rigid stationary under plate does not wear and the profile being finished is 'highly deterministic'. The DC electromagnet generates a non-linear field and yield stress distribution in a typical fluid comprising 36% carbonyl iron, 55% water, 3% anti oxidising and stabilising agents, plus 6% cerium oxide polishing abrasive. A 10Angstrom surface can be obtained in minutes of polishing, depending on the abrasive, of which nano-diamond is a preferred material for the larger micro circuit substrates etc which are similary finished.

Being liquid bourne particles, the process takes place in a cooled and load distributed fashion without leaving much in the way of remnant stress concentration in the work-piece. Since the electronic-hydrodynamic process is both fast, continuous and repeatable material can be removed precisely and a 'lens' contoured via a programmed demand with continuous measurement and adjustment of field excitation taking place. The process is competitive, with novel flexibly operable machines costing hundreds of thousands of dollars each.

5.3 Analytical Aspects

Recently the behaviour of a yield stress fluid flowing through a converging channel of stationary parts has been studied [2]. Though the flow is pressure gradient driven (Poiseuille type) and not the more complex Couette plus Poiseuille combination present, e.g. in the QED machine, the work serves as a warning to the pitfalls awaiting the unwary. These are caused by the presence of plugs of unsheared liquid due to the yield stress and their apparent incompatibility with reality.

Another analysis [3] considered the flow between axi-symetric curvilinear surfaces of revolution - with and without relative rotation. This general approach to the problem of ESF modelling in difficult geometries included the yield stress dependence on excitation strength in a Shulman non-linear model in order to allow various types of viscoelectric continuum to be appraised. Main conclusions include a suggestion that the use of the method of averaged inertia shows that momentum terms may be large and that the pressure distribution between the surfaces can be Reynolds number dependent.

Both of the papers cited above are seemingly associated with the 'forced tribology' - optical machining process but are not supported by experimental verification. Their importance in this chapter is to set into context the approach that follows viz [4]: to use a relatively simple model which may be verified (or not) by experiment in order to answer the question of whether the plastic viscosity of the ESF mixture or its base oil property is more appropriate to the formation of a hydrodynamic film and how fast will it respond to excitation.

5.4 Hydrodynamic - Rayleigh Step Appraisals

The Rayleigh step apparatus shown in Figure 5.2, is chosen as the theoretical and experimental model for ESF hydrodynamic flow along with the Bingham plastic constitutive equation on the grounds of simplicity, and because this combination is as likely to throw light on the key facets of the general problem as any other. At the fully loaded condition, Figure 5.3, the film thickness and through flow rate are respectively equal to the step height and zero. Also, the plug is (at its simplest), Figure 5.4, complicated enough to indicate any limitations on the continuum analysis approach.

The theory assumes the flow to be steady, incompressible, laminar, fully developed, one dimensional and isothermal. Initial unexcited tests on a well known T46 Shell Newtonian oil and the plane bearing theory validated the procedure.

For the excited ER fluid use is made of the Buckingham solution for flow rate $\dot{q}$ through parallel plates of separation h and unit width, one of which is moving at uniform speed U:

$$\dot{q}/Uh = P\left[\left(\frac{y_1}{h}\right)^3 - 2\left(\frac{y_2}{h}\right)^3 + 3y_2/h\right]/6$$

where $P = h^2\,(dp/dx)/\eta_p\,U$

y_1 and y_2 are defined in the velocity profile Figure 5.4 and η_p is the plastic viscosity.

In short duration tests to avoid complications of time dependency on unexcited ERF experimental and theoretical pressures had been compared on the basis of P values . In the absence of authoratitive yield stress values the experimental data and theory above were used to calculate apparent values for the yield stress, in the fully loaded bearing (Figure 5.5), the Newtonian viscosity from the unexcited tests being used in calculations.

It was shown that the fluid and flow properties, when combined with the Bingham plastic analysis gave a reasonable prediction of the pressure generated - for design purposes, provided isothermal conditions were maintained. Little hysteresis (loading/unloading direction differences) was encountered on the 30% volume fraction ER Lipol fluid, the wall slip problem was not seen to be important over the range of film thickness $\frac{1}{4} < h < \frac{1}{2}$ mm, and the effects of temperature generation were shown to be secondary in comparison to any problems encountered via time dependent non-Newtonian features. Though flow development and width of the finite film (thickness ~ 50 diameters of the ~10 μm particles) were excluded as factors in the low Reynlds number flow, the possibility of particles entering the $h_2 \rightarrow 0$ part of the apparatus was of some concern (was h_2 effectively zero?)

Whilst it was concluded that the mixture viscosity was more important in design calculations than base oil viscosity - for transport and film thickness calculation, what the case would be in a very thin film remains a matter of some speculation. The same applies to inconclusive time response tests [5] - (Figure 5.6) where the rise time exceeds those seen in other ESF flows by a wide margin.

5.5 EHL Problems

Investigations in the electohydrodynamic regime have mainly taken place on ERF since some MRS particles can scratch the expensive glass disc (see later) used for observation of the phenomenon in interferometric apparatus [6]. In the case of rolling ball on disc tests preparations carried out on silicone based ER fluids (Figure 5.7), it was found that the majority of particles avoided the contact zone with the result that predominantly base fluid was introduced there and consequently, due to its low viscosity, a full film could not be formed. At low speeds some particles did enter the contact zone but, many particles were charged into the region in front of the ball. The film formed was time dependent (Figure 5.8), the entry mechanism was dominant.

These results show the difficulty in achieving adequate lubricating films at realistic contact speeds (typically a rolling bearing might operate at speeds of 0-5 m/s). Lubricating films should ideally be thicker than the combined surface roughness of the contacting elements (typically 0.1 - 1 μm for a ground surface) to minimise surface contact and hence friction and wear. Thus to achieve adequate elastohydrodynamic lubrication with an ER fluid a viscous base is required. This raises a dilemma. The contribution of this base viscosity to the overall ER fluid mixture viscosity would probably result in excessive heating and power loss at high speeds.

An MRS fluid comprising some 32% by volume of nominally soft iron particles carried in a mineral oil fortified with anti-coagulant, wear and separation additives was used in EHL tests on a ball on disc machine. Though some similarity was seen with the ERF tests - see [6] and Fig 9 stable running was not usually achieved. The MRS base oil seems very thin - the 'white oil' viscosity is 18 mPa.s. It was again observed from pure sliding tests that particles accumulated in front of the ball, reached a critical amount then they entered the contact causing about a 0.3 micron variation in film thickness.

At the end of tests the ball and disc surfaces were covered in particles. The passage of particles can be seen from video photos shown as Figure 5.10. The unsteady situation described could have arisen from the trapping of the hardened particles or, in the light of subsequent experience of too high a speed. However, since scratching of the expensive optical plate was taking place, the tests were postponed pending future developments.

A tapered bearing/line contact EHL optical rig had been verified, performance wise, on lubricating oil. In the present instance various high concentrations of polyurethane particles (50 to 100%) produced circumstantial evidence of self lubrication properties of the ERF particles (in silicone oil) in EHL. Essentials of the rig are shown in Figure 5.11. In principle of operation it is similar to the ball on disc rigs with the important difference that the length of line contact inhibits the avoidance of the contact by the particle though side leakage is not infinitesimal. In the event little new perspective was gained into the particle entry process; particles seemingly piling up in the inlet zone without the formation of an adequate film. Photographic evidence from the tests is given in Figure 5.12.

5.6 Friction and Wear Phenomenon

As the intention of early tests was to alert scientists to the problems which could be caused by the choice of base fluid on the grounds of yield stress, low parasitic shear stresses and conductance alone, relatively standard wear and friction apparatus (NCT) could be used - see Figure 5.13. Both 'pin on disc' and 'disc on disc' tests were adopted, the latter being utilised only on fluid B - silicone base.

Wear and friction results indicated that ERF wear rates were much greater than with the mineral oil and that at high loads fluid B (silicone based ERF) caused more wear than its base fluid alone but the position was reversed in the apparent low load EHL region. In the other fluid (A) the wear rate by the mixture and base alone oil was much the same. The detrimental effect of lack of polar additives is clearly demonstrated and a considerable amount of the load must be carried by the particles: the reduction in friction by particle addition gives confirmation that a substantial number of particles pass through the contact zone and perhaps aid the process by virtue of a rolling action - the particles have a greater nominal size than the film thickness.

In the pin surface inspection fluid A produced but fine grooves - 5 μm deep, suggesting mild abrasion by grinding of asperities, incident particles were 2 to 3 μm and of various shapes. By comparison with the base oil the lithium salts increased the severity of the abrasion but there is little evidence of metallic adhesion. It is concluded that the PCB base fluid must be a reasonable boundary lubricant.

Whilst the solid lubricant can enhance friction performance it is likely to cause an earlier onset of boundary lubrication and as deduced from SEM and optical inspection of wear surfaces an unacceptable level of wear. It has to be said that unsteady operation of the wear mechanism was observed from time to time. The presence of large numbers of spherical ~3μm wear particles in the waste oil, Figure 5.14 suggests that adhesion and subsequent cooling of molten debris occurred in silicone based fluids. More particles were present in the base oil B tests than in the B mixture tests run off. Figures 5.15 and 5.16 give an example of results and [7] a bibliography for these series of tests. The disc apparatus is much the same as a block or pin apparatus save that a mild steel disc replaces the stationary pin. Use of discs confirmed the pin result.

Health regulations have caused the withdrawal of PCB fluids. This factor with the higher potential yield stresses was enough to cause attention to widen to include the other ESF - magnetorheological suspensions which could prove easier to stabilise and treat for boundary lubrication. Preliminary tests on Lord MRF 132 suggests better wear performance than ERF though some particles are hard enough to cause pin damage [8 and 9], Figs. 17, 18 and 19.

Sliding friction tests on Lord MRF 132 which utilised an Instron model 4204 test machine [10] have indicated that at a sled velocity of 2.6 m/s and compressive forces between 10 and 20N on the 76mm square lubricated surface produced a factor of 2 to 3 times less fiction than on dry iron to iron and nylon to iron surfaces. No wear rate data is given.

The effects on friction and wear of applying an electric field of ~ 4 V across boundary lubrication situations involving mineral oil and sliding motion can be considerable [11]. Equivalent results are not available for ERF or MRS though the paper does stimulate thoughts towards their provision.

5.7 Summary

The size and hence number of particles in an ESF mixture of a given solid volume concentration is determined by the required yield stress and zero volts viscosity, material availability and other operational considerations. A survey of hydrodynamic film generation using a Bingham plastic fluid indicates that it is unlikely that particle sizes will be as small as the size of a flow regime outside or inside of any plug formed or indeed smaller than the film thickness itself. Much the same problem applies in the EHL regime where the film would likely at best be intermittent in its formation: due to the entry mechanism and the avoidance of the contact zone by the particles the resulting (base fluid) viscosity may prove to be too small for full film formation. Any application of excitation would probably cause an increase in pressure in all films with an increased scraping effect of the particles on the surrounding surfaces, as presumably occurs in the QED machine.

In terms of the limitation of friction and wear, boundary and mixed lubrication regimes probably deserve the most attention. Properties of the base fluid are important and need not be sacrificed in the formulation of MRS whereas they may have to be secondary to polarisation, conductance and breakdown of an electrostatic nature in an ESF. Hence, the harder (presumably) particles of an MRS can cause less damage than an ERF, especially as the correct surface additives plus extreme pressure agents etc can be introduced with relative impurity. However, processes are not entirely ideal and some hard particles may find their way through the production phase or the particles can work harden when between bearing surfaces. This could be a very important factor, especially in reciprocating/sliding regimes where the particles have limited scope for circulation.

If ESF are to approach nearer to the tribo quality of mineral oil blends for the same purpose further work needs to be done. Not least on this horizon is the need for systematic design data similar to the pressure - velocity graphs for solid lubricants and/or a more individual means of quantification of the boundary performance of the fluid/lubricant. In the end in prototype tests will give the most realistic friction and wear versus duty results.

For completeness it should be added that ESF surface machining can be achieved by the use of low frequency excited ERF. This causes reciprocating motion of the particles between the electrodes [12]. Generally surfactant covered diamond ERF particles are used with silicone oils. This work is in its infancy and, like squeeze film bearings, is hardly relevant to the present work. There is considerable divergence between theory and practice (the polishing effect is dependent on E, its frequency and the materials involved) and is not dwelt on further at this juncture.

References

1. Kordonski W and Golini D, Magnetorheological Suspension - Based High Precision Finishing Techniques, Jnl. Intel. Mat. Syst. and Struct., Vol. 9, No 8, (Aug. 1998), pp 650 - 654.
2. Ganz R F, On the Flow of a Yield Strength Fluid Through a Contraction, Jnl. Non-Newtonian Fluid Mech. 81 (1999), 183 - 195.
3. Walicki E and Walicka A, Reynolds Number Effects in the Flow of an Electrorheological Fluid between Fixed Surfaces of Revolution, Jnl. Intel. Mat. Struct. and Syst., Vol. 9, No. 8, Aug (1998), pp 662 - 666.
4. Leek T H, Lingard S, Bullough W A and Atkin R J, Hydrodynamic Pressure Generation with an Electrorheological Fluid. Part I - Unexcited Fluid and Part II - Excited Fluid, Proc. 3rd Int. Conf. ER/MR Fluids held Feldkirch, (1993), pp 608 - 624 and 625 - 642.
5. Leek T H et al, The Time Response of an Electrorheological Fluid in a Hydrodynamic Film. Int. Jnl. Mod. Phys. B, Vol. 10, Nos 23 and 24 (1996), pp 3343 - 3355.
6. Dwyer-Joyce R S, Bullough W A and Lingard S, Elastohydrodynamic Performance of Unexcited Electrorheological Fluids, Int. Jnl. Mod. Phys. B, Vol. 10, No 23 and 24 (1996), pp 3181 - 3189.

7. Lingard S, Bullough W A and Shek W M, Tribological Performance of an Electrorheological Fluid, Jnl. Phys. D: Appl. Phys. 22 (1989), pp 1639 - 1645.
8. Bullough W A and Wong P L, Tribology of Electrostructured Fluids, Proc. IV Minsk International Forum to be held May 2000, Acad. Nauk, Minsk.
9. Wong P L and Bullough W A, Tribological Performance of a Magnetorheological Suspension, submitted to Wear.
10. Jolly M R, Bender J W and Carlson J D, Properties and Applications of Magnetorheological Fluids, Proc. SPIE 5th An. Int. Symposium on Smart Materials, held San Diego (1998), pp 1 - 15.
11. Yamamato Y, Ono B and Ura A, Effect of Applied Voltage on Friction and Wear Characteristics in Mixed Lubrication, Lubrication Sci. 8-2, (Jan 1996), pp 199 - 207.
12. Yakagami Y, Ogasawara Y, Nisimura B, Jeyadevan B and Fujita T, Characterisation of Particle Motion for Polishing and Texturising Under AC Field Using Particle Dispersion Type ER Fluid, Jnl. Intel. Mat. Syst. and Struct., Vol. 9, No. 8 (Aug 1998), pp 666 - 675.

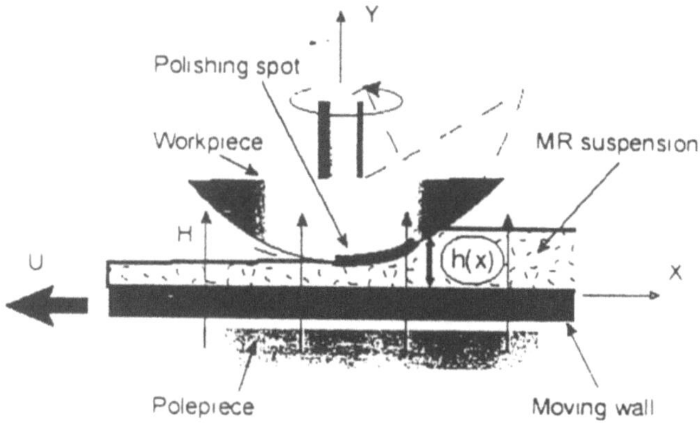

Figure 5.1: Schematic of MR lens polishing process.

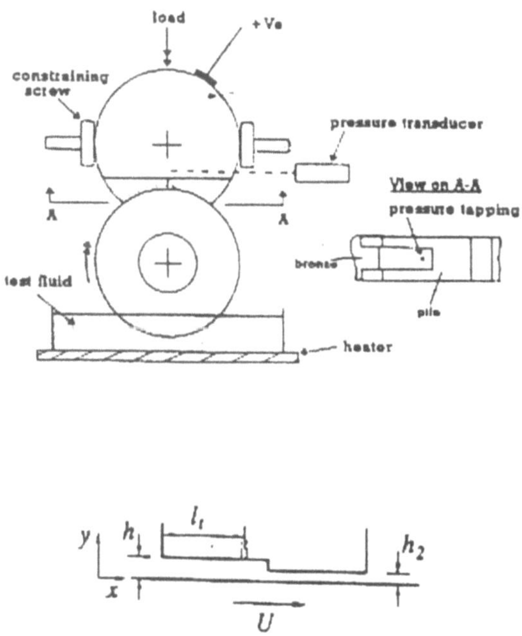

Figure 5.2: Rayleigh step arrangement: experimental above, analytical below.

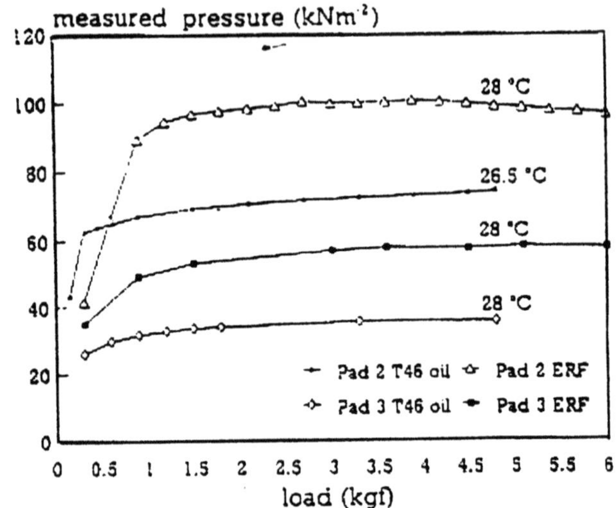

Figure 5.3: Example of generated pressure v loading in Rayleigh step: fully loaded to right.

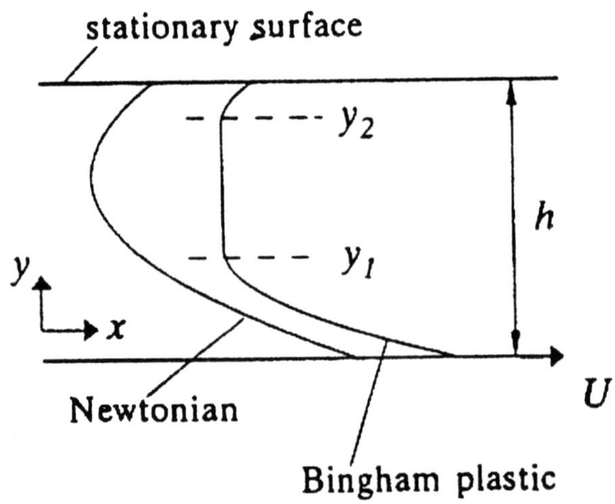

Figure 5.4: Velocity profile of Bingham plastic flow in the Rayleigh step at zero flow rate

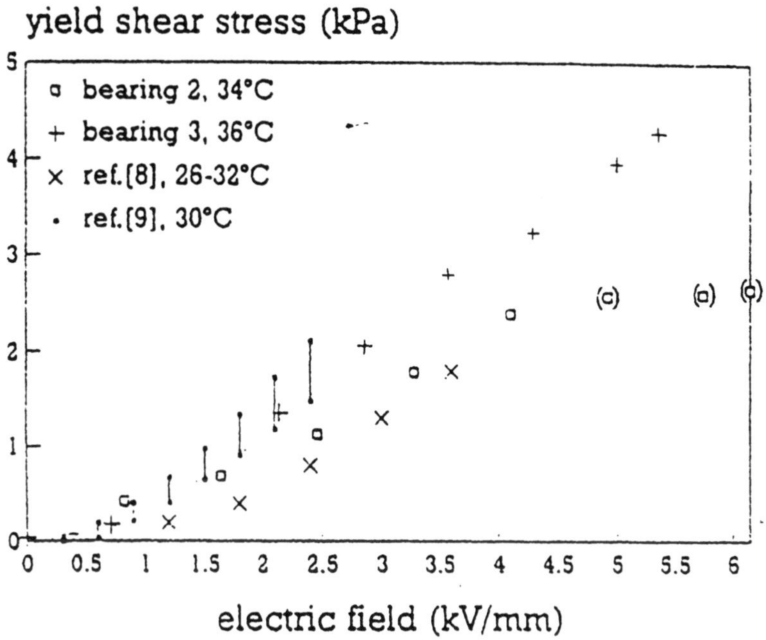

Figure 5.5: Bingham plastic yield stress values extracted from step bearing at zero flow rate and from valve and clutch test data. Plastic viscosity is taken from corresponding apparatus when unexcited. Brackets indicate isufficient loading toproduce limiting pressure.

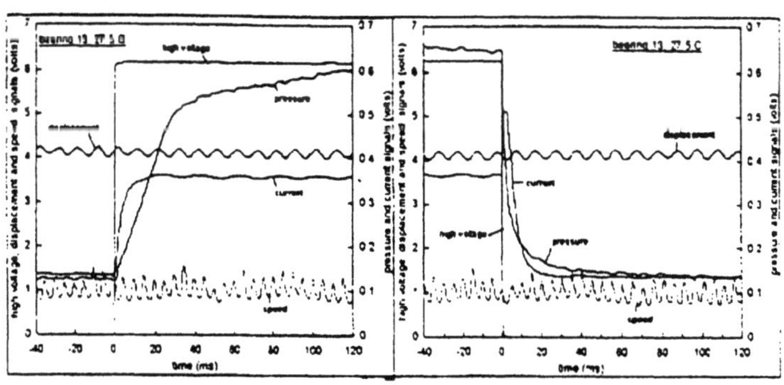

Figure 5.6: Examples of time response data signals from Rayleigh step.

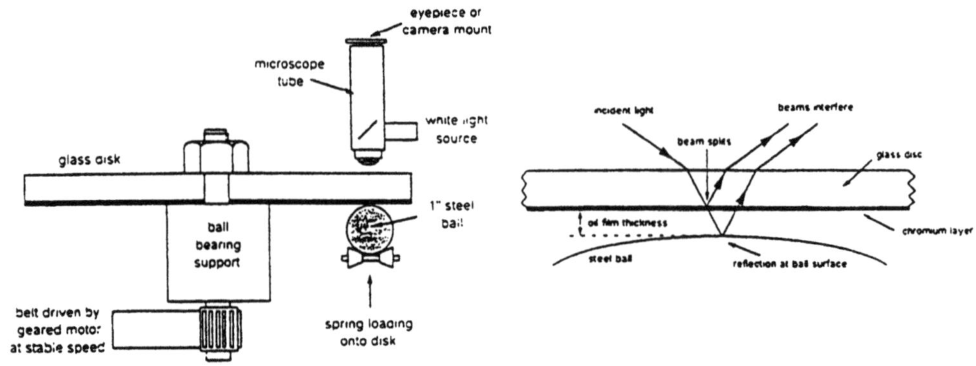

Figure 5.7: Schematic diagram of ball on disc EHL test equipment.

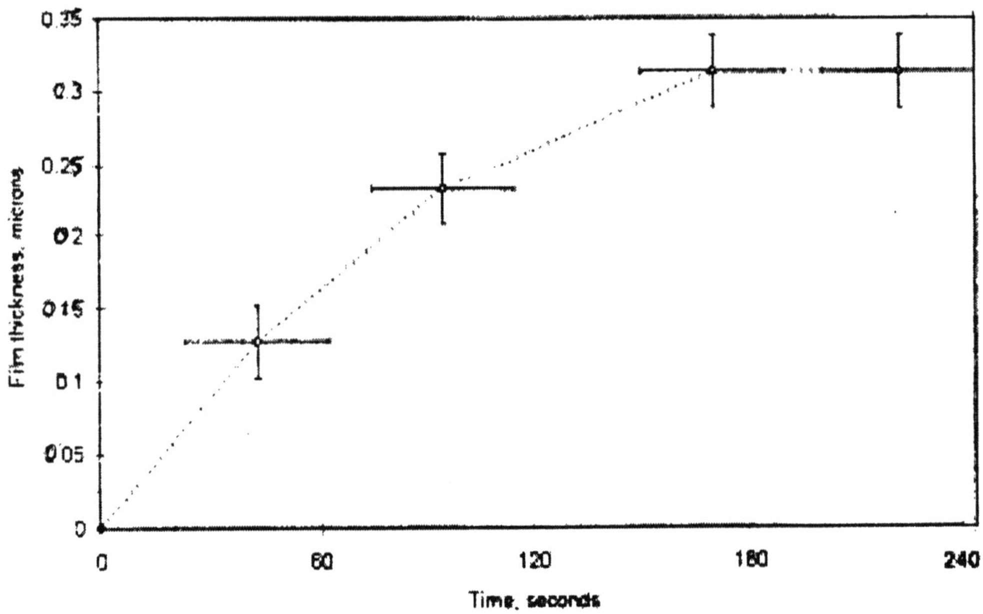

Figure 5.8: Variation of film thickness with time for silicone based ERF polymer 60% mix, on 'ball on disc' machine at 3 mm/s rolling speed.

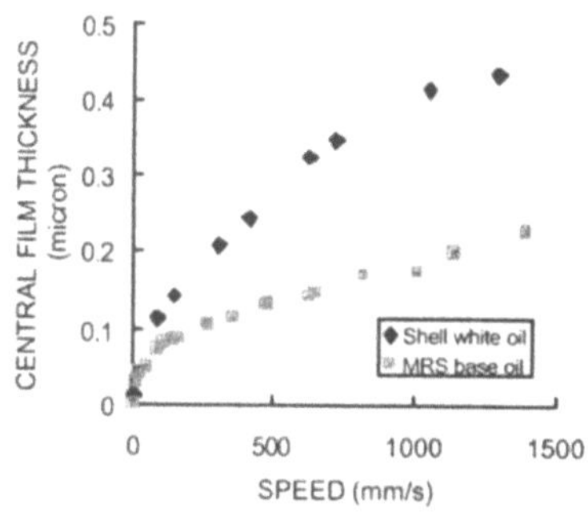

Figure 5.9: EHL ball on disc result for MRS base oil

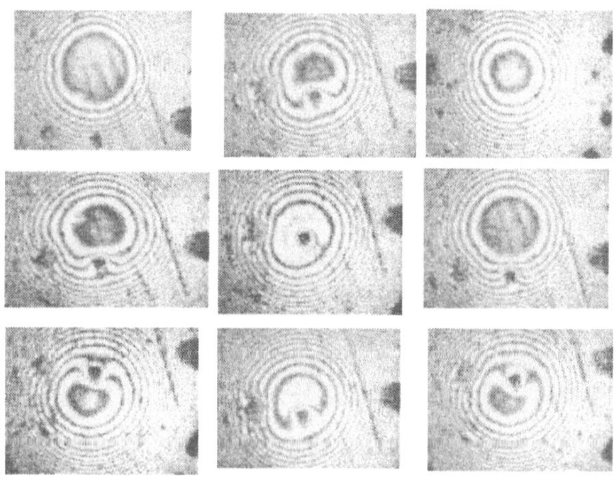

Figure 5.10: MRF particle approaching, passing through and leaving contact zone. Ball on disc EHL tests. Film thickness ~ 0.1μm.

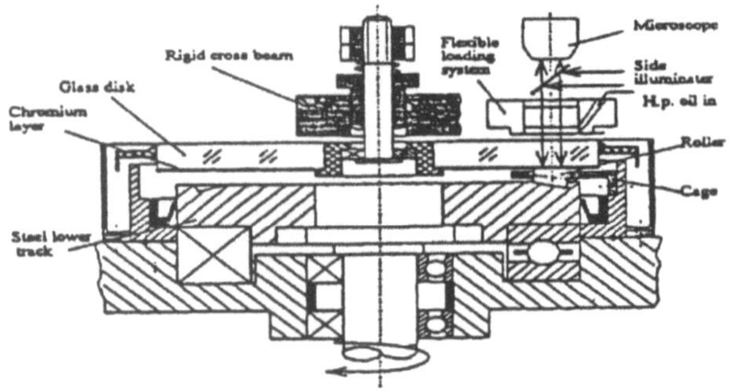

Figure 5.11: Tapered bearing EHL rig - line contact

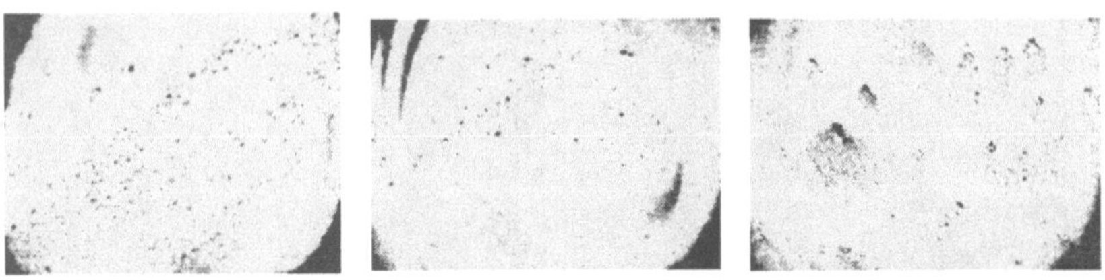

Figure 5.12: Typical rolling track appearance for (a) slow and (b) high speed tests on apparatus of Figure 11 with ERF and (c) larger magnification of slow test - respectively, showing particles flattened and adhering to track.

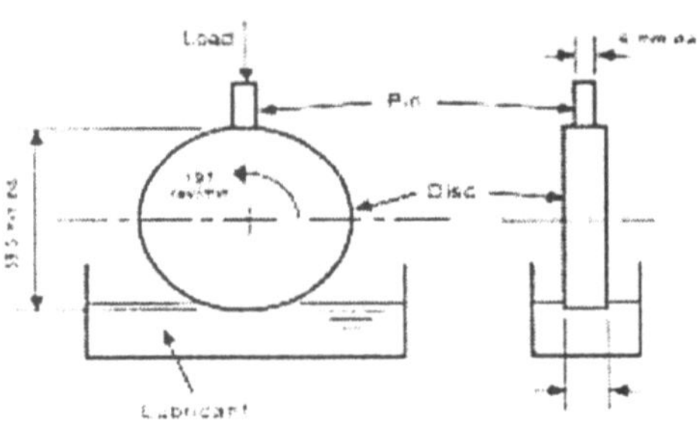

Figure 5.13 Pin on disc test apparatus (schematic).

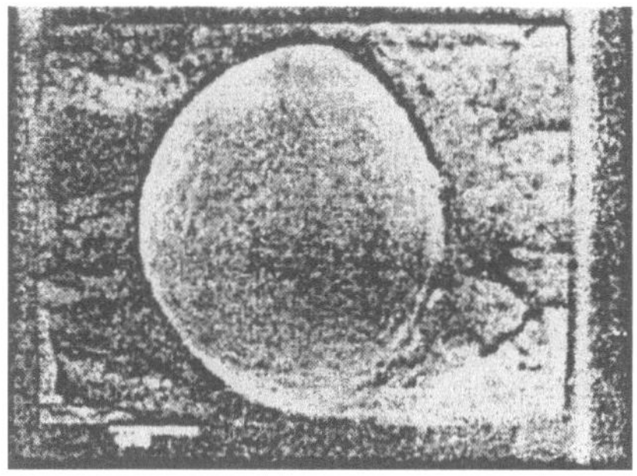

Figure 5.14: Spherical ferrous particles from used ER base fluid 4 to 12 μm dia, 10^5 per litre.

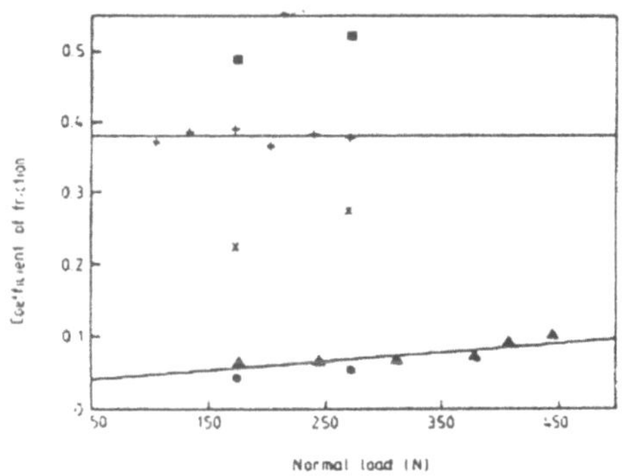

Figure 5.15: Coefficient of friction, pin on disc test +, mild steel pin, ER fluid; ■, mild steel pin, ER base fluid; ▲, mild steel pin, Tellus 46; x, brass pin, ER fluid; ●, brass pin, Tellus 46.

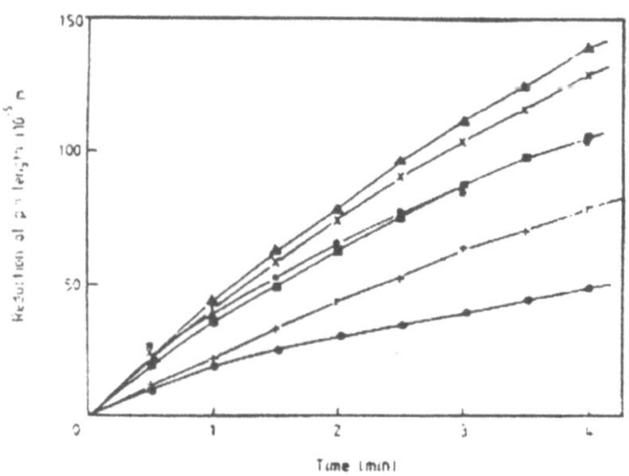

Figure 5.16: Reduction in pin ;length with time, pin on disc test, mild steel pin, ER fluid. ●, 105 N; +, 135 N; *, 173 N; ■, 203 N; x, 2409 N; ▲, 270 N.

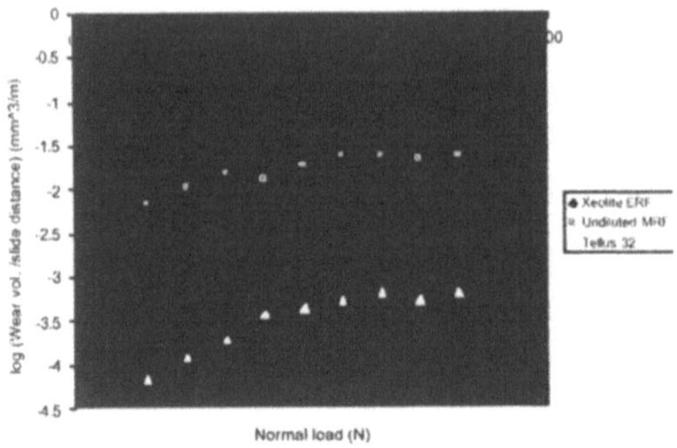

Figure 5.17: Comparative MRS wear tests.

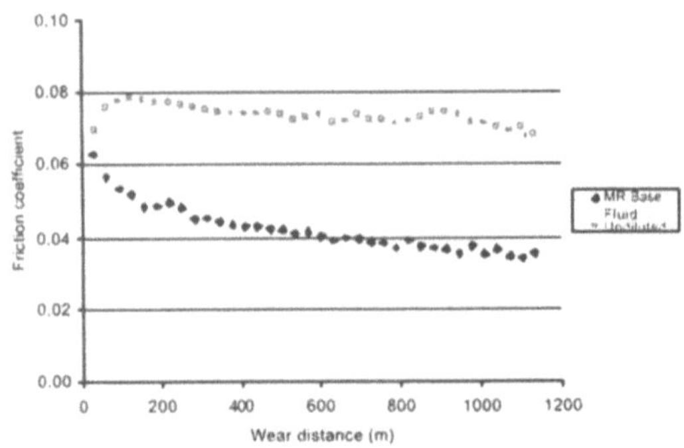

Figure 5.18: Comparative MRS friction coefficient test results.

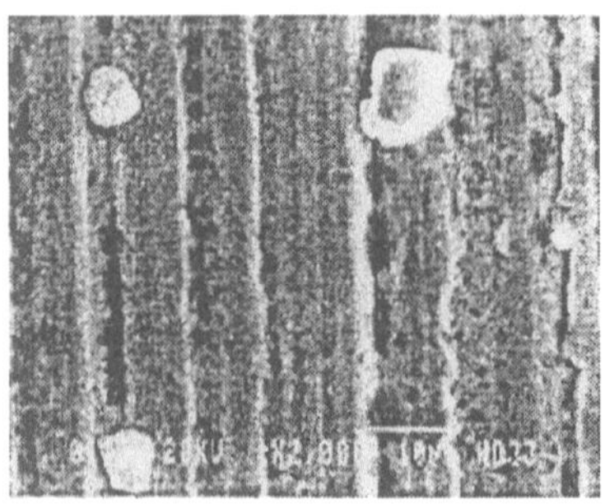

Figure 5.19: Ploughed grooves formed by hard abrasive particles embedded in block surface with rolled and ploughed plastic deformation.

CHAPTER IV

MICRO ELECTRO MECHANICAL SYSTEMS

1. Microelectromechanical Systems
2. Small and large: scaling
3. Silicon micromachining
4. Principles of measurement mechanical quantities: transduction of deformation
5. Case studies

M. Elwenspoek
MESA+ Institute
University of Twente
P.O. Box 217
7500 AE Enschede
Tel: x31-53-4893845
Fax: x31-53-4893343
Email: m.elwenspoek@el.utwente.nl
http://www.el.utwente.nl/tdm/mmd/mmdhome.html

1. Microelectromechanical Systems

M Elwenspoek and R. Wiegerink

Faculteit der Elektrotechniek, University of Twente, Enschede, The Netherlands

Abstract The technology for miniaturisation develops from a number of fabrication methods: We mentioned IC-fabrication methods, but there are many groups at universities and companies that aim at the fabrication of small systems using technologies that derive from more conventional machining, such as cutting, drilling, sand blasting, spark erosion, embossing, casting, mould injection etc. In this chapter we concentrate on silicon micromachining and on the mechanical microsensors that can be fabricated by this method.

1.1 Miniaturisation and Systems

In many technical systems there is a strong trend for miniaturisation. This trend results on one hand from the fact that small components and systems perform differently: small systems can perform actions large systems cannot (example: minimal invasive surgery). In many cases a miniaturisation makes the systems more convenient (example: GSM telephone). On the other hand technology derived from IC-fabrication processes allows the production of miniature components in large volumes for low prices (examples: pressure sensors for automobile applications, ink jet printers)

The technology for miniaturisation develops from a number of fabrication methods: We mentioned IC fabrication methods, but there are many groups at universities and companies that aim at the fabrication of small systems using technologies that derive from more conventional machining, such as cutting, drilling, sand blasting, spark erosion, embossing, casting, mould injection etc. In this chapter we concentrate on silicon micromachining and on the mechanical microsensors that can be fabricated by this method.

There are two basic notions used to indicate the science of miniaturised mechanical components and systems: Micro systems technology (MST; this notion was coined originally in Germany) and Microelectromechanical Systems (MEMS, invented in the USA). While MEMS is more specific to mechanical components, MST includes also microoptical systems, chemical sensors, analysis systems etc. However, in all microsystems mechanical and electromechanical functions must be realised, therefore in practice, both notions have a very large overlap. The important word in both notions is *system*. Here, system is used in contrast to *component*. Microthings are necessarily systems because of the small details in the systems. One cannot assemble a microsystem from components from the shelf. Example: if a factory wants to build a car, which is a pretty complex system, the whole car is designed using (in principle) available components (motor, transmission, doors, windows, wheels, etc.). These are purchased or fabricated following the specification of the designers and assembled. Such a procedure is impossible with microthings. First of all, the components are too small to be assembled for reasonable costs. Further, there are no components that fit together. In designing a microsystem, the components have to be designed during the design of the whole system. There is no design phase of the system separate from the design phases of components: the design of a microsystem is an integral process. Even more: the fabrication of the microsystem is an integral process. This process is being provided by silicon micromachining. We shall illustrate the design- and fabrication process a few times.

Referring to microsensors, we have leaned by a painful process that the sensing element, the electronic interfacing and the package (if you wish, the mechanical interface) must be designed as a whole. Designing one part of the system after the other in most cases will lead to sensors which cannot be packaged, or of which the package is too expensive.

A second systems aspect plays a major role in the fabrication process. Micromachining is very complex and not developed to a standard technology. The process space is still largely unexplored. When designing a microsystem, the fabrication process of the system must be designed, too. Thus system design and process design are integrated in a single design process. Microsystems designers therefore must have a large number of skills and a rather broad experience. This is quite different from IC-design and processing. For IC-processes there are strict and clear design rules which guide the circuit designer, and he does not have to worry how the circuit is actually made. Microsystems designer must be able to design both: the system and the fabrication process.

A third systems aspect is the following: Usually microsystems have complex functions. The functions have roots in different physical domains. For a sensor this is clear: a sensor must transmit a signal from a particular domain (chemical, mechanical, thermal) to the electrical domain. A sensor designer must of course know the domains to which the sensor has ports. As an example, in a pressure sensor a membrane is deformed and the deformation is the measure for the pressure. The sensor designer must know the mechanical properties of membranes. The deflection of the membrane can be done by optical interferometry. In this case, the designer must know optics as well. The designer must know how to treat his (usually) small electrical signal, how to amplify it, how to realise the interfaces to a computer, so he must have knowledge of electronics, too. Sensor designers therefore are Jacks of all trades. In practice, they have to work in groups of engineers of different colours, and understand enough of other disciplines in order to be able to communicate.

1.2 Examples for MEMS

Since this chapter is devoted to microsensors, there will be no extensive discussion of other microsystems such as actuators, microstructures, microrobots etc. We give a few examples here. We find this important because the technology - silicon micromachining - enables us to fabricate complex systems and simple systems for more or less the same price. So there is the chance (and the challenge) to integrate many functions in a single piece in an integrated production process.

Using silicon it is obvious to think of the combination of mechanical and electronic functions. Much work has been done to develop integrated sensors, or "smart sensors". A few are on the market (e.g. accelerometers from Bosch and Analog Devices, a pressure sensor from Bosch), however it seems that the development of the production process for the integration of electronic and mechanical functions is extremely expensive. The integration of several mechanical and electromechanical functions seems to be much simpler and economically more feasible.

1.3 Bubble jet

The ink jet printer is one of the successful microsystems now. One of the ways to dose drops of ink on a piece of paper is to squeeze the liquid ink close to an opening - the nozzle - in a way that a single drop with a reproducible volume and a certain speed will leave the nozzle. This can be accomplished by a piezoelectric element; however, the first successful technology uses thermal expansion of ink vapour. Schematically the ink jet printer head looks like as shown in fig. 1.1.

The thermal ink jet printer head consists of a reservoir, a channel, a nozzle and a heating element. When the heating element heats the ink, its temperature is raised above the critical point, so a bubble is formed without a nucleation barrier (which would result in a too slow and not reproducible formation of a bubble). The pressure in the bubble is used to displace the ink, and because the flow resistance to the nozzle is smaller than to the ink reservoir ink is ejected. When switching off the heating power the ink cools and the vapour bubble collapses and an ink bubble (in fact, often many bubbles) breaks off. This sequence is shown in fig. 1.2.

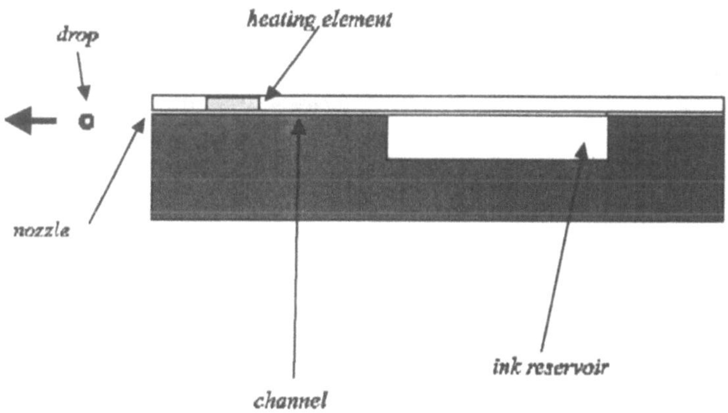

Fig. 1.1 Schematic of an ink jet printer head.

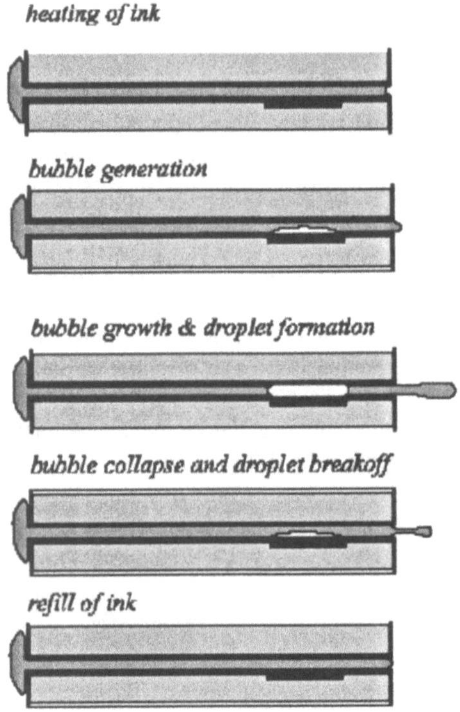

Fig. 1.2. Cycle of drop formation

Thé ink jet printer is a very nice example of a microsystem in a number of respects. It is a system. In fact, what we have drawn here is only a very small portion of the system. There must be some electric wiring. The printer must be able to communicate with a computer. so there must be an electric interface. When designing the channel and the nozzle, one must know what type of ink will be used: the viscosity is a very important parameter and is strongly temperature dependent. When choosing the ink, one must know on what type of paper will be printed. One needs to know the size of the droplets. The ink jet printer as an example par excellence for multidisciplinary work: thermodynamics (critical points and phase transitions), hydrodynamics (very complex because of the drop formation), and electronic engineering for the head, mechanical and electric engineering. The distribution is changed by convection and measured. The ink jet printer head has a heater, so micromachining allows to integrate more heating elements into the channel, some can be used as actual heaters, and some as temperature sensing element e.g. by exploiting the temperature dependence of the electrical resistance of the materials used.

1.4 Actuators

Microactuators can be used for positioning of small masses. An example of such an actuator is shown in fig. 2.3. This is a so-called comb drive actuator, in which interdigiting teeth contract under a voltage difference. The maximum force these actuators can deliver are quite small (in the order of 10 pN!), therefore one needs many of them in parallel. Future applications of comb drive actuators are in positioning mirrors or tips in read-write heads in data storage devices. Currently, much effort is taken to develop microresonators for telecommunication purposes. Connecting a comb drive to a spring gives a resonator, which can be used in an electromechanical high quality factor filter.

Currently there is a search for actuators, which are fast and strong. Electrostatic actuators are fast but the maximum force, even using clever designs, seems to be limited to about 1 mN. Therefore considerable effort is undertaken to fabricate piezoelectric actuators. In general, thermal actuators tend to be strong but slow.

Comb drives have important applications in mechanical microsensors. Since they represent a capacity the value of which depends on the overlap of the fingers, they can be used as position sensitive devices. Deflection of seismic masses in inertia sensors (see chapter. 7) and shear stress in fluid flow sensors (see chapter 8) is measured using comb drive actuators. In both types of devices comb drives are also applied in servo systems where the force is the sensor output, needed to keep the fingers in a fixed position.

1.5 Micropumps

A great promise in MEMS comes from micro fluid handling systems (MFS), after the first components for MFS have been demonstrated. Among them a silicon micropump. Historically the first silicon micropump was a peristaltic pump, designed by Jan Smits during the first half of the eighties. Smits' work was published not before Transducers '89 (Smits 1990), two years after the follow up of this work was presented at Eurosensors 87 by van Lintel, van de Pol and Bouwstra (van Lintel 1988), see fig. 1.4, which in fact was the first working silicon micropump ever published.

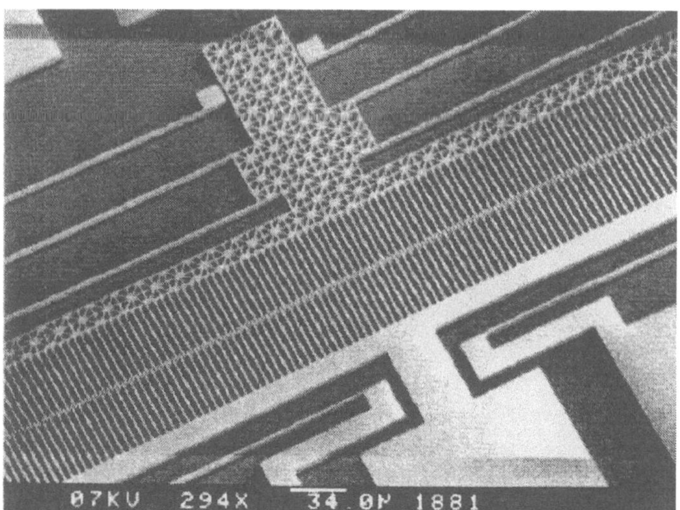

Fig. 1.3. Micromechanical comb drive actuator. The gap between the teeth is app. 2 μm. In the upper half there is a shuttle.

Van Lintel used a piezoelectric disc glued to a glass membrane; under a voltage difference the piezoelectric disc changes its lateral dimension which results in a bending moment in the dimorph. The original prototype was able to produce a maximum pressure of 100 cm water and had a maximum yield of ca. 10 μl/min at 1 Hz block wave actuation.

Mainly from aesthetic reasons van de Pol suggested an alternative for the piezoelectric actuation which he called thermopneumatic actuation. By heating a gas or a liquid under its vapour pressure the pressure is increased and a membrane can be deflected. This pump, when using ambient air as expanding medium in the actuator, was able to produce a maximum pressure of 40 cm water and had a maximum yield of ca. 30 μl/min. Maximum yield occurred at 1 Hz block wave actuation.

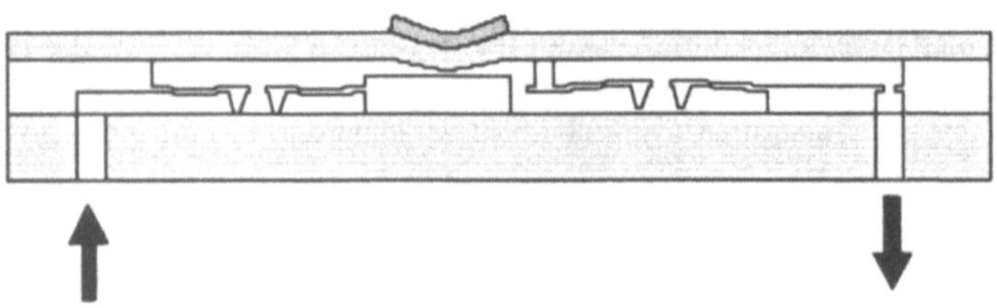

Fig. 1.4. Schematic of van Lintels membrane pump (van Lintel 1988). In the centre the pump chamber and the pump membrane are drawn. If the membrane bends down, the right valve opens and the left valve is forced to close; this state is shown. If the membrane is released or bends upwards, the left valve opens. Note that this pump is asymmetric. Reproduced from 'Silicon Micromachining', M. Elwenspoek and H. Jansen, 1999, by permission of Cambridge University Press.

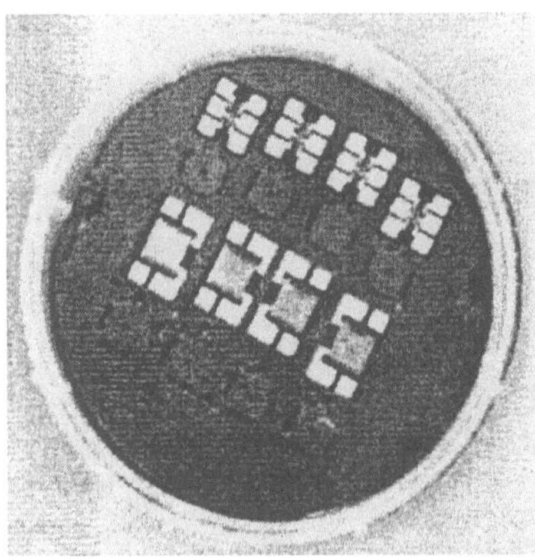

Fig.1.5. A dosage system with a micropump and a flow sensor integrated on one chip. Photograph of the wafer containing four dosage systems (Lammerink et al. 1993). Reproduced from 'Silicon Micromachining', M. Elwenspoek and H. Jansen, 1999, by permission of Cambridge University Press.

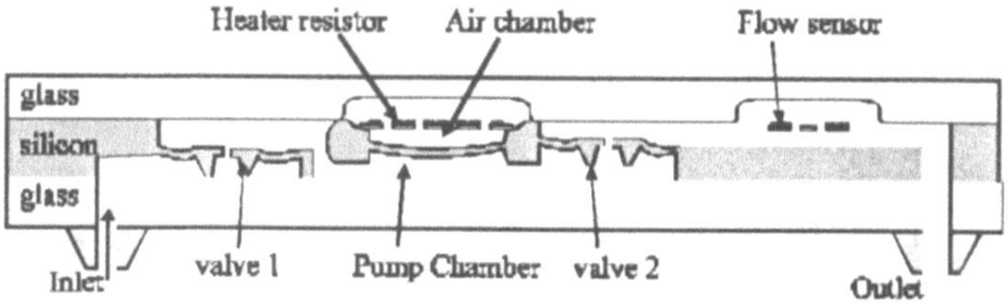

Fig. 1.6. Cross-section of the dosage system. Reproduced from 'Silicon Micromachining', M. Elwenspoek and H. Jansen, 1999, by permission of Cambridge University Press.

There are many groups now active in the design of micropumps (examples: Forschungszentrum Karlsruhe, Universtät Freiburg, Institut für Mikrotechnik Mainz (IMM), Hahn-Schickart Gesellschaft, Institute for Micromachining and Information Technology, HSG-IMIT, in Villingen - Schwenningen), and the performances has been increased considerably during the past years. A recent overview can be found in (van Kuijk 1996). The latest pumps are self-priming and resistant against the formation of gas bubbles in the system. (H. Linnemann 1998, K.-P. Kämper 1998).

The integration of the thermopneumatic pump and the flow sensor is rather straightforward. Fig. 1.5. shows an optical photograph of the wafer with four dosage systems, and in fig. 1.6. we show the cross section of a dosing system. It has been described first at MEMS '93 in Ft. Lauderdale, USA (Lammerink 1993). Once having demonstrated the feasibility of integrating several functions of a MFS on one wafer the progress is obvious: tedious and expensive assembly of micro components can be avoided. Flow sensors are important components in microliquid handling systems and will be described in detail in chapter 4.

Micro fluid handling systems will possibly replace chemical sensors. In chemical sensors the sensing material reacts directly with the chemical whether or not in the ambient. MFS can be used as chemical laboratory on a chip, which includes functions such as sample pre-treatment, calibration and possibly a chemical reaction if the detection of a reaction product is easier that the that of the chemical its self.

Below we give a list (far from complete) of applications of MEMS. A recent overview can be found in Fluitman 1996.

Medical applications:

- hearing and seeing aids, prosthesis for eye and ear
- systems for nerve stimulation
- dosage systems for medication
- implanted dosage systems
- minimum invasive surgery

- implanted drainage systems
- endoscopy with active endoscopes
- surface structuring of implanted systems

Consumer applications:

- microsystems for information storage
- CD-players in the size of match boxes, CD's in the size of coins
- ink jet printers
- control instrumentation (sensors) for products such as cars, houses. copymachines. telecommunication
- (optical) communication systems

Applications in environmental control, chemistry, pharmacy, agriculture:

- (bio-)chemical sensors, analysis systems, robots
- on-line monitoring and control of waste production and gas emission
- fight vermin by microrobots (in place of hazardous chemicals)
- noise control (microphones. soundintensity sensors)

Applications in fabrication:

- production facilities for chips in the size of an average living room
- control instrumentation in industry
- industrial robots
- microrobots for product control

Applications in science:

- micro instrumentation for:

* wind tunnels (pressure flow patters drag forces)

* aeronautics (sensors and actuators)

* space travel, "micro" and "nano" satellites
* cell micro manipulation
* distributed sensing for the study of e.g. turbulence
* sensors for friction and wear

- materials science
- scanning microscopy

1.6 Available fabrication technology

Technologies based on lithography

In photolithography light impinges a transparent plate with opaque regions, called a mask. Below the mask a photosensitive material is placed, which can be "developed" after illumination. In this process, the illuminated regions (or the non-illuminated ones) are dissolved, and the underlying material is accessible to further processing. Fig. 2.11 shows an example.

Silicon micromachining

The basis of silicon micromachining is photolithography. Photolithography defines regions on a silicon wafer where the machining is done. The machining includes etching (in a liquid, plasma, gas or by high energy ions or particles), doping and deposition of thin films. It is clear then that micromachining allows only the fabrication of three dimensional structures with complex forms only in two dimension, everything in the third dimension is in some way extended from two dimensions. Some extension to really three-dimensional structures can be accomplished by means of wafer bonding. These technologies will be described in greater detail later in chapter 3.

The main difference of silicon micromachining with most other machining techniques is that the first technology is suitable for batch processing, similar to the fabrication techniques of integrated circuits. The technology is quite expensive, mainly because a clean room is required, but the price of components is not proportional to the number of fabricated pieces. In the limit of very large production volumes, the production costs of the whole production is independent from the number of components fabricated. The other technologies described below (with the exception of the LIGA process, which is in some respects similar to silicon micromachining) do not have this property. Here each component must be made piece by piece. Low prices for large production volumes are the result of mechanisation.

LIGA

LIGA is an acronym based on the German words Lithography, Galvanik (electroplating) and Abformung (moulding). The technology originated at the Kernforschungszentrum Karlsruhe in Germany, where attempts were made to machine microfluidic elements for isotope separation. The process involves illumination of a thick X-ray sensitive resist by high-intensity radiation from an X-ray source. Since X-rays are practically not scattered or diffracted at the dimensions of interest (μm), this first illumination step results in structures with tremendous high aspect ratios and very flat sidewalls. Illumination is done through an X-ray absorbing mask; to prepare this mask, often the same X-ray source - a synchrotron - must be used. The resist is deposited on top of a metal to allow electroplating into the opened features. The resulting metal structure then is used as a mould insert. Polymer microcomponents fabricated this way are on the market now.

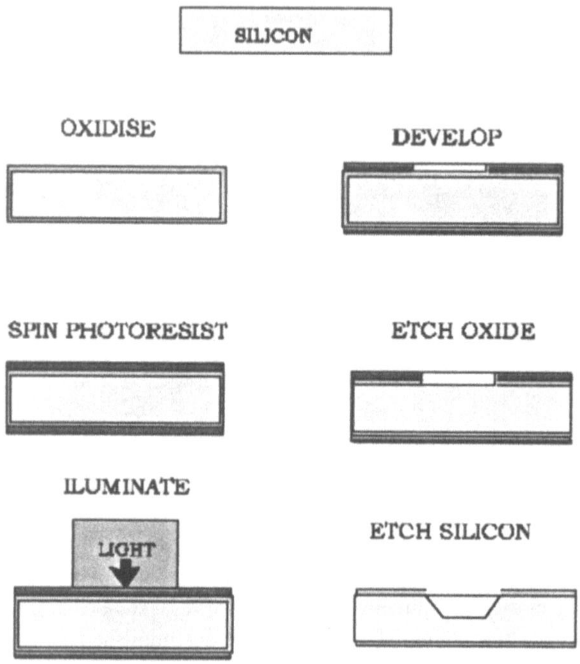

Fig. 1.7. Basic photolithographic process Reproduced from 'Silicon Micromachining', M. Elwenspoek and H. Jansen, 1999, by permission of Cambridge University Press.

Obviously the illumination step is the most expensive one in the process. The development of microsystems using LIGA is very expensive, because many masks must be made before a process is stable and the system has the desired properties. LIGA is economically only feasible for very large production volumes. Therefore alternatives have been looked for, with some interesting results.

One of the alternatives is to use conventional photolithography to illuminate thick mask layers (10-30 μm), of course with all the trouble caused by light scattering. A new photoresist (SU-8) performs much better in this respect (B. Eyre 1998).

A second alternative is based on silicon micromachining by first fabricating the high aspect ration structures with the help of reactive ion etching (RIE). This step replaces the illumination step in the LIGA process.

1.7 Miniaturisation of Conventional Technologies

The most important difference with photolithography based fabrication technologies is that using conventional technology the components must be made piece by piece. Finest details that can be machined are in the range of 10 to 100 μm, one to two orders larger than what photolithography makes possible. One of the greatest advantages is that one is not restricted to flat substrates.

Cutting, turning, drilling, milling and *shaving* can be done using highly specialised precision machines. An example and a first guide to the literature can be found in (Y. Yamagata 1996).

Ultrasonic machining is a well known technique to fabricate quartz tuning forks, but the technology is being extended to other materials and the precision increases (example: Xi-Qing Sun 1996).

Also techniques such as *sandblasting, laser ablation* and *spark erosion* are miniaturised. Powder-blasting has been used to fabricate features in glass in the size range of 100 μm and larger for flat panel displays at Philips Research (Slikkerveer 1996). The aspect ratio seems to be limited by 1:1 and the minimum feature size by fracture mechanics (10 - 20 μm) and by the particle size (powders with particles below 10 μm are very dangerous). Laser ablation works by heating the substrate locally so that the material evaporates. The technique suffers from heat production. If the heat is not carried away by the gas but diffuses into the substrate it deforms plastically due to large temperature gradients and thermal expansion. Light can also be used to enhance etch rates locally and to enhance the nucleation and growth of materials. Mark Madou gives a rather recent overview in his book (M. Madou 1997).

Acknowledgements

Contributions of Han Gardeniers, Meint de Boer, Gijs Krijnen, Remco Wiegerink, Erwin Berenschot and Theo Lammerink are gratefully acknowledged.

2. Small and Large: Scaling

M Elwenspoek and R. Wiegerink

Faculteit der Elektrotechniek, University of Twente, Enschede, The Netherlands

Abstract There are a few matters one needs to keep in mind when miniature technical systems are designed. Many children imagine worlds on the top of a pin. Linear dimensions shrink slower than surfaces and surfaces shrink slower than volumes. So, friction associated with surfaces becomes relatively more important than masses: weight and inertia tend to be negligible in the microworld. But also stiffness, electric forces, magnetic forces, adhesive forces, the character of convection, the importance of boundary layers, surface finish, characteristic times associated with diffusion and resonance all scale in their own way.

2.1 Introduction

There are a few matters one needs to keep in mind when miniature technical systems are designed. Many children imagine worlds on the top of a pin. There is an impressive movie (The fantastic voyage, written by Isaak Asimov) in which a submarine is shrunken to the size of a few hundred micrometers and injected into the human body for microsurgery. From a scientific point of view, these thoughts are ridiculous. Shrinking the linear dimensions does not shrink forces in the same way. A few things are quite obvious. Linear dimensions shrink slower than surfaces and surfaces shrink slower than volumes. So, friction associated with surfaces becomes relatively more important than masses: weight and inertia tend to be negligible in the microworld. But also stiffness, electric forces, magnetic forces, adhesive forces, the character of convection, the importance of boundary layers, surface finish, characteristic times associated with diffusion and resonance all scale in their own way.

It is instructive to analyse scaling by comparing forces and effects by studying the exponent of the length scale of linear dimensions l. We then have:

Linear dimension: $l \sim l^1$.

Surfaces: $A \sim l^2$.

Volume: $V \sim l^3$.

2.2 Electromagnetic forces

The electrostatic force between capacitor plates is given by the derivative of its electric energy W with the gap distance x:

$$F = -\left(\frac{\partial W}{\partial x}\right)_Q \tag{2.1}$$

where the charge Q must be held constant. W can be found in the easiest manner by the energy density w:

$$w = \frac{1}{2}ED = \frac{\varepsilon_0}{2}E^2 \tag{2.2}$$

E and D are electric field strength and the displacement current, respectively. We assumed that the dielectric between the capacitor plates is air (ε_r=1). It follows that

$$W = \int w dV = wAx \tag{2.3}$$

with A the surface of the electrodes and x their distance. For w = constant, meaning E = constant we obviously have

$$F_{el} = -wA = -\frac{\varepsilon_0 E^2 A}{2} = -\frac{Q^2}{2\varepsilon_0 A} \tag{2.4}$$

We have used that the electric field of a plane is proportional to the charge density Q/A. Note that taking the derivative in (2.4) at constant Q corresponds to constant E or constant w. If we are able to shrink the whole capacitor at constant energy density, meaning at constant electric field or constant charge density (this sounds reasonable), the energy of the capacitor scales with the volume and the force between the capacitor plates like the surfaces:

Electric Energy: $W_{el} \sim l^3$.

Electric force: $F_{el} \sim l^2$.

Consequently, electrostatic forces *increase* relative to the volume when the systems shrink.

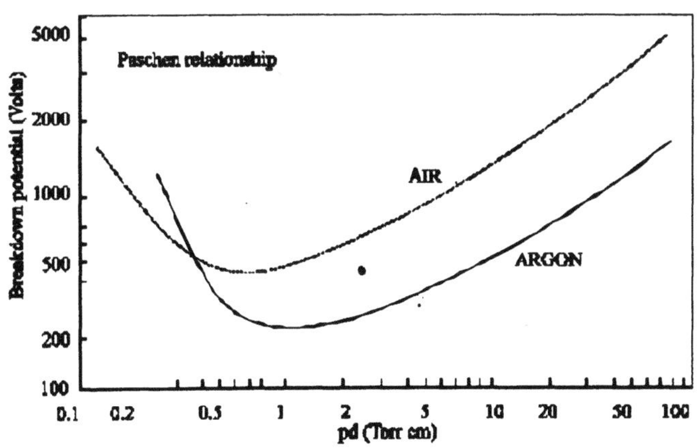

Fig. 2.1. The DC breakdown voltage as a function of the gas pressure p and the electrode spacing d for plane parallel electrodes in air and Argon. Such curves are determined experimentally and are known as Paschen Curves

The maximum field in air is not independent of the distance of capacitor plates if they approach closely. This has been observed first by Paschen. In Fig. 2.1 we show the breakdown voltage in air and Ar as a function of the product of plate separation and pressure. If the separation between the plates is roughly of the order of the mean free path of molecules in air, the maximum field increases.

Magnetic forces can be calculated in a analogous way. The magnetic field energy density is given by (with B, H and μ the magnetic flux density, magnetic induction and the permeability, respectively)

$$w_{mag} = \frac{1}{2}BH = \frac{B^2}{2\mu} \tag{2.5}$$

so one is tempted to assume that the energy in a gap of a magnetic circuit scales with the volume and the forces with the surface, but this conclusion is too quick. The reason is that the magnetic field in the gap is given by

$$B = \frac{1}{\mu_0}\frac{n}{L}I \tag{2.6}$$

with n the number of turns of wire wound around a magnetic jock, L the total length of the coil and I the current through the coil. When shrinking the coil and the magnet, it may be possible to keep n constant, but the cross-section S of the wires must shrink, so that it will be impossible to keep I constant. It is more reasonable to keep the current density constant, but this means that B~S: B scales like l, see eq. (2.6)! The energy density then scales like l^2. We have:

Magnetic energy: $W_{mag} \sim l^5$.

Magnetic force: $F_{mag} \sim l^4$.

Some care is necessary here: The maximum current density is limited by the dissipation, accordingly by the maximum temperature the system can stand. Small systems loose their heat faster than larger systems: Heat flow is proportional to the temperature gradient, which is proportional to l^{-1}. Taking this into account, one would end up with $F_{mag} \sim l^2$. However, it depends very much on the details of the design if one really can use the large temperature gradients. Furthermore, the assumption that the number of wires is independent on the system size becomes dubious if the coils shrink below 1 mm^3. In practice, magnetic actuators smaller than 1 mm^3 are hardly feasible.

Any comparison between magnetic motors and electrostatic motors will end up that at the small end the electrostatic motors are stronger, while at the large end the magnetic motors are stronger.

If one is able to use materials with a high dielectric constant (e.g. PZT - load circonate titanate has ε_r=1300) the energy density and the total outputforce increases by the amount of the dielectric constant. This increases the maximum force of electrostatic micromotors to about 1 N.

2.3 Coulomb friction

In the macroscopic world friction forces are independent of the contact area. The physical reason is that two rough bodies without load touch each other only at three points. In silicon micromachining, this seems to be different in some cases because silicon is extremely smooth (the micro surface roughness of silicon wafers is of the order of 1 nm, rather less). The adhesive forces can be very large, which is being used in wafer bonding and it leads to process problems in surface micromachining, as we will show in the next chapter. These surface forces scale with l^2, and become large in small systems.

2.4 Mechanical strength

Small mechanical constructions are stiff. One can see this easily in models of sailing ships: the form of the cables does not follow what gravity tries to impose, but it is the internal stress, which results in their arbitrary shape. The differential equation for a cantilever beam loaded by its own weight per length P is

$$\frac{\partial^4 \zeta}{\partial x^4} = \frac{P}{EI} \tag{2.7}$$

with E Young's modulus, I the moment of inertia, $I = bh^3/12$, and $\zeta = \zeta(x)$ the deflection. b and h are width and thickness respectively. See fig. 2.2. This equation will be derived and discussed in chapter 4. Obviously with $P = \rho bL$ (ρ = density, L = length of the beam)

$$\zeta \sim L^4 bh \,/\, bh^3 = L^4 \,/\, h^2 \sim l^2 \tag{2.8}$$

Deflection under own weight $\zeta \sim l^2$

A ten times smaller (thinner and shorter) beam bends 100 times less due to its own weight.

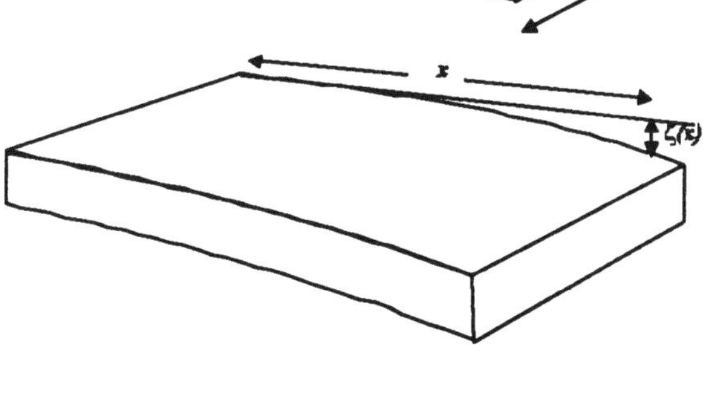

Fig. 2.2. Geometry of a bent cantilever beam

A beam under a lateral compressive stress buckles if a critical load is exceeded. This load is given by

$$T_{cr} \sim \frac{EI}{L^2} \tag{2.9}$$

Again if we assume that the beam is loaded by its own weight, T = $\rho gLbh$, (g = acceleration) and we find that the critical length is given by

$$L_{cr}{}^3 \sim \frac{EI}{\rho g} = \frac{E}{\rho g}\frac{bh^3}{12} \tag{2.10}$$

from which we may conclude that

stable length $\qquad L_{cr} \sim l^{4/3}$.

2.5 Dynamic properties

The time a particle needs to travel a distance L by diffusion is given by:

$$\tau = \frac{L^2}{\alpha D} \tag{2.11}$$

where D is the diffusion constant and α is a numerical value two times the dimension of the diffusion process (e.g. $\alpha = 6$ for diffusion in three dimensions). Self diffusion coefficients in normal liquids at room temperature (water, mercury) is $D = 10^{-5}$ cm^2/s. In gases D is 1 cm^2/s. We see that diffusion tends to be very fast in small systems:

diffusion times $\qquad \tau \sim l^2$

This equation is also valid for thermal diffusion. However, in liquids heat diffusion coefficients are much larger than particle diffusion coefficients.

A mass subject to viscous friction moves after a certain time with constant velocity, which depends on the force acting on the mass. Taking again the mass's weight as the force, the velocity of a spherical mass (radius r) is given by

$$v_\infty = 18\eta r / 4\rho g r^3 \tag{2.12}$$

with η the dynamic viscosity. We find that viscous forces heavily damp any motion. Shrinking a ball falling though water by a factor of 10 leads to a steady state velocity 100 times smaller. The transient time also varies like l^2.

Drift velocity $v_\infty \sim l^2$.

Transient time $\tau \sim l^2$.

The electrical resistance of resistance increases when scaling down. We have for the resistance R of a wire with specific conductivity σ, length L and cross-section S

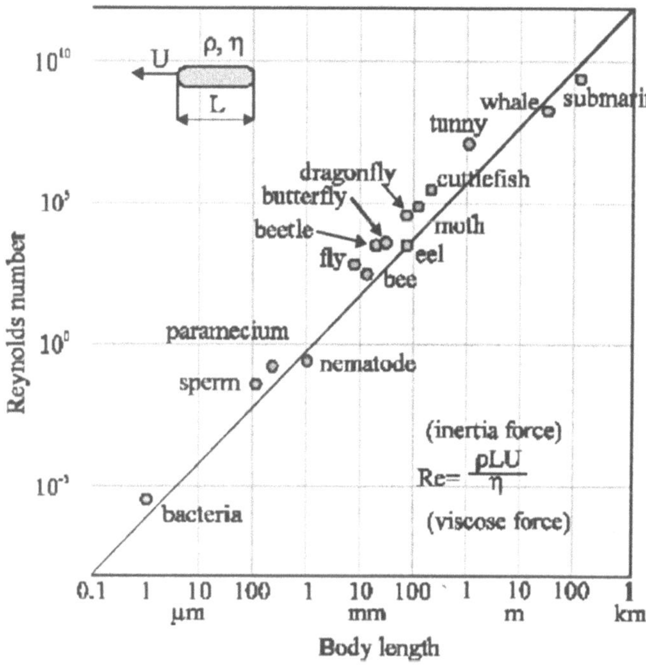

Fig.2.3. Reynolds number as a function of size (Hayashi 1994) Reproduced from 'Silicon Micromachining', M. Elwenspoek and H. Jansen, 1999, by permission of Cambridge University Press.

$$R_{el} = \frac{1}{\sigma}\frac{L}{S} \tag{2.13}$$

electric resistance $R_{el} \sim l^{-1}$.

The hydraulic resistance of a fluid streaming through a pipe with length L and cross-section S increases much faster. The physical reason is that at the walls of the pipe the velocity of the fluid vanishes. The hydrodynamic resistance is proportional to L/S^2,

$$R_{hy} = a\eta L / S^2 \tag{2.14}$$

hydraulic resistance $R_{hy} \sim l^{-3}$

With high velocities the hydrodynamics become unstable and the convection becomes turbulent. The Reynolds number Re is the criterion: If in pipes Re > 70, there is a chance for vortices, and the transition to turbulent flow is around Re = 2300. The Reynolds number is given by

$$\mathrm{Re} = \frac{\rho U r}{\eta} \tag{2.15}$$

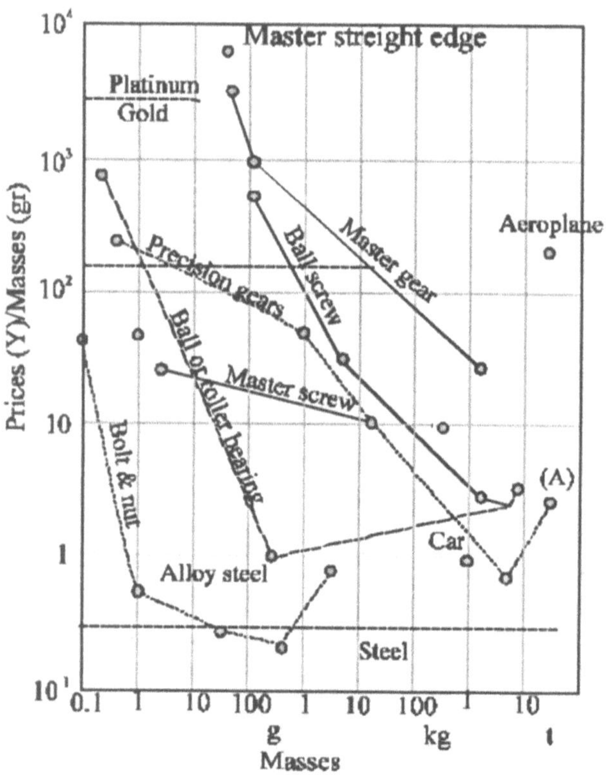

Fig. 2.10. Price per mass of machines as a function of the size (data from Hayashi (1994)) Reproduced from 'Silicon Micromachining', M. Elwenspoek and H. Jansen, 1999, by permission of Cambridge University Press.

If the velocity U scales linearly with the size, we have

Reynolds number $\mathrm{Re} \sim l^2$.

If we take the drift velocity, the Reynolds number scales even like l^3. There is no turbulence in microsystems in which liquids flow. There can be turbulence in microsystems with gas flow.

In fig. 2.3. we show the Reynolds number of animals swimming or flying as observed in nature. The empirical exponent is very close to 2, indeed (solid line in the figure). Note that also man-made "animals" (the submarine) follows the trend.

Resonance frequencies become large for small systems. The easiest way to see this is by studying the eigen frequencies of a string or a tube. The lowest eigen frequency corresponds to a state where the length of the string equals half the wavelength λ. From the well-known relation c = λν (with c the wave phase velocity, ν the frequency) it is immediately seen that eigenfrequencies scale like l^{-1}.

Resonant frequency $\nu \sim l^{-1}$

This relation is also valid for the eigen frequency of a beam where the restoring force is due to stiffness. The resonant frequency is given by

$$\nu_r = \frac{\alpha}{L^2}\sqrt{\frac{EI}{\rho S}} \tag{2.16}$$

α is a number, the value of which depends on the mode and the boundary conditions. A beam with a rectangular cross-section ($S = bh$) has $I = bh^3/12$, so

$\nu_r \sim h/L^2. \sim l^{-1}$

We conclude this section with commenting on the price of machined mechanical parts and systems. In fig. 2.10 the price of system or component per gram is plotted as a function of the mass. It can be seen that prices generally increase steeply if the systems become small. This indicates that it is difficult to machine small components.

Table 2.1 Scaling of physical properties.

Linear Dimension	$l \sim l^1$
Surfaces	$A \sim l^2$
Volume	$V \sim l^3$
Electric Energy	$W_e \sim l^3$
Electric Force	$F_e \sim l^2$
Magnetic Energy	$W_{mag} \sim l^5$
Magnetic Force	$F_{mag} \sim l^5$
Deflection Under Own Weight	$\zeta \sim l^2$
Stable Length	$L_{cr} \sim l^{4/3}$
Diffusion Times	$\tau \sim l^2$
Drift Velocity	$v_\infty \sim l^2$
Transient Time	$\tau \sim l^2$
Electric Resistance	$R_{el} \sim l^{-1}$
Hydraulic Resistance	$R_{hy} \sim l^{-3}$
Reynolds number	$\mathrm{Re} \sim l^2$
Resonant Frequency	$\nu \sim l^{-1}$

Acknowledgements

Contributions of Han Gardeniers, Meint de Boer, Gijs Krijnen, Remco Wiegerink, Erwin Berenschot and Theo Lammerink are gratefully acknowledged.

3. Silicon Micromachining

M Elwenspoek and R. Wiegerink

Faculteit der Elektrotechniek, University of Twente, Enschede, The Netherlands

Abstract The arsenal of technologies for silicon micromachining comprise photolithography, thin film deposition, doping, etching by wet chemicals and plasmas and waferbonding. Mainly for the sake of clarity and to describe the notions necessary to understand the limits and possibilities of micromachining we give a brief account of these fabrication methods.

3.1 Photolithography

The basic photolithographic process includes a drawing, which defines transparent and opaque areas on a mask. More information on silicon micromachining can be found in the literature mentioned in the text and in a few recent textbooks (Büttgenbach 1991, Elwenspoek 1999, Heuberger 1989, Kovacs 1998, Madou 1997, Menz 1993, Muller 1991, Muller 1999, Ristic 1994, Sze 1994, Tabib-Azar 1998).The mask can be produced by direct writing on a glass plate in the desired size, but it can also be drawn much larger and subsequently reduced by photographic means.

Since for pattern transfer light is used, the minimum feature size feasible is defined by the wavelength of the light.

The patter transfer from the mask to the substrate requires the following steps:

1. -resist spinning
2. -prebake (depending on the resist, typically 10 minutes at 90°C)
3. -illumination in a mask aligner. The mask aligner allows precise alignment of the mask with features on the substrate and the crystallographic orientation of the silicon wafer
4. -postbake (depending on the resist, normally 20 minutes at 120°C)
5. -development

Usually, a mask is expensive. Typical prices are in the range of 1000$ per mask. Emulsion masks are considerably less expensive, but subject to wear during use.

Basically there are two types of photoresist: negative and positive.

Positive resist becomes soluble after illumination as a result of depolymerisation. Negative resist becomes insoluble after illumination. It is much harder to work with negative resist, due to competing chemical reaction of the material with ambient air and a poorer adhesion. Furthermore, the achievable minimum linewidth is greater (2-3 μm) than the properties of light in principle allows. The resolution of positive resist is much better (0.5μm).

Practically all processes need a wafer cleaning before the process actually is carried out. Standard in micromachining is RCA-cleaning.

In order to be able to fabricate small features a dust free environment is required. The maintenance of a clean room is very costly.

3.2 Thin Film Deposition and Doping

It is impossible to make a sensor only from silicon. One has to be able to etch silicon, therefore materials are required which are inert against the etching agents used to machine silicon. Sensors have some kind of electric function, so resistors and conductors are needed. Finally one needs special materials for transduction. These materials typically are used in the form of thin films deposited on top of silicon wafers. By doping (electrical) properties of silicon are altered. Here we shall give a very brief account for the most important techniques

Silicon dioxide

Silicon dioxide (SiO_2, in jargon "oxide") usually is used as a mask material for wet chemical etching. SiO_2 etches slowly in KOH solutions and in practically not at all in EDP (see below).

The growth rate of oxide is strongly dependent on the temperature. The rate controlling process seems to be the diffusion of oxygen atoms or molecules through the growing layer of silicon dioxide, once the films are thicker than a few nm. The growth rate of wet and dry oxide is given in fig. 3.1. In the "wet" process, steam is added to the oxygen that flows though the tube. A standard way to add steam, is to direct the oxygen through (nearly) boiling demineralized water. As seen from fig. 3.1. water vapour increases the growth rate quite considerably.

The thickness of the oxide film required depends on a number of things. For a following etch step, one has to think of the etch rate of the oxide, which is not insignificant in KOH. This etch rate will be discussed later, but to give a figure: etching through a 300 μm wafer requires an oxide layer of 2 μm.

The density of silicon dioxide differs of course from that of Si. The oxidised silicon has a volume larger than the silicon itself. Upon oxidation a oxidised layer occupies a volume larger by 46%. The result is a compressive stress[1] in the layer with respect to the underlying bulk silicon. Thermal oxide therefore is useless for most applications. It is under a compressive stress in the order of $2 - 4\times10^{10}$ Pa. Membranes and microbridges buckle; this material is only useful, when buckling is aimed at.

[1] The stress of a film grown on top of substrate is indicated with respect to the underlying material. An expanding layer is then said to be under compressive stress, a contracting layer under tensile stress.

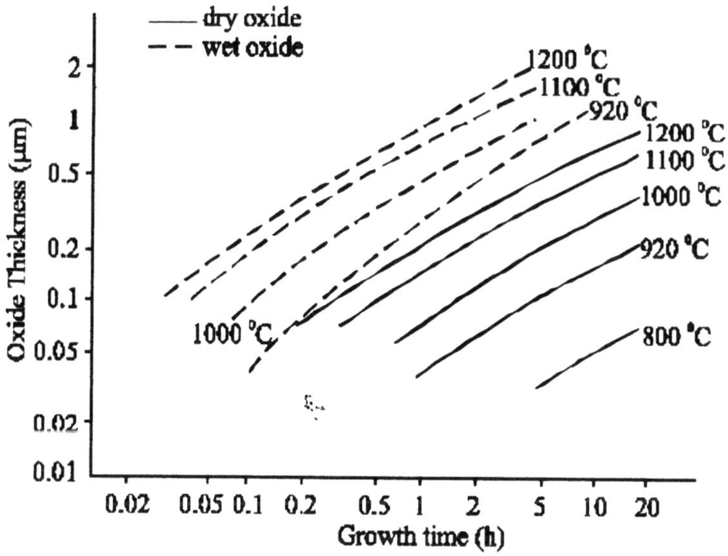

Fig. 3.1. Thickness of a thermally grown oxide layer as a function of the growth time for several temperatures. solid line: dry oxide. broken line: wet oxide (redrawn from Büttgenbach 1991)

Chemical vapour deposition

Chemical vapour deposition (CVD) is a very important deposition method in micromachining. Prominent materials are silicon in the polycrystalline form ("polysilicon" or simply "poly"), silicon nitride and phosphor silicate glass (PSG). The latter is mainly used as a "sacrificial layer" in a process called "surface micromachining" (see below), while the former two are often used as structural materials. Poly has properties comparable to single crystalline silicon, and silicon nitride is a very hard, chemically inert and strong - though brittle material with very small thermal conductivity.

In the CVD process basically a gas is fed through a pipe which contains wafers at elevated temperature. The gas is cracked close to the hot substrate, and the reaction products are deposited. We distinguish between APCVD (atmospheric pressure CVD), LPCVD (low pressure CVD) and PECVD (Plasma enhanced CVD). The main difference between APCVD and LPCVD is that in the latter usually the step that determines the growth rate of the film is the integration of atoms in the surface, while in the former diffusion of gas to the wafers is dominant. On says that the former process is controlled by surface kinetics, the latter is controlled by volume transport kinetics. This has important consequences for micromachining. If surface integration governs the kinetic processes one gets a conformal coverage of the surface by the thin film. In particular, steps are covered (c.f. fig.3.2.) using LPCVD much better than using APCVD. Also the mechanical and chemical quality (in terms of impurities, pinholes and density) of LPCVD films usually is much batter that APCVD films.

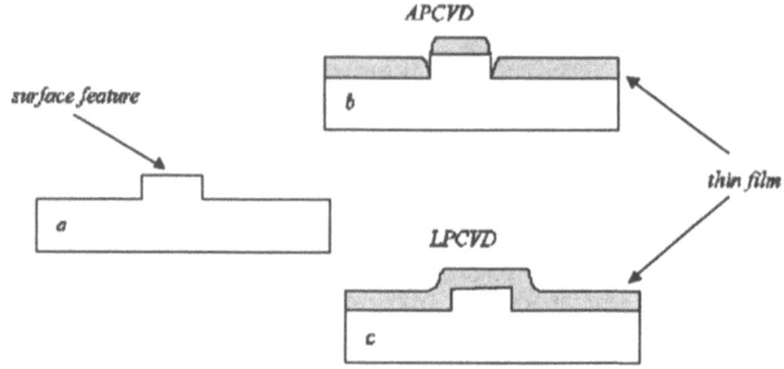

Fig. 3.2. Good and bad step coverage. A shows a step on the substrate surface. This step is covered poorly by a thin film deposited in a process governed by volume transport in is the case for APCVD (b). If surface kinetics control the process, a good step coverage results (LPCVD, c)

Polysilicon is generally deposited by Low Pressure Chemical Vapour Deposition (LPCVD) from pyrolysis of silane (SiH_4) at temperatures around 600°C and pressures of several hundred mTorr. Both compressive and tensile stresses of large magnitude are known to exist in polysilicon. Deposition temperatures below the silicon crystallisation temperature (ca. 605 °C) are generally found to produce tensile films, while a higher deposition temperature generally results in compressive films. Polysilicon films, deposited near 600 °C have been shown to grow initially in the amorphous state, and subsequently recrystallise during further deposition. As the density of amorphous silicon is smaller than the density of polysilicon, it contracts against boundary constraints, inducing a tensile stress. Tension occurs only in films that are deposited in the amorphous state and subsequently recrystallise. Annealing at high temperature induces recrystallisation, which allows intrinsic stresses in the silicon films to relax. In Fig. 3.3. the strain in polysilicon as a function of the annealing temperature and time is shown for undoped and boron doped samples (Guckel 1988).

Silicon nitride is deposited from a mixture of dichlorosilane (DCS, SiH_2Cl_2) and ammonia (NH_3) at elevated temperature (800 - 850 °C). The resulting film has properties that depend on the composition of the gas fed into the reactor. Stochiometric silicon nitride (Si_3N_4) deposited on silicon is under tensile stress in the range of 10^9 - 10^{10} Pa. Resulting structures can be very fragile because the strain in the film is close to the fracture strain. Deposited nitride at larger DCS flow compared to ammonia flow result in films having much smaller stress, and it is even possible to deposit films under compressive stress. Low stress nitride films can be deposited in an LPCVD reactor at 835 °C, 0.15 mbar pressure and a flow ration DCS/ammonia of 40:8 (Bouwstra 1990).

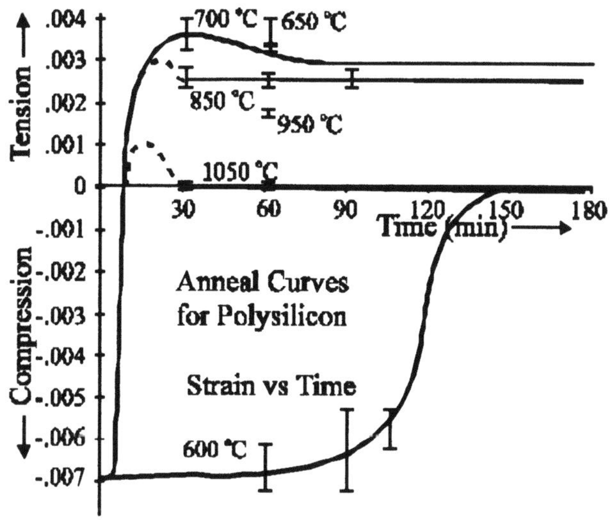

Fig. 3.3. Internal strain for LPCVD polysilicon films deposited on oxidised silicon substrates as a function of post-deposition anneal time and temperature. These films were deposited using the pyrolysis of 100 % silane at a temperature of 580°C and a pressure of 300 mTorr (Guckel 1988). Reproduced from 'Silicon Micromachining', M. Elwenspoek and H. Jansen, (1999), by permission of Cambridge University Press

Nitride can also be deposited in a PECVD reactor (PE = plasma enhanced) at much lower temperature (400 - 500 °C). These films however suffer from pinholes and a high H concentration in the film. They are less suitable as structural material.

Evaporation

Materials with sufficiently high vapour pressure can be evaporated by resistive heating. In practice the target (this is the material to be evaporated) can be heated up to 1200 °C at most so this defines the range of materials that can be evaporated by resistive heating. Evaporation is largely restricted to deposition of elements since the composition of deposited compounds is difficult to control. If evaporation of compound is done it is advisable to use separate targets for each element. Evaporation can also be done by RF heating, laser ablation or electron beam heating. E-beam heating is more common than resistive heating due to a number of advantages, among which less contamination, a better process control and a more efficient heat transfer to the target, so that practically everything can be evaporated using E-beam heating.

Evaporation is a fast process if the vapour pressure is large enough. Growth rates up to 1 μm/min are

Sputterdeposition

In deposition processes using sputtering high energy ions hit the substrate and sputter material from the target. This materials is deposited on the substrate.

Usually one uses Ar ions generated in a plasma between the target and the substrate. The target is negatively charged, so the Ar ions impinge of the target and not on the substrate which is grounded or floating.

There are some important advantages of sputterdeposition compared to evaporation: practically all materials can be sputterdeposited, there are more parameters for process control, the deposition may be more homogeneous over the wafer, one can use the sputtering for cleaning the target before deposition starts and the technology is suitable for upscaling. An important advantage is that the adhesion of the sputteredeposited films is generally much better. This is due to the high kinetic energy of the sputtered atoms when they reach the substrate. The atoms penetrate a bit into the substrate.

Doping

The change of electric properties of semiconductors lies at the basis of integrated circuits. But not only electrical properties of silicon but also electrochemical, chemical and mechanical properties can be altered. Therefore doping is an important process in micromachining and micromechanics. Wet etch rates of silicon depend on the voltage difference between the silicon substrate and the etching solution, and the dependency is different for p-doped and n-doped silicon. We shall see later how this fact can be exploited. When etching electroless, the etch rate in certain anisotropically etching solutions can be reduced drastically by doping silicon with boron. Here we briefly describe the processes used for doping.

In general there are two possibilities for incorporating impurities: Ion implantation and indiffusion.

With *ion implantation* high energy ions from an accelerator hit the substrate and penetrate to a certain depth into the substrate. The penetration depth depends strongly on the ion energy. One needs keV for implantation to reach a few 100 nm. After implantation annealing helps to increase the thickness of the implanted layer, of course at the expense of reducing the concentration. Annealing is also required for restore the upper layer, which is damaged by high energy, impacts. Finally, annealing changes the electrical properties of the doped film by allowing the impurities to occupy lattice sites.

Indiffusion can be done in two ways. Solid source indiffusion is done in tube ovens in which one places silicon wafers next to boron or phosphorous wafers. The boron or phosphorous sublimates and the atoms diffuse into the silicon. During this process also an oxide layer grows, which must be removed before one continues the process. Boron silicate glass dissolves very slowly in HF, but the dissolution process can be speeded up considerably after placing the wafers in an wet oxidation oven. Indiffusion can also be done by depositing a boron or phosphor glass and anneal.

3.3 Wet Chemical Etching

Wet chemical etching is the oldest silicon micromachining technology. We distinguish between isotropic and anisotropic wet chemical etching. The chemistry of the etching is - in spite of great effort - still debated and poorly understood, and we shall not dwell into these discussions here but refer to (Elwenspoek 1999). In fig. 3.4. we show the result of etching silicon through a mask using anisotropically and isotropically etching solutions.

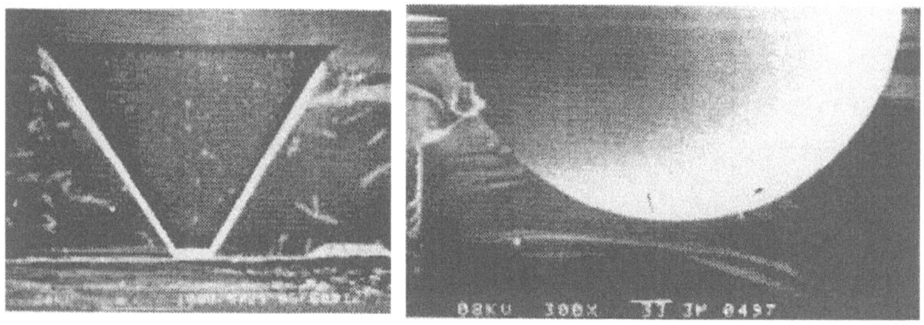

Fig.3.4. Etchpits formed by etching silicon through a mask opening using (top) EDP (ethylenediamine, pyrocathecol, water mixture) and bottom) HNO_3:HF:CH_3OOH. Reproduced from 'Silicon Micromachining', M. Elwenspoek and H. Jansen, 1999, by permission of Cambridge University Press

The wet etching process in total is quite complex and requires considerable skill from the operators. We start here with brief descriptions of the isotropic processes, and pay more attention to anisotropic etching because it is much more important. The anisotropy of the etch rate is due to crystallographic properties of silicon, and a rather precise control of the geometry of etched structures is possible - within the limits of the crystallography. The shapes resulting from isotropic etching on the other hand result from concentration fields in the etching solution, so generally control is difficult.

Anisotropic Etching

The etch rate of silicon in OH-containing solutions is anisotropic. There are deep etch rate minima in the crystallographic <111> directions, and - depending on the solution - shallow minima in <001> (KOH) and <110> (EDP, KOH with isopropyl alcohol). The process temperature is at 60 °C or higher, and maximum etch rates are a few µm/min. The minimum etch rate in <111> direction is smaller by a factor typically of 100 or more, the etch rate in the direction of the shallow minima is smaller by a factor of 2 – 5 compared to the maximum etch rate. In certain cases, the shallow minima have been used for micromachining (Rosengren 1994, Bäcklund 1992), but generally, the <111> minima play the dominant role in micromachining using wet anisotropic etching.

In Fig. 3.8. we show the definition of the three crystal planes, which are important here.

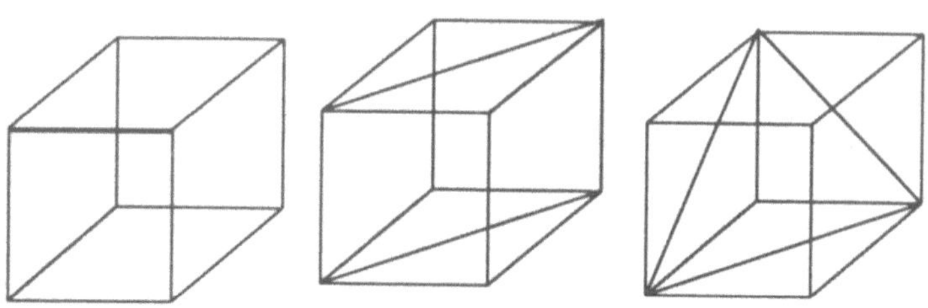

(010) (110) (111)

Fig. 3.8. The family of planes (from left to right) (010), (110) and (111) in cubic lattices

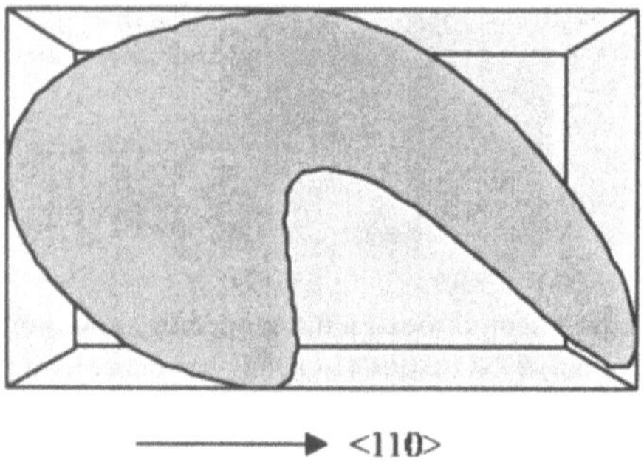

Fig.3.9. Etch pit under a mask opening of arbitrary shape

Silicon wafers can be purchased with various orientations. Here we only consider <001> orientation. On the <001> wafers the slowly etching <111> planes are inclined by an angel of ca. 54°, and the edges run along the <110> direction on the wafer. This direction is indicated by a flat portion of the side of a wafer, the so-called wafer flat. Because of the deep minimum in the <111> direction, a pit etched through a mask opening on a <001> oriented wafer will ultimately be bounded by <111> planes, see fig. 3.9.

Mask materials are silicon dioxide thermally grown in a wet oxidation oven or silicon nitride grown in an LPCVD reactor. While Si_3N_4 is practically not attacked by KOH, SiO_2 is. The etch rate ratio of Si<111> to SiO_2 is shown in fig. 3.10. as a function of the KOH concentration.

Silicon etch rates depend on the temperature and the concentration. There is a maximum of the etch rate around 20 wt%, and this concentration is used normally because small changes in concentration do not affect the etch rate. Temperature and concentration dependency is shown in fig. 3.11. and 3.12.

Anisotropic etching as such allows a good control of the lateral dimensions of features[2], but without additional measures the etch depth can be controlled only by the etch time. Normally thin structures must be machined which are made by etching right through wafers, leaving only a thin membrane or beam, much thinner than the wafer. There are some means to optimise the control of etch depth by optically monitoring the etching (Minami 1994), however in the research laboratory the fabrication of thinner structures than 10 μm is not practical, *a fortiori* for production.

The last point we have to comment on is compensation of convex corners. Anisotropic etching stops at concave corners of two (111)-faces. Convex corners however are etched. During etching higher index planes appear such as (411). Often it is required to etch mesas, structures bound by (111) planes but convex, and it is necessary to prevent the etching of convex corners. This can be

[2] Although sometime we hear that the underetching depends on the batch KOH: sometimes the reproducibility is bad [Moldovan 1998, Steckenborn 1998]

achieved by giving the etch mask an appropriate form. An example is given in fig. 3.13. The width of the bars must be approximately equal to twice the height of the mesa. Programs that simulate the structure one gets by etching with a given mask are commercially available. For an overview and more details and references, we refer to (Elwenspoek 1999, Madou 1997).

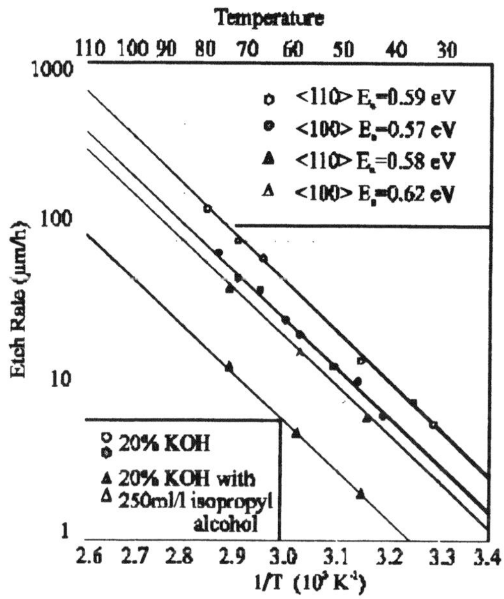

Fig. 3.11. Temperature dependence of etch rates in 20% KOH solutions in <100> and <110> directions, with and without the addition of isopropyl alcohol Reproduced from 'Silicon Micromachining'. M. Elwenspoek and H. Jansen, 1999, by permission of Cambridge University Press (Seidel 1990)

Etch stop

There rest three basic ways to stop the etching at a well controllable thickness. In the section on doping we referred to boron (p-doping) and phosphor (n-doping) for micromachining purposes. This is explained here.

In fig. 3.14. we show how the etch rate of silicon in EDP (ethylendiamide, pyrocathecol pyridine and water) type S[3] depends on the boron concentration. It is seen that around 2-3×10^{19} atoms/cm^3 the etch rate starts to decrease, and close to 10^{20} atoms/cm^3 the etch rate is smaller by a factor of 100. By indiffusion or ion implantation (large dose!) it is possible to reach this high concentration (close to the solubility of B in Si). Membrane thickness between a few 10's of nm to 20 μm have been achieved using this etch stop technique (van Huffelen 1991). The problems associated with B+-etch stop are high tensile stress and high conductivity of the material. Further, after indiffusion the wafer is too rough to bond (see below) a wafer on the B+ layer by silicon fusion bonding. It is also important to

[3] Water 133 ml, ethylenediamine: 1l, pyrocatecol: 160 g, pyrazine: 6g

keep in mind that areas of indiffusion are a bit oxidised, after removal of the oxide there is a shallow recess compared with the undoped silicon.

The B+ etch stop does not work so well in KOH. One needs either a small KOH concentration (below 10 wt%), or one has to add isopropyl alcohol to the solution. Both leads to very rough etched <001> surfaces with many pyramids, small ones (~5 μm) in the first case, fewer but large ones (50 μm) in the latter case.

The etching can also be stopped in an electrochemical set up. The etch rate of silicon n- or p- doped as a function of a bias is shown in fig. 3.15a. The potentials indicated in the curves by "PP" denote the "passivation potential". At this potential there is a competition between etching silicon and the growth of an oxide layer. The oxide layer finally passivates the silicon against the etchant. This effect is exploited in the electrochemical etch stop.

3.4 Waferbonding

Waferbonding refers to the fixation of wafers on top of each other. It is an indispensable technique for silicon micromachining. Waferbonding not only allows us to fabricate constructions with rather complex structures in three dimensions, but it also plays a role in packaging. Because whole wafers can be bonded containing many single components (here: sensors), tedious and expensive assembly can often be circumvented by wafer bonding.

Silicon is very smooth. Wafers, when contacted, do not touch each other at three small spots, as normally is the case, but on extended regions. Wafers, when clean and flat (not bent too much), spontaneously adhere to each other. This tendency can be enhanced using external forces, such as a weight on top of the wafer package or - much more efficient - electric fields. Bonding without the help of external forces between silicon wafers with or without thin films on top is called "direct silicon bonding", DSB, or "silicon fusion bonding", SFB (although the process has nothing to do with fusion). Using electric forces requires additional materials (glass), and is referred to as "anodic bonding".

Anodic Bonding

Anodic bonding was discovered in the sixties (Wallis 1969). The technology involves the bonding of a metal (electrical conducting) plate and a glass wafer, which becomes conductive at a high temperature. The electric field is then confined to a thin depletion layer in the glass and in the gap between the wafers. If the wafers are not too rough (maximum roughness must be below 1 μm), the resulting pressure is enough to deform the material at the interface so that the wafers join smoothly. Furthermore, the process temperature must be high enough for oxidation of the metal, so that covalent bonds form at the metal-glass interface.

A suitable glass for silicon anodic bonding is Pyrex, but also Corning 7070 is OK (it is practically the same material). It is also possible to sputter a thin glass film from a Pyrex target on a silicon wafer and proceed with anodic bonding with a second silicon wafer (Hanneborg 1991).

A schematic set up is shown in fig. 3.16. The wafer package is placed on a hot plate (process temperature 400-450 °C), and the glass wafer is connected to a high voltage source (required voltage in the range of 1000 V). The process can be monitored by the naked eye - the colour of the bonded region looks grey, quite different from unbonded regions. When using a thin sputterdeposited Pyrex film, the voltage must be much smaller (20-50 V). Bonding has been demonstrated with Pyrex thin films ranging in thickness between 50 nm and 3 μm (Berenschot 1994). Both processes are quite reliable, provided the wafers are dust free.

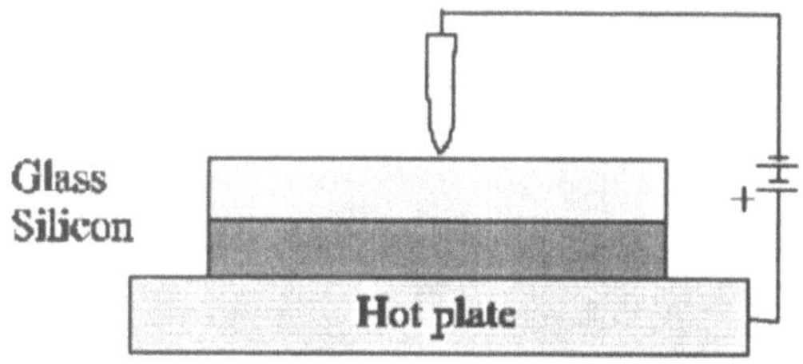

Fig. 3.16. Schematic of an anodic bonding set-up (Hanneborg 1991)

Silicon Fusion Bonding

The interfacial energy between two bare silicon wafers covered with native oxide (hydrophilic wafers) is due to hydrogen bridging Stengl 1989). The resulting forces are orders smaller than forces that can be achieved with electric forces as in anodic bonding. The result is that direct silicon bonding is a much more difficult process. The microroughness of silicon wafers must be below 1 nm.

The oxide on the wafers can be removed (in HF) making the wafers hydrophobic. The interfacial energy of the contact is then due to van der Waals forces (Bäcklund 1992), which are still on order of magnitude smaller than forces due to hydrogen bridging. Nevertheless bonding is possible, though difficult, and this process even has advantages.

Normally one uses hydrophilic wafers, as they come out of all wet cleaning procedures. The wafers are contacted under clean room conditions. The adhesion is large enough to allow normal handling of the wafer package. The package is annealed in a wet oxidation oven at 700 °C (to our own experience, this is the optimal process temperature (Sanchez 1997)) or higher. Between 300 and 700 °C voids form, which result from water.

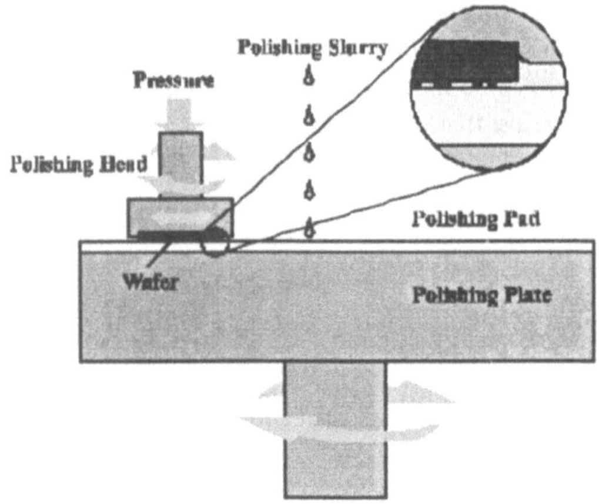

Fig. 3.18. Schematic view of a CMP machine (Gui 1998)

Prior to bonding the wafers can be processed (Harendt 1991), but each process step increases the roughness of the wafers. The process yield declines after too many steps. Some steps make DSB absolutely impossible, prominently wet etching of the whole surface (it is possible to etch cavities), deposition of too thick Si_3N_4 (more than 200 nm) or boron doping by indiffusion.

There is however a remedy. The wafer manufacturers polish silicon to the fantastic smoothness using a process called chemical-mechanical polishing (CMP, fig. 3.18 (Gui 1998)). In CMP the wafer is placed on a rotating polishing head and moved under a certain pressure over a polishing pad. A slightly alkalic slurry containing nm sized silica particles is added. The combined action of wear and etching yields the smooth surfaces. This process can also be applied to other materials among others silicon nitride, silicon oxide and poly silicon (Gui 1998). It appears that all materials that can be polished down to a roughness better than 1 nm are bondable (Haisma 1994).

In fig. 3.19 we show a silicon nitride surface (1 mm thick stress free S LPCVD nitride) before and after polishing and in fig. 3.20 we show the bonding interface (Gui 1998).

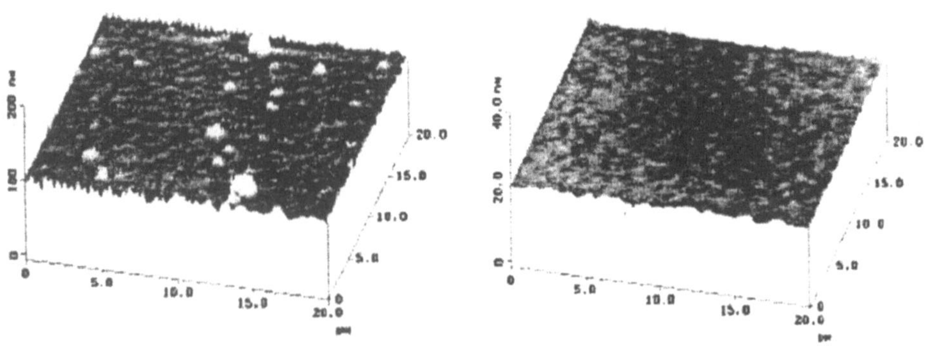

Fig. 3.19. left: LPCVD $Si_{3-x}N_4$ surface before CMP. Root mean square roughness 3.6. nm. Right: LPCVD $Si_{3-x}N_4$ surface after CMP. RMS = 0.4 nm (Gui 1998)

3.5 Plasma Etching

Wet chemical etching is confined either by the <111> planes of single crystalline silicon or by poorly controllable diffusion fields. Both techniques severely limit the possible design space for mechanical construction. Plasma etching offers much more. In plasma etching ions impinge on the substrate, and neutral particles reach the substrate by diffusion. The combined action - which is quite complex, as will be seen - leads to anisotropic etching. The anisotropy here is not the result of crystallographic properties but a result of the direction of the ion flux towards the substrate. In particular reactive ion etching (RIE) and related techniques have been developed during the past years for micromachining. Originally developed for IC processing (etching of thin films, pattern transfer) for comparatively shallow etching (etch depth not more than 1 - 3 µm), MEMS has a need of great structure height, so etching must be deep, with neat sidewalls, vertically straight through the wafer. Several processes have been developed to make this possible.

Plasma

An overview of the main dry etch techniques is given in fig. 3.21. This overview includes also the chemical etching from a vapour and laser assisted etch methods. Ions are created either in a glow discharge or an ion source. In the latter case we speak of ion beam techniques.

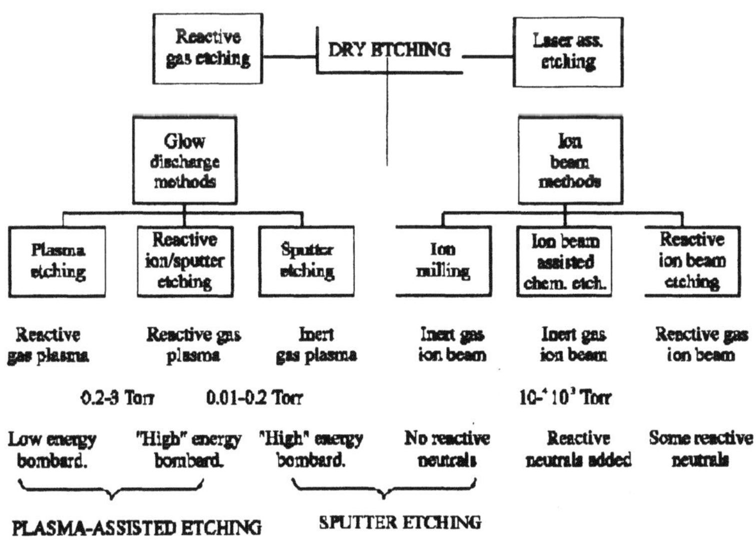

Fig. 3.21. The relationship between the various dry etching techniques (Lehmann 1991)

Chemical etching is isotropic in general. The etching is only possible if the reaction products are volatile. Further, because it is a chemical reaction, the etch rate depends strongly on the material to be etched: the etching is selective. In a plasma the etching species are radicals: atoms or parts of molecules with an unpaired electron. Radicals are very reactive. An important example for chemical plasma etching is the stripping of photoresist in an oxygen plasma, a process called "ashing". In fig. 3.22. we show an example for the result of isotropic etching of silicon in a fluor based plasma.

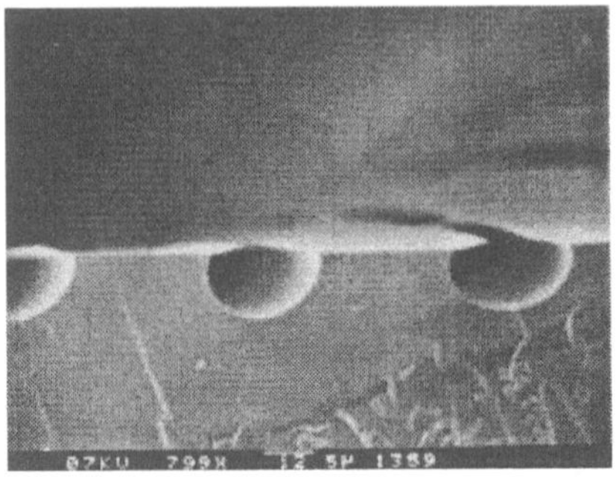

Fig. 3.22. Isotropic profile by reactive ion etching in SF_6/N_2 (Spiering 1993)

If there are only ions hitting the substrate while neutrals have no chemical effect the etching is said to be "physical". Using an ion beam we speak of "ion beam milling" (IBE), using a glow discharge we speak of "sputter etching". Both processes are physically identical. The main features are:

- the etching is anisotropic, as a result of the ion flux
- any material can be etched
- etch rates do not depend very much on the type of material
- the etched material is redeposited anywhere in the etch equipment
- maximum etch rates are in the order of 1 atom removed per impinging ion, about 1 keV is required
- there is a threshold energy for etching (tens of eV)
- the sidewalls are inclined: trenches become narrower the deeper they are etched

The combination of chemical etching and physical etching makes plasma etching so interesting for micromachining. We call this mode "synergetic". In ion beam methods one speaks of Chemical Assisted Ion Beam Etching (CAIBE), in discharge reactors this is called reactive ion etching (RIE). The main difference is that in RIE the chemical etching is due to radical formed in glow discharge, while in CAIBE a reactive gas is fed into the reactor. Because a chemical reaction is involved in the process the etching is selective, and the etch products must be volatile.

Anisotropic Plasma etching modes

There are two distinct modes for the anisotropy in CAIBE and RIE. In one mode the anisotropy is due to a modification of the chemical-physical state after impact of an ion. This mode is called the ion energy induced mode. The temperature may be raised locally, chemical bonds may be ruptured, the physical site in which the atoms are may be altered, in one way or the other the chemical reaction rate is greatly enhanced by the ion - substrate interaction. Only where ions hit there is an appreciable etch rate. Ions do not hit the sidewalls in a trench (in a first approximation), so only the bottom of the trench is hit. The etching of silicon dioxide in CHF_3 plasmas follows this mode.

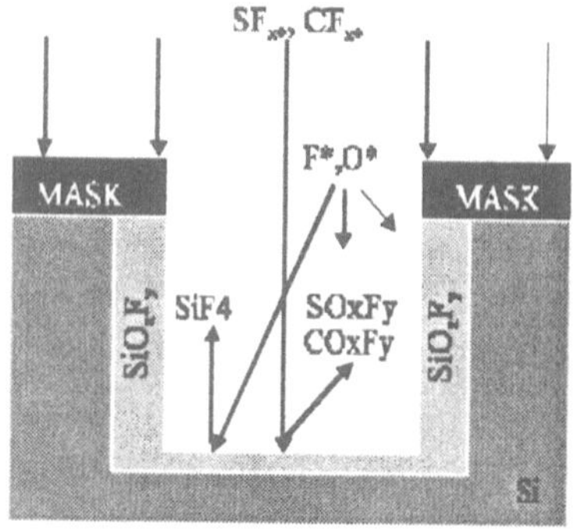

Fig.3.24. Schematic overview of processes in a SF_6-O_2-CHF_3 plasma Reproduced from 'Silicon Micromachining'. M. Elwenspoek and H. Jansen, 1999, by permission of Cambridge University Press.

The other mode is more complex. If one of the reaction products is not volatile, inert against the radical or chemicals responsible for the chemical reaction, but are etched by ions, this material will be deposited everywhere in the reactor, in particular on the sidewalls of trenches and protect the sidewalls against chemical etching. The material is also deposited on the bottom of the trench but it is removed by ions. A schematic drawing of this process is shown in fig. 3.24. In 3.25 we show a result in silicon using this method. This mode is called the inhibitor mode.

Configurations

A typical plasma etch configuration is shown in fig 3.26. The plasma is created by an RF source coupled capacitively via a tuning network to the reactor chamber. The typical glow in the plasma is due to the relaxation of atoms in an excited state after collisions of electrons with atoms or molecules. Since the electrons have a much smaller mass than the ions, they move faster. Further in collisions with atoms the electrons do not loose much energy, so the electron gas is much hotter than the ions and neutrals. Table 3.1. gives the possible reactions in the plasma, together with an example, where we also give the energies ΔE involved in the reaction. The energies must be compared to the kinetic energy of the electrons, E_{kin} which is typically a few eV. The density of the various reaction products is proportional to $\exp-\{\Delta E/kT\}$. The electrons have a certain mean free path. Electrons that hit the reactor walls are absorbed, so the plasma will be positively charged and there will be a voltage difference between the reactor walls and the plasma. The voltage drops over a region called dark space or sheath close to the reactor walls; this is a region of an extension of the order of the Debeye length, typically 0.1 mm, where only few electrons hit atoms. Since the plasma is conductive, there is no electric field inside the plasma, only in the sheath. Ions are accelerated in the sheath. In fig. 3.27 we show a schematic drawing of the distribution of ions, radicals and electrons close to a substrate.

Fig 3.28 shows how the electrical potential varies in the reactor. It is seen that the capacitively coupled electrode is on a large negative potential. Placing a substrate on this electrode will result in a high energy ion flux, and we have the possibility to control the voltage drop and the ion energy. The most important parameters for the ion energy are the pressure, because it will control the mean free path of the ions, and the power of the RF source. A low pressure gives ions more time to pick up energy from the field, so the ion energy, and the voltage drop across the sheath will very roughly proportional to 1/p. The mechanism of etching (physical, chemical or synergetic) depends mainly on the RF power but also on pressure. Low pressure tends to result in a large sputtering component: poor selectivity, redeposition. High pressure means high selectivity, isotropic etching. In between, typical at 1 - 100 mTorr, we have the synergetic mode.

Table 3.1. Examples of Electron Impact Reactions (Mucha 1983).

Reaction	Energy
Excitation (electronic, vibrational, rotational): $e + A_2 \rightarrow A_2^* + e$ $e + CF_4 \rightarrow CF_4^* + e$	4 eV
Ionisation: $e + A_2 \rightarrow A_2^{+*} + 2e$ $e + CF_4 \rightarrow CF_4^+ + 2e$	15.5 eV
Dissociation: $e + A_2 \rightarrow 2A + e$ $e + CF_4 \rightarrow CF_3 + F + e$	12.5 eV
Dissociative attachment: $e + A_2 \rightarrow A^- + A^+ + e$ or $A^- + A$ $e + CF_4 \rightarrow F^- + CF_3^+ + e$ or $F^- + CF_3$	7.3 eV
Dissociative ionisation: $e + A_2 \rightarrow A^+ + A + 2e$ $e + CF_4 \rightarrow CF_3^+ + F + 2e$	14.8 eV

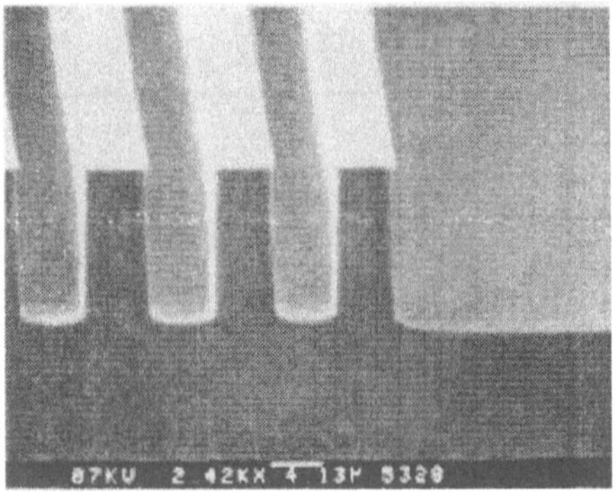

Fig.3.25. Anisotropic profile by reactive ion etching in SF_6/O_2 Reproduced from 'Silicon Micromachining', M. Elwenspoek and H. Jansen, 1999, by permission of Cambridge University Press.

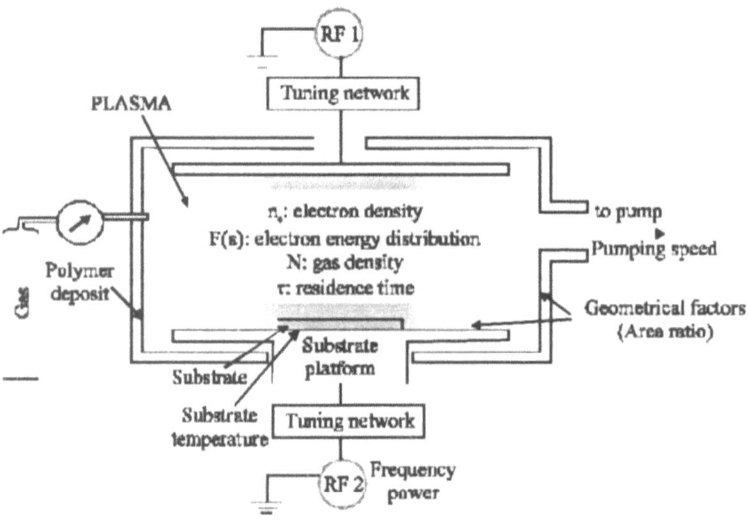

Fig.3.26 Schematic cross-section of a parallel plate reactor showing experimental and fundamental plasma chemical parameters. Usually RF-voltage is applied either to the bottom electrode (in Reactive Ion Etching, RIE) or to the top electrode (in Plasma Etching, PE) (Lehmann 1991).

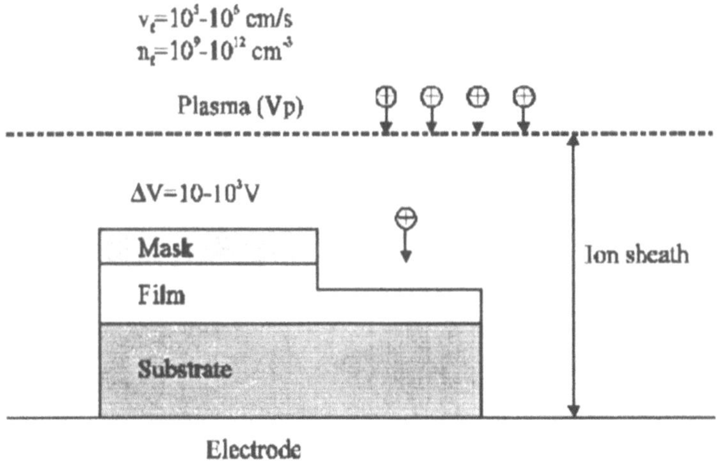

Fig.3.27. Ion motion is random in the central glow but when a positive ion drifts to the sheath boundary (dotted line), the perpendicular electric field accelerates it towards the wall and wafer surfaces. Typical parameters are shown (Lehmann 1991).

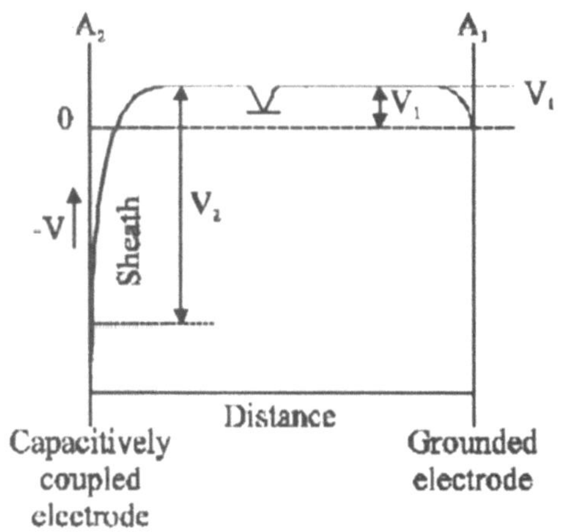

Fig.3.28. Potential distribution in a parallel plate etcher with the grounded surface area larger than the area of the powered electrode. -V: RF-induced DC voltage (self-bias), V_p: plasma potential, $V_2 - | V + V_p |$ determines energy of ions striking the surface, V_f: floating potential (Mucha 1983).

Black Silicon Method

In the inhibitor mode there is a competition of deposition and ion-induced etching of the passivating layer. If deposition prevails the sidewalls will tend to grow and the result is an inclination of the wall toward the trench. If etching prevails, there will be some underetching and the walls will be inclined towards the material. Jansen et al. found that it is possible to change the inclination by changing the etch conditions in a gas flow mixture containing SF_6, CHF_3 and O_2. In fig. 3.30 we show structures with walls inclined in either direction. The conditions for vertical etching obviously must be found right in between the conditions resulting in the two structures shown. The interesting point now is that there is practically always some sputtering in the systems, so in particular, mask material will be redeposited. If the conditions lead to structures shown on the left hand side in fig. 3.30, the redeposited material will act as a small mask, and features etched in silicon will emerge. These features are known as "micrograss" and are shown in fig. 3.31. Macroscopically, micrograss reveals itself by absorbing light: the wafer covered by micrograss will look black. Once black wafers are seen in the reactor, the etch conditions are close to the optimal point for vertical etching. This is the basis of the "Black Silicon Method" (Jansen 1996).

The conditions leading to inclination of sidewalls to either side are shown in fig. 3.32. Adding O_2 to the flow will reduce the underetching until the sidewalls incline towards the trench, and the wafer becomes black.

It must be noted that the locus of the lines shown in fig. 3.32. are strongly temperature dependent. Since the chemical reactions are exothermic the substrate will be heated by the process and the temperature changes. The process will move away from the initial process line if the temperature is not controlled.

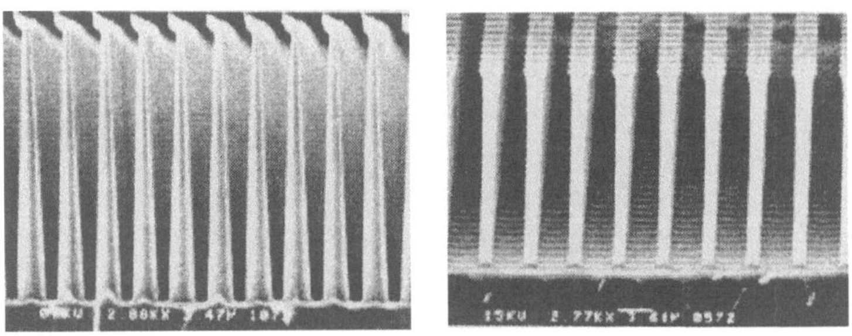

Fig.3.30. The effect of oxygen addition to a SF6 plasma. From left to right the oxygen content of the plasma is decreased, which leads to a change from a positively to a negatively tapered profile Reproduced from 'Silicon Micromachining', M. Elwenspoek and H. Jansen, 1999, by permission of Cambridge University Press.

Fig.2.31. Formation of "black silicon" (also frequently called "micro grass") Reproduced from 'Silicon Micromachining', M. Elwenspock and H. Jansen,1999, by permission of Cambridge University Press.

3.6 Surface Micromachining

Surface micromachining is a process to fabricate free standing and freely moving microstructures in a large two dimensional design space. The process idea is shown in fig. 3.33. The trick is to deposit and pattern two thin films of materials that can be etched away selectively with respect to each other. The substrate now only plays a role as a mechanical carrier; the structure is made from a thin film material.

Although many combination of materials is possible for this process, the technology has developed to a high standard for the combination of silicon dioxide or PSG as the sacrificial material and polysilicon as the structural material. Here we restrict ourselves to this combination.

Users of the process must master two basic problems, thin film stress and sticking.

Thin film stress

In section 3.2. we commented on thin film stress of polysilicon films. We saw that annealing is the essential step. During processing one must incorporate test structures to monitor the stress of the structural layers. There are basically two types of stress. We have tensile or compressive lateral stress and there are stressgradients in the films. Test structures for lateral stress work on monitoring the buckling of structures. A buckled microbridge is seen in fig. 3.34.

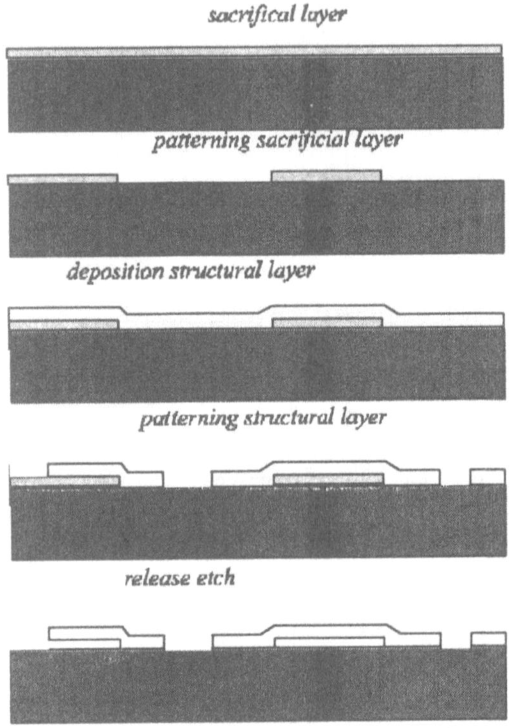

Fig. 3.33. Process scheme of surface micromachining.

Fig.3.34. Buckled polysilicon structures demonstrating compressive stress of unannealed thin films (Howe 1988)).

Sticking

The second problem that has to be mastered is sticking. When etching the sacrificial layer the free-standing structures have the tendency to touch the substrate and stay there. The cause is thought to be the surface tension of the liquid during evaporation, see fig. 3.35. The liquid forms a droplet during drying between the microstructure and the substrate, and if the drying liquid wet both the droplet has an underpressure, and the result is a collapse of the microstructure, if it is not stiff enough. There are solutions to the problem. If possible, the structures should be made stiff enough so that no sticking occurs.

Often the structures must be designed to that they are larger than the maximum length. Then other rinsing procedures can be used, such as critical point drying (Mulhern 1993) or freeze drying (Guckel 1989). Both methods avoid the free liquid surface. Another alternative is to introduce a surface roughness of the microstructure, because in the end sticking is just a waferbonding process. Special designed structures can be designed so that the surface which comes into contact with the substrate is quite small (Tas 1996).

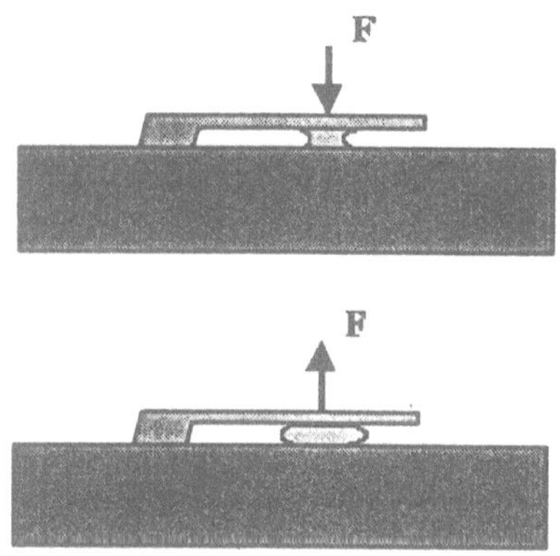

Fig. 3.35. Capillary forces during sacrificial layer etch

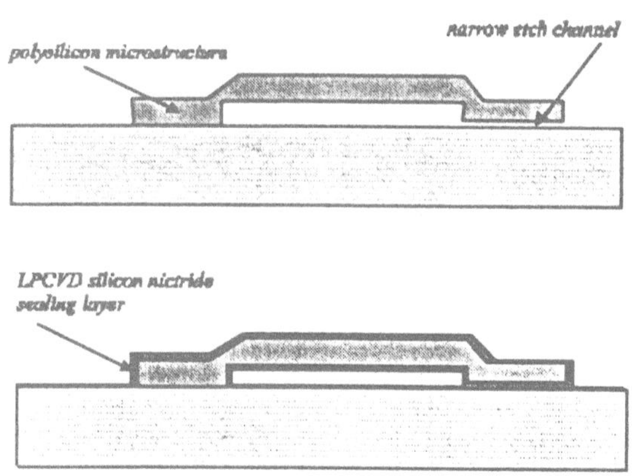

Fig. 3.36. Reactive sealing process

The surface micromachining process enables us also to seal structures in a vacuum. This special process is called reactive sealing, and due to Guckel (Guckel 1986) and Tilmans (1993). In fig. 3.36 we show the basic process. A narrow channel is left under the structural sealing layer to give access to the fluid for the sacrificial layer etch. After this process a silicon nitride is grown. The material grows also in the cavity until the channel is closed by the growing film. A fairly low pressure can be achieved in the cavity. The process has applications in absolute pressure sensors and in resonant strain gauges.

Acknowledgements

Contributions of Han Gardeniers, Meint de Boer, Gijs Krijnen, Remco Wiegerink, Erwin Berenschot and Theo Lammerink are gratefully acknowledged.

4. Principles of measurement mechanical quantities: transduction of deformation

M Elwenspoek and R. Wiegerink

Faculteit der Elektrotechniek, University of Twente, Enschede, The Netherlands

Abstract Historically the first transducers for the measurement of mechanical deformations are metal strain gauges, and we start the discussion with this subject.

4.1 Metal Strain Gauges

The electrical resistance of a piece of metal depends on his size and shape. For a rod with cross-section S and length l the resistance R is given by

$$R = \frac{l}{\lambda S} \tag{4.1}$$

with λ the specific conductivity. If as a consequence of an external load the resistor changes its dimensions the resistance will change. We have for the change of $R(l,S)$ therefore

$$\mathrm{d}R = \frac{\partial R}{\partial l}\mathrm{d}l + \frac{\partial R}{\partial S}\mathrm{d}S = \frac{\mathrm{d}l}{\lambda S} - \frac{l\mathrm{d}S}{\lambda S^2} \tag{4.2}$$

If the rod is subject to an axial stress $\sigma = F / S$ the change of the length and the cross-section change both lead to an increase (tensile stress) or a decrease (compressive stress) of the resistance. An illustration is given in fig. 4.1.

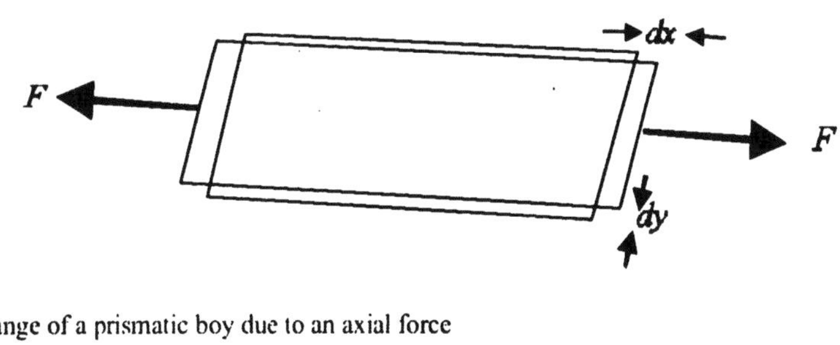

Fig. 4.1. The change of a prismatic boy due to an axial force

The reason is that the change of the length and of the cross-section has different signs. We have $\mathrm{d}l = l\varepsilon = lE\sigma$ and

$$dS = (y + dy)(z + dz) - yz$$
$$\approx ydz + zdy = yz(\frac{dz}{z} + \frac{dy}{y}) = -2\upsilon\varepsilon S = -2\upsilon E\sigma S$$

z denotes the thickness of the rod (not drawn in fig. 4.1). Using these equations in (4.2) we find for the relative change of the resistance

$$\frac{dR}{R} = (1 + 2\upsilon)\varepsilon \quad (4.3)$$

The ratio of the relative change of the resistance per unit strain is called the gauge factor G. Typical values of υ are 0.1 - 0.4, so for most materials the gage factor is between 1 and 2. The yield strain of metals mostly is well below 1% (this is about the yield strain of high quality steel), so changes in resistance of metal straingauges is maximal 1 to 2%.

4.2 Semiconductor Strain Gauges

The dominant effect in metal strain gauges is the change of the geometry. The materials property λ is independent of the strain. This is different in semiconductors. The effect is called piezoresistivity.

Conductivity, and therefore all effects connected to conductivity such as piezoresistivity in crystals are anisotropic. The piezoresistive coefficient π is defined as the relative change of the resistivity per stress. There is a longitudinal and a transversal effect: the resistivity changes in a direction parallel to the strain and normal to the strain.

$$\pi = -\frac{1}{p}\frac{\Delta\lambda}{\lambda} = \frac{1}{p}\frac{\Delta\rho}{\rho}$$

The effect is observed in single crystalline silicon and in polysilicon thin films. In single crystalline silicon the effect is exploited by doping p-regions in n-type silicon microstructures. Polysilicon thin films can be deposited on top of microstructures, doped and pattered.

Piezoresistive Effect in Single Crystalline Silicon

The gaugefactor in silicon depends on the crystallographic orientation and on the doping. The maxima of the coefficients are:

$$p-Si: \pi_{l<111>} = \quad 93.5\times 10^{-11}\,Pa^{-1}$$
$$n-Si: \pi_{l<100>} = -102.2\times 10^{-11}\,Pa^{-1}$$

If the electrochemical etch stop is used, the deformed member (a membrane in a pressure sensor, or a cantilever beam in an accelerometer) will be n-silicon, and p-type resistors will be indiffused or implanted into the upper layers of the member. Therefore, the larger π for n-silicon is of no use for micromachined sensors using this etch stop. Further, membranes or cantilever beams are machined in <100> wafers, so there is no <111> direction available. On a <001> oriented p-silicon wafer the maximum piezoresistive effect is a little bit more than $70 \times 10^{-11} Pa^{-1}$, both, for the transversal and longitudinal effects. The orientation of the maximum is in <110>, there are minima in <100> directions.

The piezoresistive gage factor can be found easily from π. If we include the change of the specific resistivity ρ ($\rho = 1/\lambda$) in eq. (4.3), we have

$$\frac{dR}{R} = (1+2\upsilon)\varepsilon + \left(\frac{d\rho}{\rho}\right)_l \varepsilon_l + \left(\frac{d\rho}{\rho}\right)_t \varepsilon_t$$
$$= (1+2\upsilon)\varepsilon + \pi_l E\varepsilon_l + \pi_t E\varepsilon_t \qquad (4.4)$$

the suffixes l and t refer to longitudinal and transversal piezoelectric coefficients.

The gaugefactor of $p\text{-}Si_{<110>}$ is 133, much larger than the geometric effects. Change in geometry can be neglected.

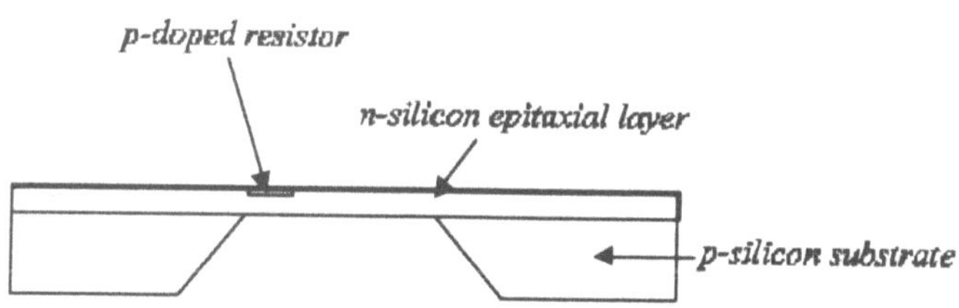

Fig. 4.2. Examples of a membrane with indiffused or ion implanted doped resistors

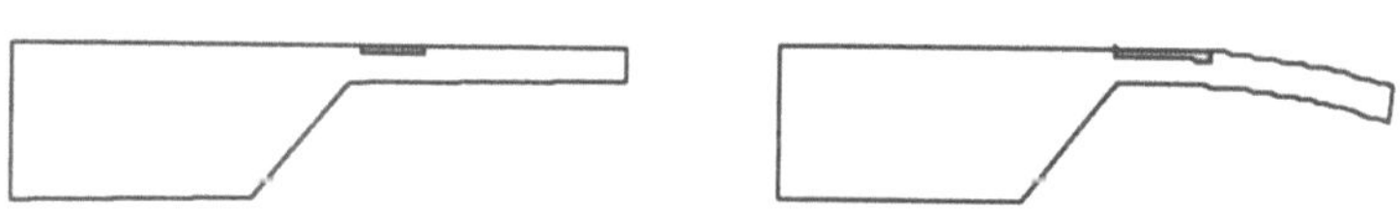

Fig. 4.3. Diffused resistors in silicon flexible structures

It is important to note that the piezoresistor must be placed in the upper part or on top of the flexible structure. Bending results in a strain distribution across the structure, and the strain will change its sign in the centre. A piezoresistor right across the flexible structure would not change its resistance upon bending.

The electrical isolation of the doped region relies on junction isolation. So the voltages are limited and there is some current leakage.

Piezoresistive Effect in Polysilicon Thin Films

Polysilicon thin films deposited by LPCVD processes on top of flexible structures can also be used as piezoresistive strain gauges. They have a few advantages compared to indiffused or implanted resistors: The technological possibilities allow a greater freedom of design and the choice of materials for membranes is greater, and the strain gauges do not rely on junction isolation, so there are less problems with leakage currents and greater voltages can be used. Electrical isolation from the substrate is done by means of a dielectric thin film (preferably stress free silicon nitride). Electrical contact can be made either by high-doped polysilicon leads, or better, because of higher conductivity, by metal leads.

Similar to single crystalline silicon, the piezoelectric coefficient in p-doped silicon is larger than in n-doped silicon. Using polysilicon it is possible to micromachine films with both types of doping, in contrast to implanted strain gauges. However, due to grain boundary effects and texture, the size of the piezoresistive effect is reduced. The maximum gauge factor in p-doped polysilicon is close to -40, in n-doped silicon to +20. Doping level must be in the range of a few $10^{19}cm^{-3}$, deposition temperature 560°C, annealing at 1000-1100°C (French 1989).

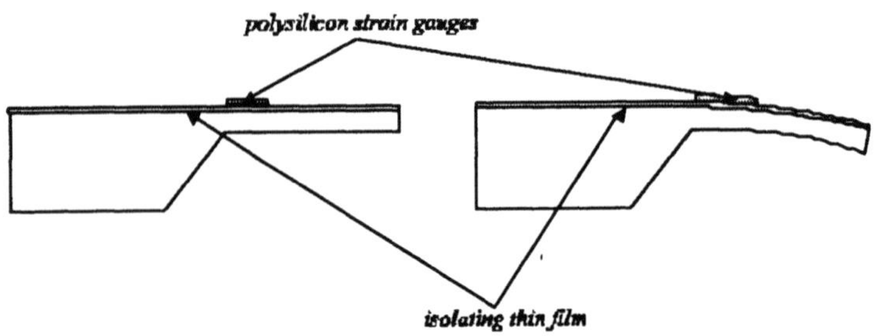

Fig. 4.4. Configuration of polysilicon this films on to of flexible silicon structures

Besides the possibility of doping by indiffusion and implantation, polysilicon can be doped during the deposition by feeding an additional gas for the doping through the reactor. Bouwstra et al. used in situ phosphorous-doped polysilicon by adding phosphine to the silane flow through the reactor (Bouwstra 1990). This way has some advantages, one of them is that the temperature coefficient of the gauge factor can me made zero. However the standard procedure is to dope the thin film after growth by ion implantation (Burger 1996).

Transduction from deformation to resistance

The piezoresistors are placed on top of flexible structures. Upon bending parts of the structures will be under compressive stress, other under tensile stress. Obviously it is important to know how the structure deforms under an external load to find the optimum positions for the strain gauges.

The strain in a bent member is

$$\varepsilon = \frac{z}{R} = zw'' \tag{4.4}$$

z is the distance normal from the neutral plane, R is the radius of curvature and w is the deflection of the neutral plane. The last equality holds for small deflections. This equation also holds locally. Generally the radius of curvature varies along the deformed construction. Equation (4.4) means that per length dx at a distance z from the neutral plane there is a change of the length due to the bending dΔx. dΔx/dx is then the local strain. A piezoresistive element at a distance z from the neutral plane sees an average strain, or a total change in length of the element. This is given by the integral over $d\Delta x$. So the average strain between two points x_1 and x_2 is given by

$$\bar{\varepsilon} = \frac{1}{x_2 - x_1}\int_{x_1}^{x_2} d\Delta x = \frac{1}{x_2 - x_1}\int_{x_1}^{x_2} \frac{z}{R}dx$$
$$= \frac{1}{x_2 - x_1}\int_{x_1}^{x_2} zw''dx = \frac{z\left(w'(x_2) - w'(x_1)\right)}{x_2 - x_1} \tag{4.5}$$

Thus, it is the derivative of the deflection, which is important for the average strain in a strain gauge.

This is simple for the cantilever beam: the sign of the curvature is constant, so the piezoresistor can be placed on top of the beam along the whole length. The total strain is easily calculated from (4.66) for a cantilever beam with length l:

$$\bar{\varepsilon} = \frac{20 P_0 h l^2}{48EI} \tag{4.6}$$

A bridge covered in its whole length by a piezoresistor would measure no effect. Since at both clamped edges the slope of the bridge is zero, there is no average strain. For an optimal effect, the strain gages must extend either from the edges to the two inflection points, or it must be placed in between them. The inflection points can be found from. (4.63), they are located at

$$x_{inf} = l\left(1/2 \pm \sqrt{1/12}\right) \tag{4.7}$$

The form of a circular diaphragm loaded by a constant pressure p_0 is

$$w(r) = \frac{3}{16}\frac{1-\upsilon^2}{Eh^3} p_0\left[r_0^2 - r^2\right]^2 \tag{4.8}$$

where r_0 is the radius of the diaphragm. It follows that the inflection line is at $r = r_0/\sqrt{3}$. A strain gage must be placed either at the edges, without exceeding the inflection line, or in the centre of the diaphragm between $\pm r_0/\sqrt{3}$. The relative change of the resistance is given by

$$\frac{\Delta R}{R} = G\bar{\varepsilon} = \frac{1}{6}G\frac{p_0}{E}\left(\frac{r_0}{h}\right)^2 \tag{4.9}$$

For square diaphragms only the numerical factor differs from this equation. For large sensitivity, the diaphragm must be thin and wide.

4.3 Capacitive Transducers

Electromechanics

A capacitive transducer is a transducer of a very different type. The capacitor can store energy, and its properties are completely described once the energy function is known. Therefore, capacitive transducers can be use for sensing as well as actuation purpose. There are two distinct ways to change the energy of a capacitor: by changing its charge, this would represent an electric work on the capacitor, or by changing the geometric arrangement, and this would represent mechanical work. The total energy change is then

$$\mathrm{d}W = u\mathrm{d}Q + \mathbf{F}_{ext}\,\mathbf{dr}\,. \tag{4.10}$$

with u the electrostatic potential and $\mathbf{F}_{ext}$ the force needed to realise the displacement $\mathbf{dr}$. Let us restrict ourselves for the moment to a situation in which the geometrical change of the arrangement of the conductors is only given by the distance of the plates of a parallel plate capacitance, as shown in fig. 4.5.

In this case, the mechanical work is $-F_{\mathrm{int}}\mathrm{d}x$ and the total energy as a function of the charge and the distance of the plates is [1]

$$W = \frac{1}{2}\frac{x}{\varepsilon A}Q^2 + K(x - x_0)^2 \tag{4.11}$$

with A the overlap surface of the plates. We have added a spring to the system, with spring constant K. The position of the plate hold by the unloaded spring is x_0 (measured from the lower plate).

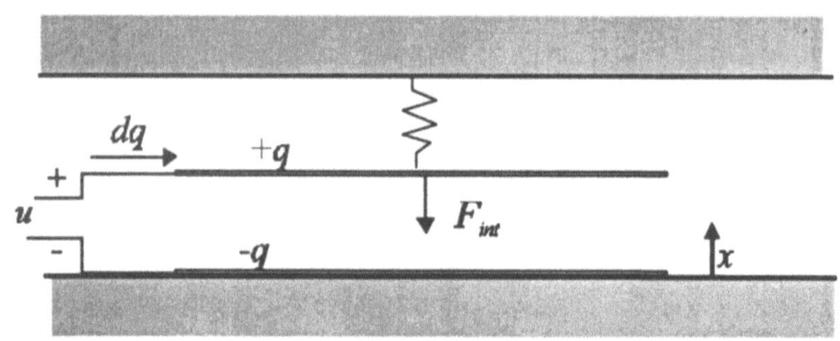

Fig. 4.5. A capacitor with a movable plate connected to a spring

The total differential of W is given by

[1] Note that in a quasistatic process $\mathbf{F}_{\mathrm{int}} = -\mathbf{F}_{\mathrm{ext}}$

$$dW = \left(\frac{\partial W}{\partial x}\right)_Q dx + \left(\frac{\partial W}{\partial Q}\right)_x dQ \tag{4.12}$$

from which we find by comparing with (4.10)

$$\left(\frac{\partial W}{\partial x}\right)_Q = F_{ext} = \frac{Q^2}{2\varepsilon A} + Kx \tag{4.13}$$

and

$$\left(\frac{\partial W}{\partial Q}\right)_x = u = \frac{Qx}{\varepsilon A} = \frac{Q}{C}. \tag{4.14}$$

We also see that the capacity and the spring constant are given by

$$\left(\frac{\partial^2 W}{\partial Q^2}\right)_x = \frac{1}{C} \equiv \frac{1}{C_x} \tag{4.15}$$

and

$$\left(\frac{\partial^2 W}{\partial x^2}\right)_Q = K \equiv K_Q. \tag{4.16}$$

In (4.13) the first term the right hand side represents the electrostatic force. Note that this force is independent of the plate separation. This might be surprising, one would expect that the force becomes smaller if the plates are farther apart. But this impression is based on the assumption that the voltage is constant when one changes the distance. In fact, the electrostatic field close to a plane surface is independent of the distance, so will be the forces on charges. Usually, in these systems the voltage is controlled. If the plate separation is changed while the voltage is constant, the charge must change. So a battery must also deliver work to the capacitance. We can see from eq. (4.10) that the derivative of W with respect to x at variable charge does *not* give the force. In fact, the battery must do a work equal to $u\mathrm{d}Q = u^2\mathrm{d}C$.

There is a mathematical trick to circumvent this difficulty, which is called the Legendre transformation. One adds to the energy function the product $-uQ$, and one gets the so-called co-energy:

$$W' = W - uQ \tag{4.17}$$

We get immediately:

$$dW' = dW - u\,dQ - Q\,du = -Q\,du + F_{ext}\,dx. \tag{4.18}$$

With the help of $Q = Cu$ we can express W' by u and x:

$$W' = -\frac{1}{2}Cu^2 + K(x-x_0)^2 = -\frac{\varepsilon A u^2}{2x} + K(x-x_0)^2 \tag{4.18}$$

and we get now

$$F_{ext} = \left(\frac{\partial W'}{\partial x}\right)_u = \frac{1}{2}\frac{\varepsilon bl}{x^2}u^2 + Kx \tag{4.19}$$

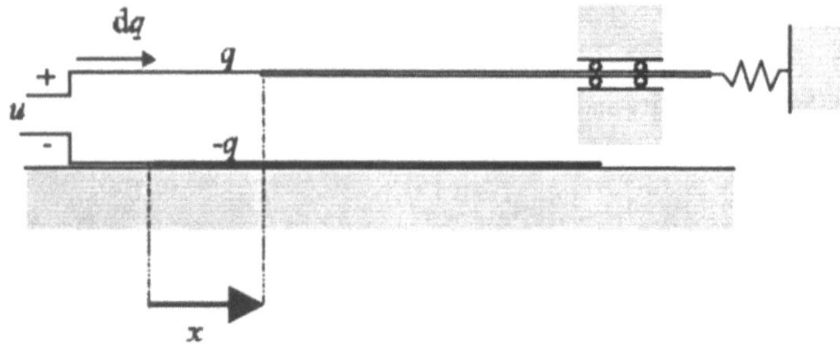

Fig. 4.6. Schematic of the capacitor with variable overlap b-x. b is the width of the plate.

Now we do have an x-dependency of the force. But note that this is exactly the same force as in (4.13), as can be seen if one eliminates u in (4.19). The important point is that the stiffness of the capacity is different. If we calculate the spring constant from (4.19) we get a voltage dependency:

$$K_u = \left(\frac{\partial^2 W'}{\partial x^2}\right)_u = K - \frac{\varepsilon A u^2}{x^3}. \tag{4.20}$$

There is a voltage u_c at which the spring constant K_u vanishes. This voltage is the so-called pull-in voltage: the capacitance collapses. When using the capacitance as a sensor for an external load, it is clear that one has to stay away from u_c because of uncontrollable instabilities. On the other hand, a small K_u means a sensitive sensor.

There is a second way to change the geometry, which is of importance for mechanical microsensors and microactuators. This is the change of the overlap of the electrodes at constant separation of the plates. We have given an example of a micromachined structure for this in chapter 2, fig. 2.3. Besides in actuators, these types of electrodes are used in accelerometers, gyroscopes and resonators. Schematically, the device is shown in fig. 4.6.

In this case the energy is given by

$$W = \frac{Q^2 d}{2\varepsilon a(b-x)} + K(x-x_0)^2 \tag{4.21}$$

Here, the surface of the capacitor plate is $A = ab$, and the overlap is $a(b\text{-}x)$. d is the separation of the plates. The force is now

$$F_{ext} = \left(\frac{\partial W}{\partial x}\right)_Q = \frac{Q^2 d}{2\varepsilon a(b-x)^2} + Kx \tag{4.22}$$

and we see that in this case the electrostatic force (first term in eq. (4.22)) is dependent on the position. It also follows that the electromechanical spring constant is dependent of the charge of the capacitor. The second derivative of (4.21) is

$$K_Q = K + \frac{Q^2 d}{\varepsilon a(b-x)^3} \tag{4.23}$$

and we do not get an instability. On the other hand, the co-energy is now

$$W' = -\frac{u^2 \varepsilon a(b-x)}{d} + K(x-x_0)^2 \tag{4.24}$$

and at constant voltage the force is independent of the overlap:

$$F = \left(\frac{\partial W'}{\partial x}\right)_u = \frac{u^2 \varepsilon a}{d} + Kx \tag{4.25}$$

and the spring constant is simply K.

In practice, the distance between the plates is kept constant by means of a flexible mechanical construction, which is much stiffer than the spring in fig. 4.6. However, when controlling the comb-drive by voltage, there is the same instability as described by eq. (4.20). Therefore, one is limited in the voltage of these electromechanical transducers. See fig. 4.7.

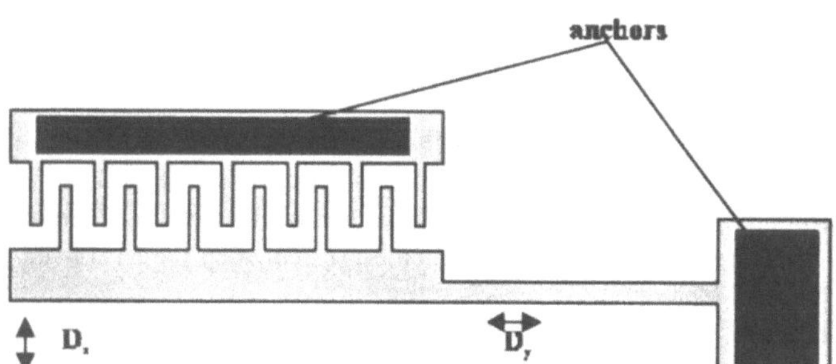

Fig. 4.7. Comb drive suspended by a spring with spring constants D_x (vertical movement) and D_y (horizontal movement). The latter spring constant is important to keep the fingers right in place. Generally, $D_x << D_y$. D_y is decisive for the largest deflection.

Diaphragm pressure sensors

There are three types of mechanical constructions for mechanical diaphragm pressure sensors. A simple diaphragm bending under the pressure load (fig. 4.8a), a partially collapsed capacitance (fig. 4.8b) and a diaphragm with a stiff portion, a so-called boss (fig. 4.8c)

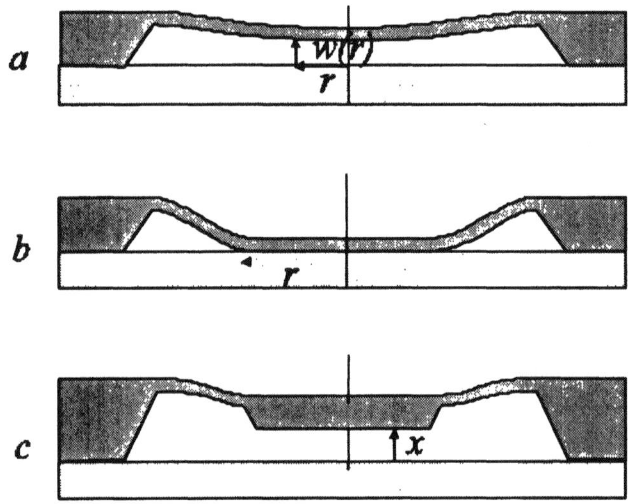

Fig. 4.8 Three configurations for pressure sensors

The configuration in fig. 4.7c is quite similar to the schematic in fig.4.5. Here the spring constant is given by di Giovanni (1982)

$$K \approx \frac{Eh^3}{\pi r_b^2 r_0^4} \frac{1}{A_p}$$
$$A_p = \frac{3\left(1-\upsilon^2\right)}{16}\left(1-\left(\frac{r_b}{r_0}\right)^4 - 4\left(\frac{r_b}{r_0}\right)^2 \log\frac{r_0}{r_b}\right) \tag{4.26}$$

Here, r_b is the radius of the boss and r_0 the radius of the membrane, and we have assumed that the pressure acts only on the boss, with is acceptable if $r_b \approx r_0$. h is the thickness of the flexible part of the structure. Keep in mind that this spring constant is the one that belongs to operation with charge control. For voltage control the spring constant must be corrected by the term in eq. (4.20).

The relative change of the capacity is in this case

$$\frac{\delta C}{C} = \frac{\delta x}{x} = \frac{F}{xK} = \frac{p}{E}\frac{r_0^4}{h^3 x} A_p . \tag{4.27}$$

For the case shown in fig. 4.7a, the analysis of the relative change of the capacitance leads to a formula similar to this one. The capacity is now given by

$$C = 2\pi \int_0^{r_0} \frac{\varepsilon}{d - w(r)} r\,dr \approx 2\pi \int_0^{r_0} \frac{\varepsilon}{d}\left(1 - \frac{w(r)}{d}\right) r\,dr \tag{4.28}$$

We assumed a small deflection. d is the plate separation of the unloaded diaphragm. The deflection of a circular diaphragm under a pressure is given by (4.8) and the integration leads to

$$\frac{\delta C}{C} = \frac{1}{32} \frac{p}{E} \frac{r_0^{\,4}}{dh^3} \tag{4.29}$$

The response of the capacity is non-linear. In the analysis here we linearised all equations to make clear the most essential features of the capacitive sensors. The example in fig. 4.7b is an attempt due to Rosengren (1992). The result of their analysis is that once the membrane has touched the bottom electrode (one needs of course an electrical isolation), the capacitive change is dominated by the change of the contact area, and this turns out to be much closer to linearity than the capacitance change in the case 4.7a. Also the linearity of the bossed membrane is better (e.g. Mallon 1990).

It is interesting to compare (4.29) to (4.9). One can see that the deformation has quite different consequences for the relative changes of capacity and resistance. Typically we have in micromachined capacities $d \approx h$, so the sensitivity of capacitive pressure sensors is proportional to (r_0/h) to the fourth power, while in piezoresistive pressure sensors the sensitivity is only proportional to the square of (r_0/h). Sine this ratio is small anyway, the fourth power has quite drastic consequences and easily outrules the gagefactor.

Acknowledgements

Contributions of Han Gardeniers, Meint de Boer, Gijs Krijnen, Remco Wiegerink, Erwin Berenschot and Theo Lammerink are gratefully acknowledged.

5. Case studies

M Elwenspoek and R. Wiegerink

Faculteit der Elektrotechniek, University of Twente, Enschede, The Netherlands

Abstract This section presents the following design case studies: pressure sensors, miniature microphones, accelerometers, mechanical flow sensors and electrostatic actuators.

1.1 Pressure sensors

The design of a pressure sensor depends strongly on the choice of the read out. In the literature there are the four suggestions for read out: capacitive, piezoresistive, resonant and optic.

Capacitive pressure sensors make use of the fact that the deflection of a membrane changes the capacitance, which is measured. The capacitance of a micromechanical structure is rather small. It is clear that C is a nonlinear function of the deflection. A typical value for a micromechanical capacitance is 10 - 100 pF, and its change by a deflection is a fraction of this value. It is clear that one has to minimize the gap spacing.

The thickness of membranes used in micromechanical sensors is of the order of 1 μm or more, practically never thinner than the gap spacing. This implies, since $\delta \ll d$, we automatically have $\delta \ll h$, and we can neglect the cubic term in (5.1.).

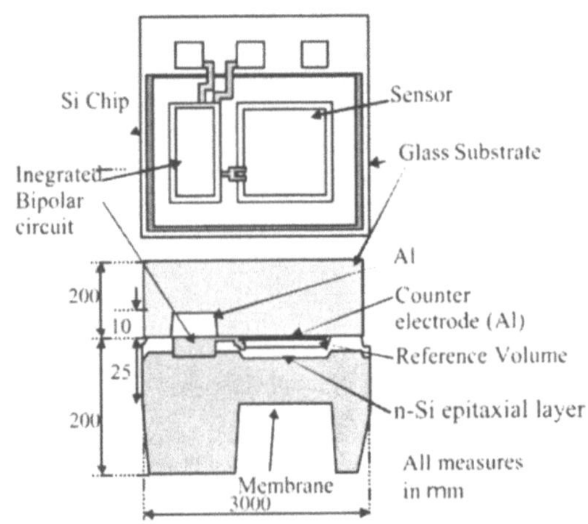

Fig.5.1. Schematic of an integrated capacitive pressure sensor (from Heuberger (1989), original work from Sander 1980)

The realization of capacitive pressure sensors requires a wafer bonding step. A schematic drawing after Sander et al. (1980) is given in Fig. 5.1. The fixed counter electrode (Al) has been deposited on a Pyrex wafer and the deflecting electrode has been realized by an n-doped expitaxially grown silicon layer. The membrane of a thickness of 25 µm has been etched from the backside by anisotropic etching using the electrochemical etchstop. In this work, a bipolar integrated circuit has been integrated on chip for a first amplification of the small signals. This particular sensor design has a sensitivity ($[\Delta C/C]/p$) of 0.35 bar^{-1}.

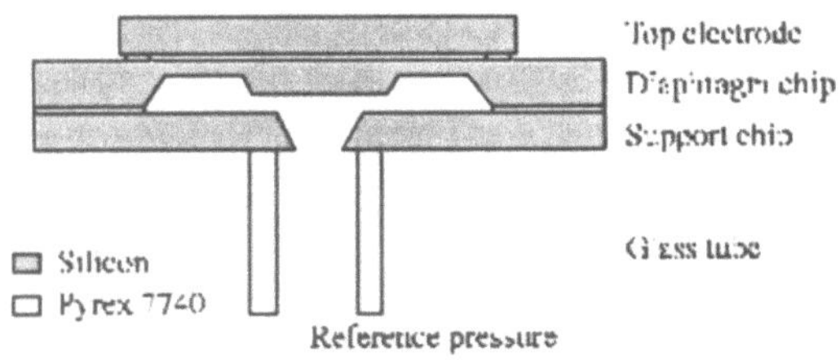

Fig, 5.2. Schematic of a capacitive pressure sensor using a bossed diaphragm and sputtered thin films of Pyrex for anodic bonding.

A similar design of a capacitive pressure sensor using anodic bonding of a silicon wafer to a Pyrex wafer has been used by Puers et al. (1990).

Hanneborg and Ohlckers (1990) proposed the use of a bossed diaphragm and anodic bonding using a sputtered thin film of Pyrex, this way minimizing thermal drifts and slow stress relief. A schematic drawing is shown in Fig. 5.2.

Fig. 5.3. shows the fabrication sequence of an alternative design using anodic waferbonding, as proposed by Chau et al (1988). They fabricated a pressure sensor small enough for use in a cardiovascular catheter, with an outer diameter of 0.5 mm. The chip with the deflecting silicon membrane was bonded on a glass substrate. It was micromachined with concentrated HF solution using a goldmask in order to realize 30 µm deep groves for lead attachment and to realize the metallization. An overall view is given in Fig. 5.3. a and b. The process sequence for the fabrication of the membrane is given in Fig. 5.3.c. After etching a shallow cavity (3 µm), in the end providing the cavity that defines the capacity, two boron indiffusion steps are involved: a deep indiffusion (12 µm) to provide the etchstop for the support of the membrane and a shallow indiffusion (1.2 µm) for the etchstop of the membrane itself. A dielectric is deposited on the membrane to protect the membrane against shorts. After this the wafer is bonded using anodic bonding. The last step is etching all undoped silicon in EDP, leaving the membrane and its support that consists entirely of heavily boron doped silicon and the dielectric layer.

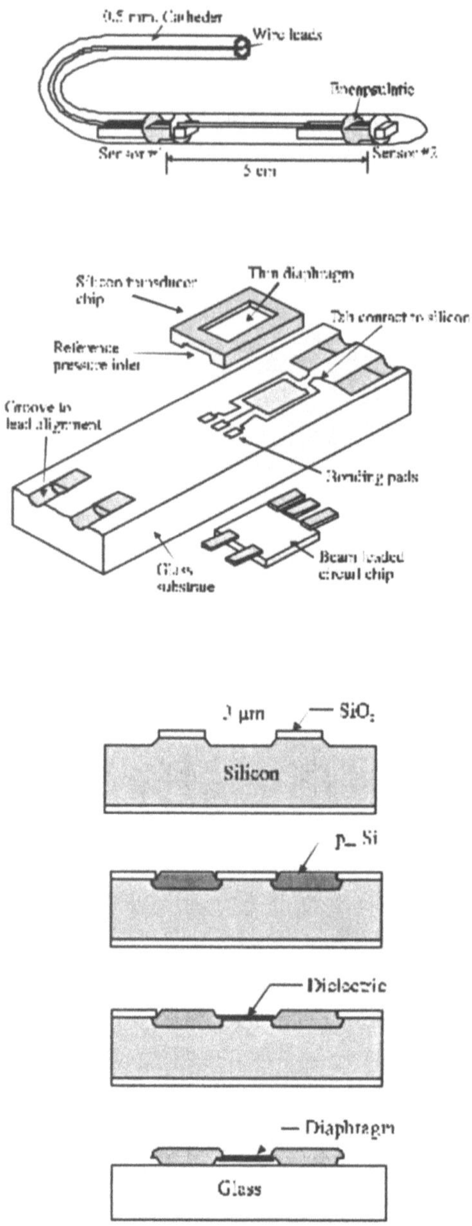

Fig. 5.3. Overall view of a the catheter (a), construction of the pressure sensor, (c) fabrication sequence of the silicon transducer chip (from Chau (1988)

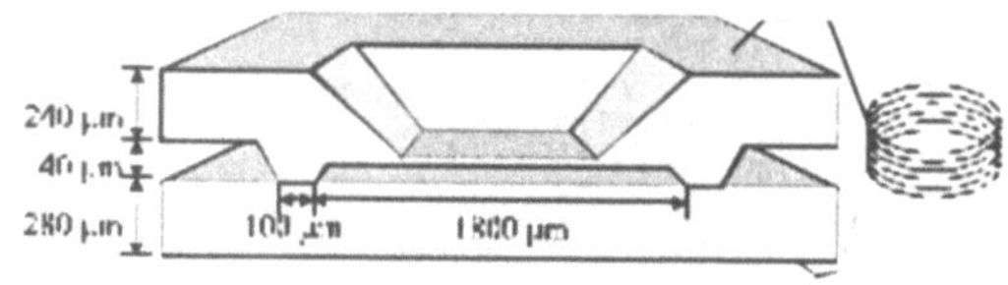

Fig. 5.4.a. Schematic cross section of the intra ocular pressure sensor

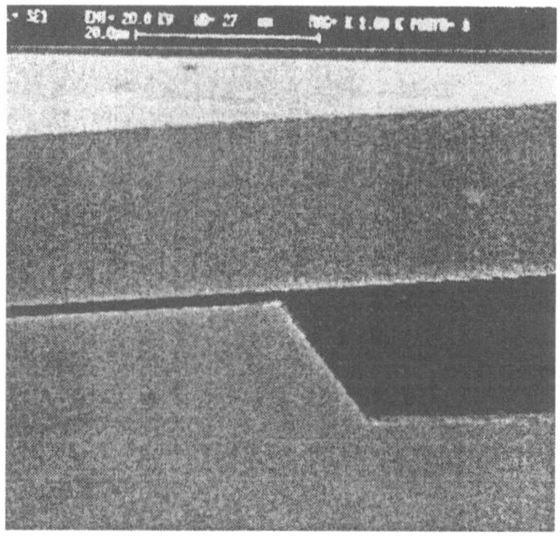

Fig. 5.4.b. SEM-photo of the membrane showing the bonded region. (The region marked with 100 µm in Fig 5.5.a, upside down)

The last example of a capacitive pressure sensor we give here concerns a device that is planned to be implanted in the human eye Bäcklund (1990. It is based on the idea that a change of the capacitance changes the resonance frequency of an LC-circuit. The implanted sensor is a passive structure, and the resonance frequency is measured remotely with an external, inductively coupled oscillator. In the fabrication of this device use was made of direct silicon wafer bonding ("fusion bonding"). The authors report on difficulties combining the boron etchstop with direct bonding. Direct bonding is only possible with perfectly flat silicon wafers, and the indiffusion process seemingly degrades the surface flatness too much. As a consequence, the authors used a time etchstop. A schematic of the device and a microphotograph is shown in Fig. 5.4. The silicon wafer containing the membrane has been etched from both sides and bonded to the second silicon wafer at the narrow 100 µm wide rim. The SEM-photo shows the bonds region and a part of the membrane.

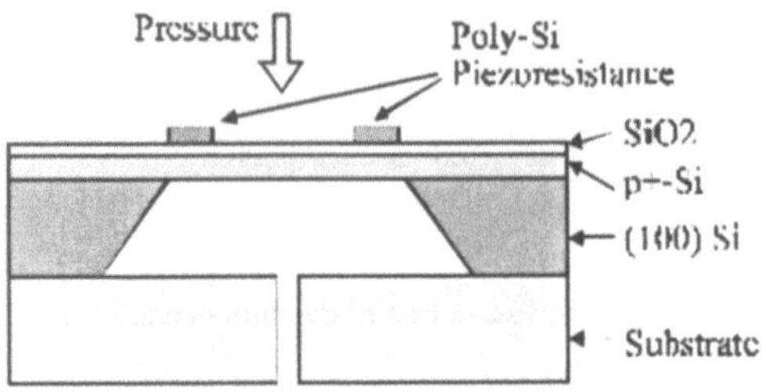

Fig. 5.5. Schematic cross section of a pressure sensor with poly crystalline silicon piezoresistive elements (from Heuberger (1989))

Since it is not necessary to have an electrode very close to the membrane, use of waferbonding technology is not a prerequisite for the fabrication of the sensor chip. However, the packaging of the membrane requires some mechanical connection that is often realized using wafer bonding. A schematic of the piezoresistive pressure sensor is shown in Fig. 5.5 .

5.2. Miniature microphones

A lot of work on silicon technology-based *miniature microphones* is done by Bergveld's group at the University of Twente. The essential parts are a charged capacity with a membrane that is deflected by the fluctuating pressure at audio frequencies and a backplate with apertures to allow for flow of air in order to reduce the viscous damping of the system. The charge of the capacity can be provided by a built-in charge (a so-called electret) or by an external voltage. First designs of silicon microphones used a mylar foil providing the flexural membrane and Teflon as an electret material.

The processing steps for the silicon electret microphone (1988) is shown in Fig. 5.6. Since the backplate of the device has to be processed from both sides, a doublesided polished wafer (<100>) is preferable, and the process must start with the definition of alignment marks on either side of the wafer. Using a doublesided maskaligner is the most convenient method, but in this particular work, there was no such instrument available. Therefore a specially designed holder was used containing the masks with the alignment marks. Wafers placed in this holder can be illuminated from both sides, providing there are alignment marks on both sides.

In step b of Fig. 5.6., the 32.5 μm deep cavities are etched in KOH. After growth of thermal oxide (leading to an oxide of 1 μm in the cavities and of 1.5 μm elsewhere), holes are etched from the backside, providing holes for the airflow out of the cavity (A) and to attach the mylar foil during one of the last steps (G). This etch stops at the oxide layers. Simultaneously, V-grooves are etched in the front side of the wafer in order to break the individual chips out of the wafer after processing. The depth of this etch is defined by the opening in the oxide mask. After this all oxide is stripped and regrown. In step e, the Teflon foil is attached to the wafer. For this procedure, which is by no means easy due to the bad adhesion of Teflon to practically all materials, we refer to Sprenkels' thesis. The Teflon foil is patterned in a dry etch (oxygen plasma, see Ch. V) and charged. For the charging procedure and the attachment of the mylar foil, we again refer to Sprenkels (1988).

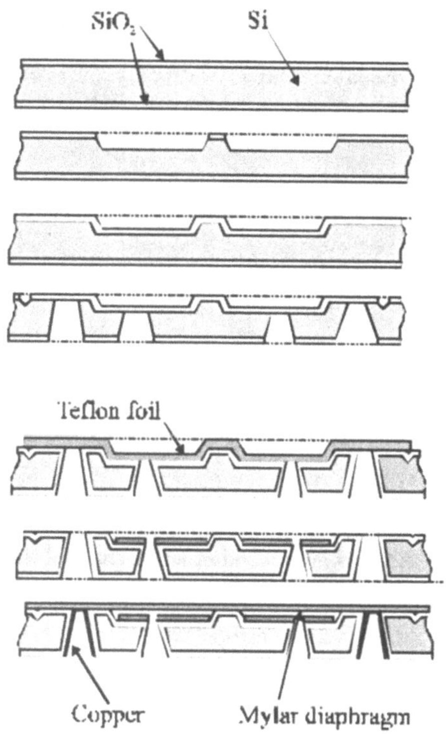

Fig. 5.6. Process sequence for fabrication of a silicon electret microphone (after Sprenkels])

5.3. Accelerometers

Accelerometers generally consist of a proof mass M which is attached to a fixed frame by one or more spring elements. Fig. 5.7. shows a simple lumped element model of such a structure. In this model, K is the effective spring constant of the spring element(s) and D is the damping factor. The operation of the device is based on Newton's second law of motion:

$$F = \frac{dp}{dt} = Ma \qquad (5.1)$$

where F is the force acting on the mass, p is the impulse momentum and a is the acceleration of the mass. From (5.1) it is clear that an acceleration results in a force being exerted on the mass. This force results in a deformation of the spring element(s) and a displacement of the mass given by:

$$d_{static} = \frac{F}{K} = \frac{Ma}{K} \qquad (5.2)$$

The subscript *static* indicates that (5.2) is only valid for slow variations of the acceleration, i.e. well below the resonance frequency of the system.

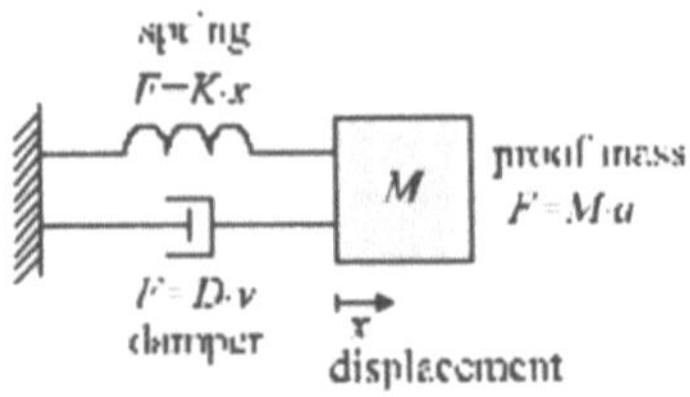

Fig. 5.7. Lumped element model of an accelerometer structure.

The dynamic behaviour of the system can be analyzed by considering the differential equation:

$$M\frac{\mathrm{d}^2x}{\mathrm{d}t^2}+D\frac{\mathrm{d}x}{\mathrm{d}t}+Kx=F_{ext}=Ma \tag{5.3}$$

where F_{ext} is the external force acting on the reference frame to which the proof mass is attached. Using Laplace transformation, the following second-order mechanical transfer function from an acceleration to a displacement of the mass is obtained:

$$H(s)=\frac{X(s)}{A(s)}=\frac{1}{s^2+\frac{D}{M}s+\frac{K}{M}}=\frac{1}{s^2+\frac{\omega_r}{Q}s+\omega_r^2} \tag{5.4}$$

where $\omega_r=\sqrt{\frac{K}{M}}$ is the resonance frequency and $Q=\frac{\omega_r M}{D}$ is the quality factor. From this equation we see that we can also write (5.2) in terms of the resonance frequency:

$$d_{static}=\frac{a}{\omega_r^2}, \tag{5.5}$$

which clearly illustrates the trade-off between bandwidth and sensitivity. For a high dc sensitivity we need a low resonance frequency. Feedback may be used to eliminate this trade-off as will be explained later.

The performance of accelerometers is limited by the thermal motion of the proof mass. According to the laws of thermodynamics, the thermal energy of a system in equilibrium is $k_BT/2$ for each energy storage mode, where k_B is the Boltzmann constant and T is the temperature. The small proof mass of micromachined accelerometers results in rather large movements. An equivalent acceleration spectral density, the so-called total noise equivalent acceleration (TNEA) can be calculated and is given by (Gabrielson 1993):

$$\text{TNEA}=\sqrt{\frac{\overline{a_n^2}}{\Delta f}}=\frac{\sqrt{4k_BTD}}{M}=\sqrt{\frac{4k_BT\omega_r}{QM}} \tag{5.6}$$

It is clear that for measuring low acceleration levels a large proof mass and high quality factor are required.

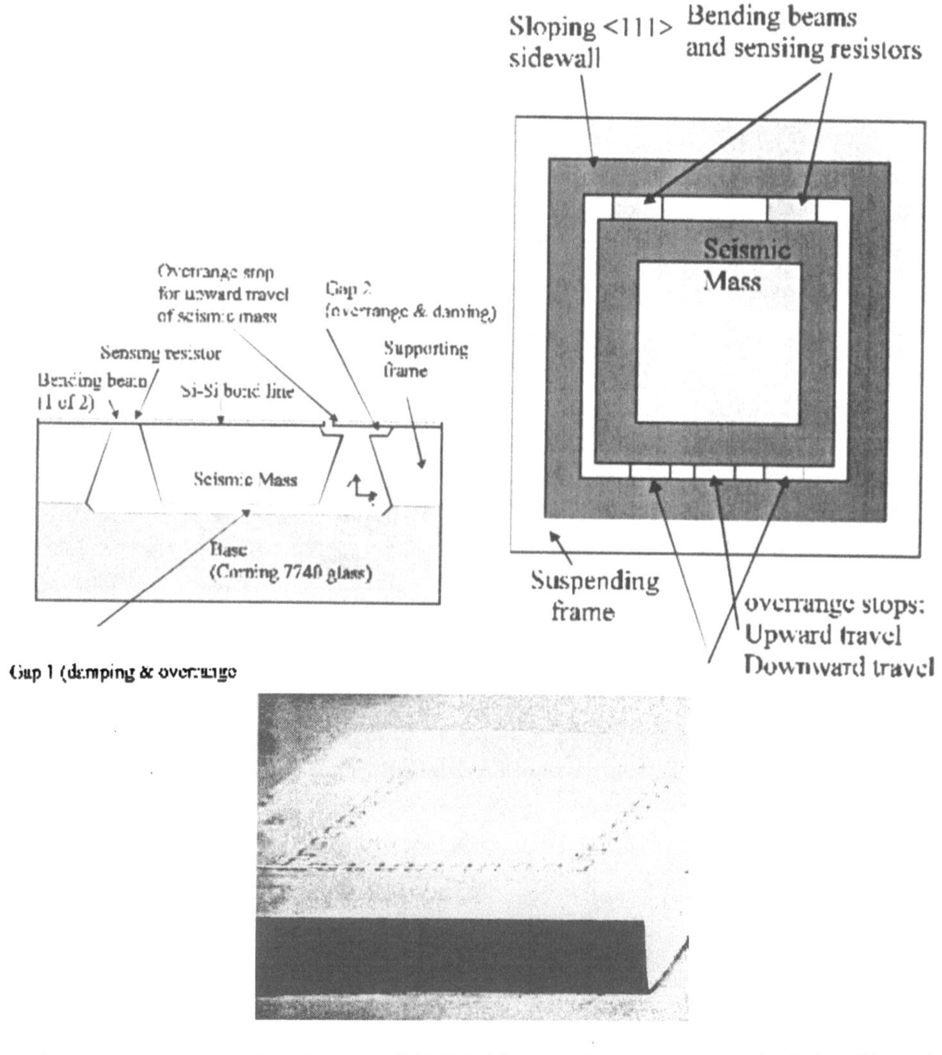

Fig. 5.8. Accelerometer according to Barth et al. (1988). (a) shows the cross section of the design, (b) the plane view. (c) is a SEM photo of the device before attaching the chip to a glass base. The interdigited fingers for overload protection are clearly visible, and the supporting beams can be seen in the far end

A design of an *accelerometer* is given in Fig. 5.8. (1988). Similar to the pressure sensor of Petersen et al. (1988), besides anisotropic etching, fusion bonding is an essential fabrication step. This technique permits great freedom in the production process. In the cross section shown in Fig. 5.8. (a) the cavities described as "gap 2" are etched before bonding. After bonding the following fabrication steps can be performed: etching the seismic mass into the lower wafer; etching a membrane by thinning the upper wafer; etching the beams out of the membrane for suspension, bending and overload protection; fabrication of the piezoresistors and the metallization. This device has two special aspects.

One is the overload protection. The central seismic mass is surrounded by the supporting frame, see Fig. 5.8.(b). The interdigited beams for overload protection are present around three sides of the mesa, and cantilever beams suspending the mass are located at the fourth side. This can be seen in the SEM-photograph in Fig. 5.8 (c).

The other special aspect is the fact that this device can be fabricated with controlled damping. The damping is viscous, and the main contributions come from the narrow gaps indicated in Fig. 5.8 a. The size of these gaps are such that the damping is critical. The overload protection of the device is better than 1000 G

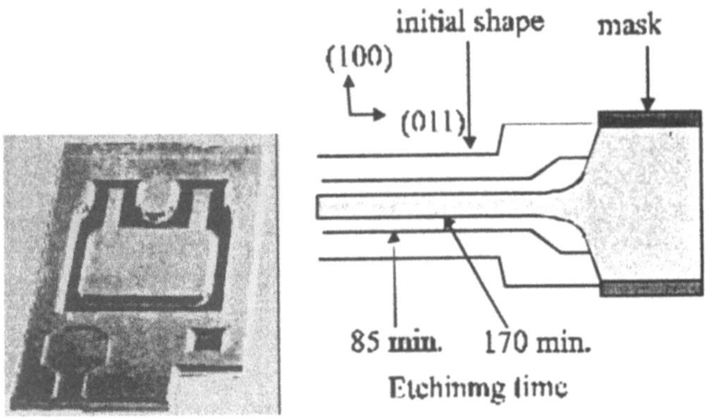

Fig. 5.9.a. (left) SEM-photograph of the silicon chip showing the mesa and the suspending beams (right) Two-step anisotropic etching process. The emergence of a rounded etch profile. the silicon chip showing the mesa and the suspending beams.

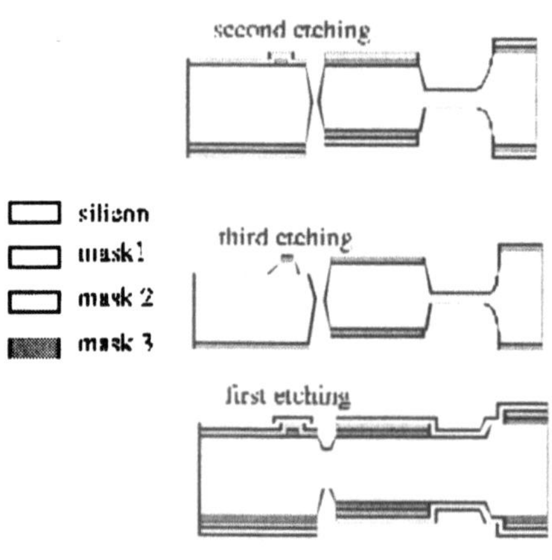

Fig. 5.9.b. How to etch rounded profiles of beams. Cross-section of a multistep anisotropic etching process.(from Koide et al. (1992)) bonding and anisotropic etching.

The readout principle is capacitive, similar to other proposals of accelerometers, e.g. (Jornod 1998). The seismic mass is supported by eight beams located at the edges of the mass. For the etching of the beams, the boron etch stop technique has been used. The high symmetry of the design serves as a small cross sensitivity, so that acceleration in the lateral direction of the chip is nearly invisible.

An interesting improvement of the mechanical strength of the suspension beams in accelerometers has been proposed by Koide et al. of Hitachi (1992). One always has stress concentration at sharp corners and anisotropic etching easily leads to sharp corners. Koide proposed using a layered mask technique to resolve this problem. By growing an oxide layer of varying thickness it is possible to perform different etch steps one after the other without the need of photolithographic steps (which are difficult to realize or impossible if there are structures on the wafer). An example is shown in Fig. 5.23. The technique has been developed for the fabrication of mesas supported by beams in accelerometers. The etching steps are performed in succession after removing the thinnest mask layers. In this etch process etch stop techniques are not used; the wafers are removed from the etch bath after a given time.

5.4. Mechanical flowsensors

Skin friction sensors

Many friction sensors are developed for the study of turbulence. Turbulence is still a phenomenon, which is subject to intense basic research, and sophisticated instrumentation is needed to further the field. Here microtechnology, MEMS, may play an important role. The elements of microsensors can be made small and fast enough to resolve phenomena in turbulent flow. This is why in the past ten years a rather large number of papers appeared addressing the problem of microsensors especially designed for turbulence studies (Yoon 1992, Liu 1994, Kälvesten 1994a, Kälvesten 1994b, Pan 1995, Jiang 1995, van Oudhuisden 1995, Padmanabhan 1995 Jiang 1996, Kälvesten 1996, Schmidt 1987, Padmanabhan 1997, Ebefors 1998, Schmidt 1988). Further more, due to the small size of the sensors, many sensors may be placed close to each other and the friction field (and the pressure and the velocity field) can be measured. Successful attempts to produce space and time resolved images of the shear stress in the presence of turbulence have been made by Jiang et al (1996) and Padmanabhan (1997). We referred to the former work in chapter 8.3.1.2. Microsensors can also be made much more sensitive because there is an arsenal of techniques for the thermal isolation of the heater (van Putten 1983, Stemme 1986, van Oudhuisden 1990a, Löfdahl 1992, Jiang 1996). This sensitivity is sufficient to detect the heat flow to a turbulent gas flow. A said earlier, the requirement in space and time resolution is app. 100μm and 100μs.

Skin friction can be measured in a number of ways. An overview on instrumentation fabricated by traditional machining can be found in a paper by Winter (1977). In micromachined versions there are only two basic types described: Thermal skin friction sensors and floating element skin friction sensors. The theory of skin friction in relation to heat transfer has been described in chapter 8.1.5. The main result is that the skin friction in a developed boundary layer is proportional to the heat flux to the fluid to the third power.

A thermal measurement is indirect, so some researchers have preference for floating element sensors. This sensor consists of a movable element attached to springs to a substrate, flush mounted into the wall on which the skin friction is to be measured. The disadvantage of these devices are the strict requirement for flush mounting and the fact that inevitably there are openings in the wall surrounding the floating element to allow a small movement. It is the latter reason that researchers often complain over the poor reproducibility when using floating elements (e.g. Winter 1977, Hanratti 1996, Löfdahl 1992). Micromachining allows however the fabrication of floating elements with very narrow gaps (2 μm is standard), and surface micromachining ensures that the element is neatly in plane with the substrate. With the integration of an actuator, a force feedback is possible which enables one to keep the gap at constant width (Pan 1995). So, in our view, at the moment the question what the best type of skin friction microsensor is, seems to be open.

A floating element consists of a movable plate attached to springs with well defined spring constants. The deflection of the plate is measured, the amount of which gives us information on the force acting on the plate. Most of the floating elements are fabricated using surface micromachining.

Ng et al. (1991) proposed a design of plates attached to springs as shown in Fig. 5.10. They used a fabrication technology involving wafer bonding and etch-back instead of surface micromachining. The element shown in Fig. 5.10 was positioned so that the tethers were parallel to the flow. The dimensions of the floating element was 120×140×5 μm^3. The tethers are then loaded such that the upstream ones are in tensile, the downstream ones in compressive stress. Their idea was to use the tethers as piezoresistive elements. The structural layer was lightly doped to get suitable electrical conductivity, and when the springs are electrically connected in form of a half Wheatstone bridge, one gets a differential signal proportional to the deflection. The authors tested their device in a liquid (Cannon N2700000) and were able to detect shear stresses in the range of a few kPa. This corresponds to a force resolution in the range of 0.1 nN.

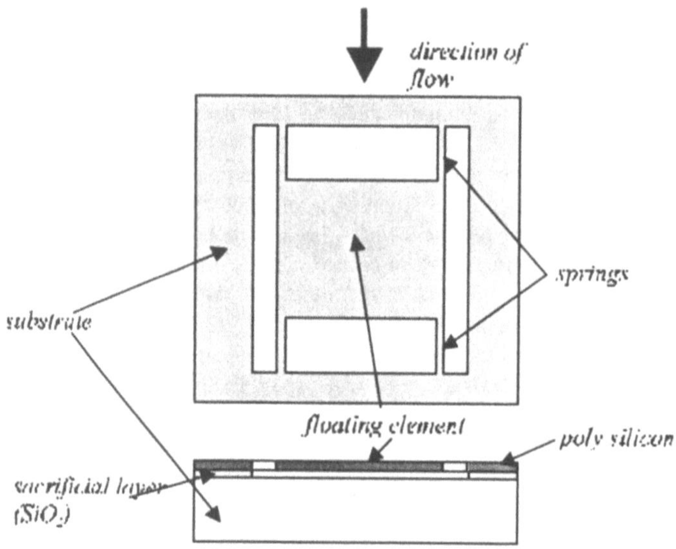

Fig. 5.10. Basic construction of a floating element

Ng's design is quite similar to earlier work by M.A. Schmidt (1987, 1988), who designed a sensor with a plate size in the 1 mm^2 range made from polyimide. Polyimide is not the most appropriate material for this purpose: it swells by absorbing water in a humid environment and it is subject to plastic deformation. Silicon is a perfect material regarding its mechanical properties for this type of sensors. It should be noted that the construction in Ng's design is mechanically overdetermined. Compressive stress in the structure may lead to buckling with the result of complete different mechanical properties. Another disadvantage of this design is that the tethers are rather stiff, which results in a relatively poor force resolution. The deflection of the plate would be much larger if the flow were directed normal to the springs, but this deformation would make the piezoresistive detection in this way impossible (piezoresistors on top of the tethers would be an outcome, see Burger 1998).

An alternative transduction mechanism is to measure the change of a capacity due to the motion of the plate. This has been done by Pan (1995) in two variants, in collaboration with Analog Devices.

The sensitivity of the skin fiction sensors described so far is limited by the way the deflection of the plate is measured. Padmanabhan (1995, 1997) introduced an optical method for this purpose. They integrated photodiodes close to the edges (1995) and under openings of the plate (1997) to achieve a stress resolution of 0.0014 Pa, two orders of magnitude smaller that Pan's sensor. Dimensions of Padmanabhan's sensor is again in the 100 by 100 μm^2 range, with a resonant frequency of 100 kHz, sufficient for the characterisation turbulent flow. The shear stress resolution corresponds to an impressive resolution of 1.4×10^{-13} N.

The high resolution of the mechanical skin friction sensors are not reached by thermal skin friction sensors: Kälvesten (1996) reports a resolution in the range of 0.2 Pa, a Figure similar to Jiang's results (Jiang 1996).

5.4.2. "Dry fluid flow" sensors

Under the head of dry flow sensors we treat all types of flow sensors where the kinetic energy of the fluid plays a role, and viscosity not. A fluid flow is "dry" if the Reynolds number is large.

In ideal liquids along a streamline (in the absence of external fields)

$$\frac{p}{\rho}+\frac{1}{2}U^2 = \text{constant.} \tag{5.7}$$

The pressure difference at points along a streamline with different flow velocities relates directly to the square root of the flow speed. Equation (8.71) tells us that the pressure is maximal at points where $U = 0$, so the velocity can be determined by

$$U = \sqrt{\frac{2\Delta p}{\rho}} \tag{5.8}$$

at a point far away stagnation from the stagnation point. Δp is then the pressure difference between this point and the stagnation point. Similarly, the pressure drop Δp over an orifice varies at large Re with the flow rate in a quadratic way, known as Toricelli's law,

$$\Phi \propto A\sqrt{\frac{2\Delta P}{\rho}} \tag{5.9}$$

Φ is the volume flow and A the area of the orifice. Equation (5.8) is very similar to (5.9), the main difference being the proportionality factor. The expression (5.9) is just a consequence of the Bernoulli equation; a discussion can be found in Feynman (1965). Toricelli's law applies if the Reynolds number is large (here the characteristic length is the length of the channel forming the orifice) if the length of the orifice is much smaller than its cross section. The hydrodynamic boundary layer is then much thinner than the cross section, which means that viscous losses are confined to a very small section of the orifice, close to the wall, where there is the velocity gradient. The proportionality constant depends on the geometry of the orifice.

The interesting aspect in equations (5.8) and (5.9) is twofold: (i) The design of a flow sensor is reduced to the design of pressure sensors and a suitable geometry to find the points where to measure the pressure. (ii) The only temperature dependent quantity defining the relation of the flow rate and the pressure difference is the density. The density depends much weaker on the temperature than the viscosity. The main draw back of the sensors discussed in this section is their insensitivity to flow at small velocities.

In the following we describe a few micro sensor designs that rely on equations (5.8) and (5.9).

Schmidt et al. (1998) manufactured a plate attached to a flexible cantilever beam in an orifice. The plate reduces the cross-section of the opening in the orifice and a force on the plate results. The deformation of the cantilever beams was determined by piezoresistive elements. At very high Reynolds numbers the pressure drop across the plate is due to kinetic energy loss and eq. (8.73) applies. A schematic drawing of the plate with some qualitative stream lines is shown in Fig. 8.11. The dimensions of the plate was between 500× 500 and 700×700 μm, a flow speed in the range of m/s, leading to a pressure drop over the orifice of a few thousand Pa.

A similar principle has been used by Richter et al. (1999) who machined an orifice in the centre of a membrane, also with piezoresistive elements to detect its deflection. In fact, they simply used a commercial pressure sensor and drilled a hole through the membrane. The drilling can be done by laser machining, reactive ion etching or powder blasting. The membrane size was in the range of 1 - 2 mm^2, with a thickness of 20μm and holes had a diameter of 30 - 400 μm, with a conical shape due to the manufacturing process. Clearly, for these relatively thick membranes the hole must be rather large. For the small hole diameter, viscous losses become important. In Fig. 5.12 we show the design and results of measurements for the largest hole investigated by Richter.

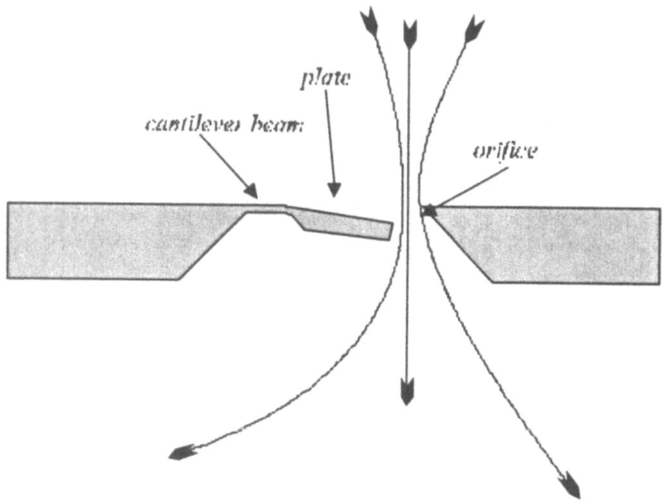

Fig. 5.11. A plate attached to a substrate by an elastic cantilever beam bends under a flow through an orifice due to kinetic pressure loss. Design by Schmidt (1998)

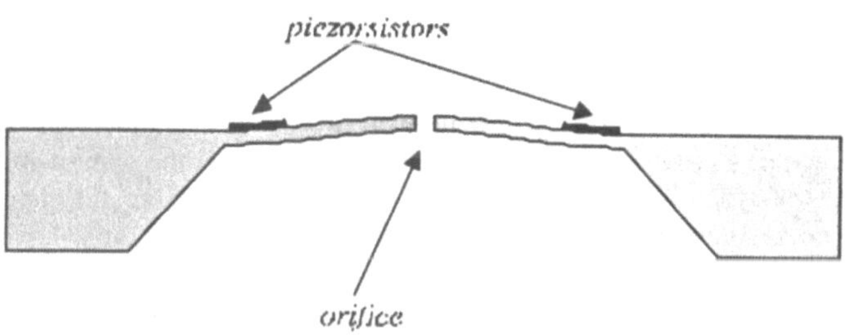

Fig. 5.12.a. Richter's design for a flow sensors based on eq. (8.71). The sensor is a commercial pressure sensors with a hole drilled through the membrane (Richter 1999)

If these sensors are used for fluid flow bubbles are of concern since they can clog the orifice. However, as shown by Richter (1999) the pressure needed to overcome the pressure associated with larger curvature is below the pressure range this design operates.

In Pandtl type flow meters, the pressure of the fluid is compared at a stagnation point where $U = 0$ and at a point far enough away from the stagnation point. According to Franke (1970) classical Prandtl flowmeters require a tube with double walls. The inner tube must have a diameter 1/3 of the outer one. The end opening of the inner tube is directed towards the flow, and there is a tiny opening (size: 1/10 of the outer tube diameter) at a distance 3 × outer tube diameter connecting the outer tube to the pressure at the opening. Any supports must be a distance at least 10 × the dimensions of the support away from the end. The inner and outer tubes are connected to barometers from which the pressure difference can be determined.

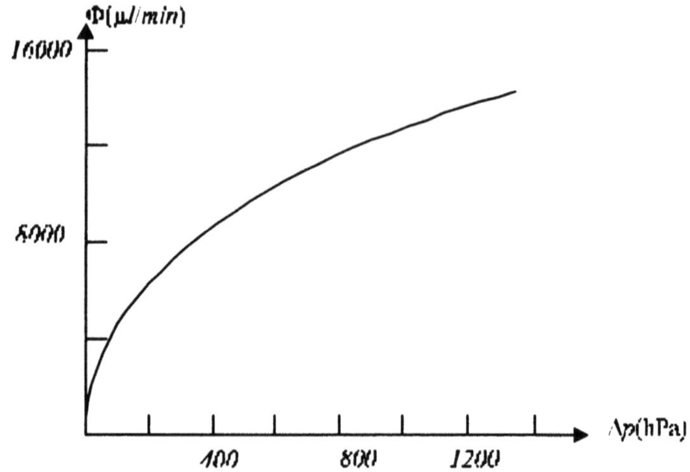

Fig. 5.12.b. Characteristic of Richter's sensor for a Honeywell 24PC (30psi) pressure sensor with an orifice of 160 μm.

Berberig suggested for a micro version of a Prandlt flow sensor the design shown in Fig. 8.13a. An opening of a capacitive pressure sensor connects the outside to the pressure sensor cavity. The sensor is arranged in the flow so that the opening is located at the stagnation point as shown in Fig. 8.13b, at the inside of the cavity the pressure will be higher than at the other side of the membrane, which experiences the dynamic pressure. If the deflection is not too large, the change in the capacity will be proportional to Δp, which in turn is proportional to the square of the velocity. Overall dimensions of Berberig's sensor is $5\times5\times1.4$ mm^3. The pressure sensor is designed to operate in linear range for air flow up to 75 m/s. Then a maximum deflection of 1 μm is reached (the gap spacing is 7 μm). Although there is some scatter in the experimental data the sensor functions according to the simple theory.

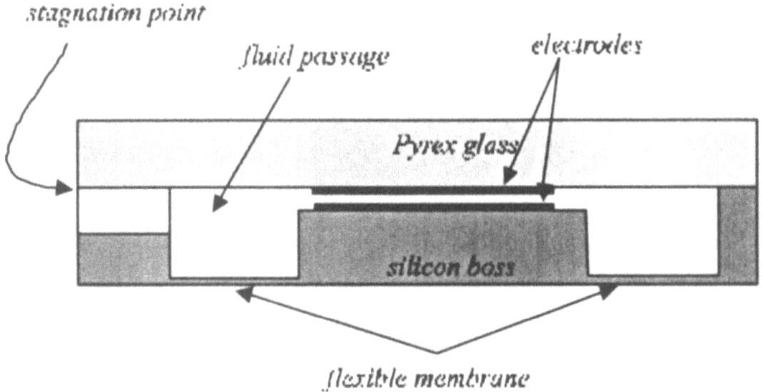

Fig. 5.13a. Design of a capacitive flow sensor using Prandtl's principle (redrawn from Berberig 1997)

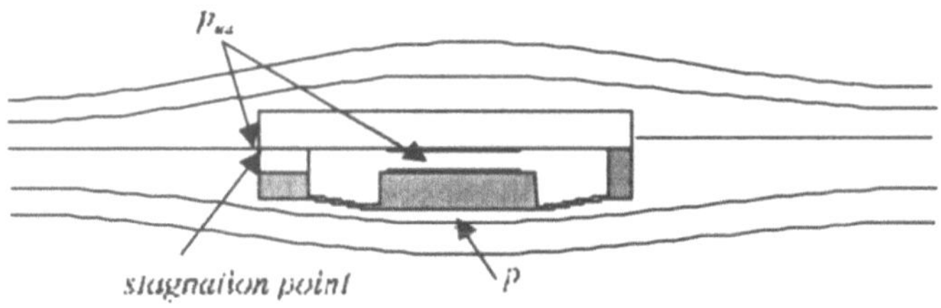

Fig. 5.13b Deflection of the Prandtl micro flow sensor when subject to a flow (Berberig 1997)

One must note that in this design the capacitor gap is connected to the medium in which the flow must be determined. So the measurement will be possibly adulterated due to corrosion and contamination. It is not necessary at all (see Oosterbroek, as discussed below) to connect the flow medium to the capacitor gap, but this requires a different design and probably a more involved fabrication process: The pressure sensor must be located inside the cavity which is connected to the stagnation point. We must also mention that when the medium is a liquid, one must expect problems with filling the device with the medium. Further, the medium has a dielectric constant (not important, of course, if the medium is a gas), which might be temperature dependent and composition dependent as in aqueous solutions and water.

"Wet fluid flow" sensors

If a viscous fluid flows through a duct there is pressure difference needed to keep the fluid flowing. Generally, the volume flow rate is proportional to the pressure difference, quite analogous to

Ohm's law for electric conductors[1]. For a cylindrical duct the relation (5) applies. The resistance is a function of geometry. In Gravesen (1993) an overview can be found on the flow resistance of ducts of various types of cross-sections. For microsensors a rectangular cross-section is the most important one. The flow resistance for a rectangular cross-section with side length a and b and a length l is given by (White 1994)

$$R = \frac{\eta l}{512} f_H{}^2 \frac{(a+b)^2}{a^3 b^3} \tag{5.10}$$

The numerical factor f_H varies from 56.91 for a rectangular cross section ($a = b$) to 96 for $b/a \rightarrow 0$.

In any case, the resistance depends on the viscosity. In liquids, the viscosity is strongly dependent on the temperature. This means that for sensors measuring the viscous pressure drop in ducts, the temperature must be measured, too, and the temperature dependence of the viscosity must be known for the temperature compensation. In gases, the temperature dependence of the viscosity is much smaller than in liquids.

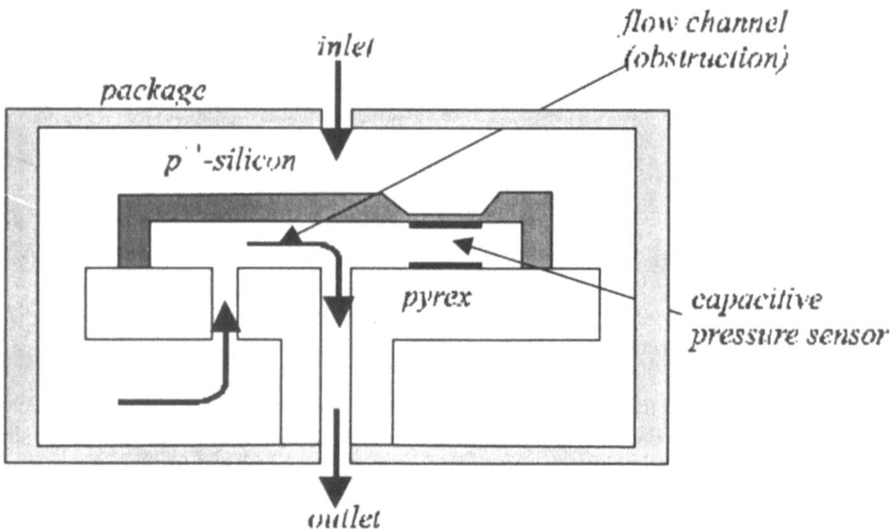

Fig. 5.14. Schematic of a flow sensor with a differential pressure sensor (Cho et a. 1992)

There are two basic designs for wet fluid flow sensors. One can measure the pressure difference between the ends of a duct, either by carrying out a differential pressure measurement using a single pressure sensor, or by two absolute pressure sensors. We begin with the first option. Here the two sides of the membrane of the pressure sensor is connected to the medium via an obstructing chan-

[1] Of course the analogy breaks down when the resistance is considered: the electric resistance is proportional to $1/A$, the hydraulic one to $1/A^2$. This is a result of the parabolic flow profile.

nel. A remark we made concerning Berberig's sensor applies also here when a capacitive pressure sensor is used: The medium has access to the gap of the capacitor, with possible problems.

Cho et al. (1992) use a capacitive pressure sensor made by the dissolved wafer process developed at the University of Michigan (see Chau 1988). This process involves shallow and deep doping of silicon by high boron concentration, anodic bonding of the doped side of the silicon wafer to a Pyrex wafer and dissolving the undoped silicon in KOH or EDP. The result is a silicon structure of a maximum height of 20 μm on top of a glass wafer. In their design for the flow sensor Cho et al. connected the cavity of the pressure sensor to the outlet of the duct and to a narrow micromachined duct over which the pressure falls. The outlet of the narrow duct is connected to the outer side of the membrane of the pressure sensor. A schematic drawing is given in Fig. 8.14. The actual channel (with a cross section of $60 \times 3.5\ \mu m^2$) for flow obstruction has been machined in the same process steps as the pressure sensor, the channel meanders over the Pyrex. The thickness and size of the diaphragm is 2,9 μm and $2 \times 2\ mm^2$, electrode separation 3.5 μm. Pressure resolution with such a pressure sensor is in the range of 10 Pa, and the flow resolution with the obstruction leads to 10^{-8} SCCM[2] with a full range of 1 mSCCM.

Flow sensors involving two pressure sensors (principle shown in Fig. 5.15) have been proposed by Boillat et al. (1995) and by Oosterbroek et al. (1997) for the measurement of liquid flow. Boillat used piezoresistive pressure sensors for flow sensors with measurement range from 1 to 300 μl/min with a pressure drop in the range of 50 hPa. A strong temperature dependence of the flow signal was observed, which turned out to be dominated by the temperature dependence of the viscosity of water. Oosterbroek machined capacitive pressure sensors (Fig. 5.16) in silicon sandwiched between Pyrex wafers. The device was designed for flow rates in the range up to 3 ml/min. Flow resolution in both designs is a few % full scale. Note a detail: In Oosterbroek's design, Pyrex has been micromachined, too. In his paper a full account is given for the models used for the design of the sensor (including membrane design for the pressure sensors, channel design for the pressure drop and considerations regarding the size of the pressure sensors to ensure that no effects of kinetic pressure affect the measurements (Oosterbroek 1999). Both designs avoid problems with contamination and corrosion of sensitive parts of the pressure sensors.

[2] SCCM = standard cubic centilitre/minute

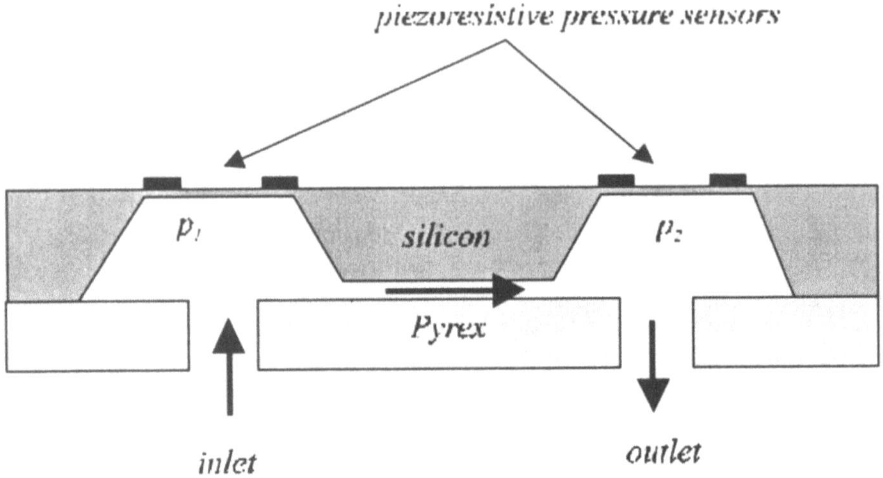

Fig. 5.15. Flow sensor based on two piezoresistive pressure sensors connected by a narrow channel (Boillat 1995)

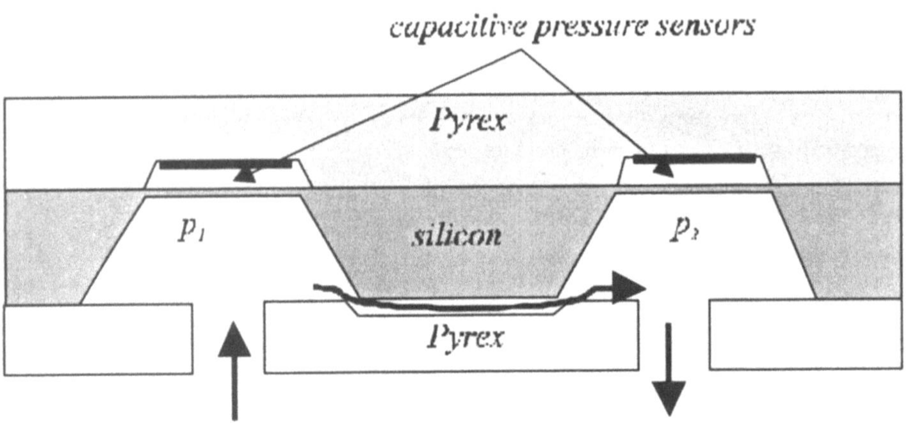

Fig. 5.16. Flow sensor based on two capacitive pressure sensors connected by a narrow channel (Oosterbroek 1997)

5.5. Electrostatic actuators

The technology of surface micromachining is now so advanced that devices fabricated by this technology are entering the market. This concerns mainly sensors for pressure, acceleration and, angular frequency, fabricated by large companies such as Bosch in Germany or Analog Devices. For a number of structures, in particular for the famous micromotors, no real application has been anticipated so far. The potential applications however are enormous. This is not restricted to surface micromachining alone but to the microtechnology as described in this book as a whole. A survey of possible applications has been described by Gabriel et al. (1989).

In the following we give a few examples of microstructures and micromachines realized with the aid of surface micromachining.

Combdrives

Combdrives are the most simple electrostatic actuators that can be fabricated by surface micromachining and related technologies. Their working relies on the fact that opposite charged capacitor plates not only attract each other but there is an additional force that drive the plates to overlap as far as possible. This is related to the energy of a charged capacitor with charge *Q*,

$$W = \frac{Q^2 d}{2\varepsilon b l}$$

with d the plate spacing, ε the permittivity, and *d*, *b* and *l* respectively the plate spacing, width and length. A force can be derived by differentiating with respect to *d*, *b* or *l*, and it is easily seen that the attractive force between the plates have a tangential component. The use of the normal component is quite limited. First of all, if this force is voltage controlled one gets an instability. Second the range of the deflection is very much limited, since appreciable forces at tolerable voltages require small gab spacing of the order of a few μm, so the normal deflection cannot exceed this Figure.

An interesting aspect is that the tangential force is independent on the overlap when the actuator is voltage controlled. To see this we perform a Legendre transformation of W, $W' = W - uQ$ (with u the voltage), and the force is then given by

$$F = -\left(\frac{\partial W'}{dl}\right)_u = \frac{u^2 \varepsilon b}{d}$$

For micromachined structures d and b (here actually the height of the capacitor plate) are of the same order, accordingly the force is of order $u^2\varepsilon = 10^{-7}$N for 100V. These actuators are weak.

In Fig. 5.17. and 5.18. we show an example for a combdrive actuator (Legtenberg 1996, 1996. These have been fabricated from a 5 μm thick LPCVD polysilicon layer grown at 590 °C, 250 mTorr and silane flow of 250 sccm. The deposition was done on a silicon wafer covered with 2 μm thick thermal oxide. The polysilicon has been doped with boron at 1100 °C for 3 hours, to obtain a sheet resistivity of 4.5.Ω/ . This procedure results in relatively small stress and stress gradient in the structure. During indiffusion a boron silicate glass (BSG) layer grows on top of the polysilicon which has been stripped in buffered HF. The etching of the BSG is much faster if the wafers are put into a wet oxidation furnace for half an hour at app. 1000 °C, as we have mentioned in Ch. II.8. On top of the doped poly silicon a 0.6 mm thick PECVD silicon oxide layer has been deposition and patterned. This layer served as an etchmask for reactive ion etching (RIE, see Ch. 9) to etch the structures in the poly silicon. The structures are released after standard cleaning for 50 min in BHF. For drying use has been made of a freeze drying procedure to prevent sticking. Finally, for backside contact a 1 mm thick aluminum layer has been evaporated.

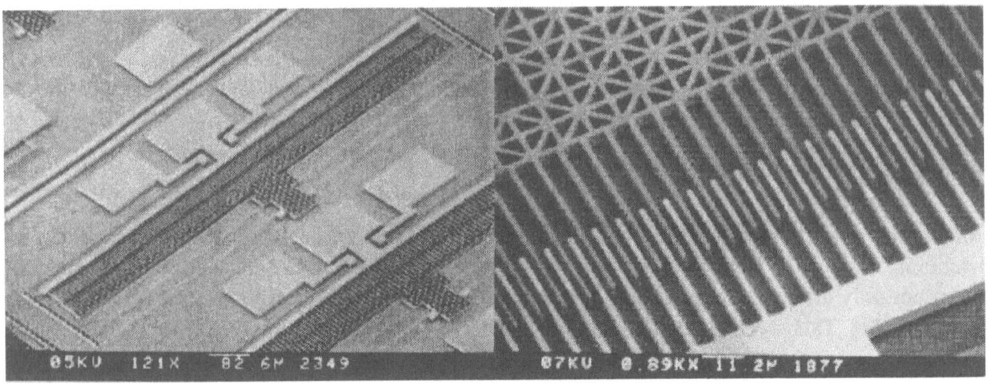

(a) (b)

Fig. 5.17. (a) Example of a combdrive actuator (R.Lechtenberg, (1996)); (b) Close up of the combdrive actuator. (R.Lechtenberg (1996))

The design of the mask is such that the anchors of the suspension beams holding the shuttle and the movable part of the combdrive (the so-called rotor) are wider than the parts that must be released. These parts are readily underetched while the anchors and the bond pads still are fixed to the substrate by the thermal oxide. This way the actuators can be fabricated in single mask process.

This has been described by Zhang et al (1991). A different design of an electrostatic drive, not based on comb actuators, has been proposed by Takeshima et al. from the University of Tokyo (1991). It is based on the deformation of tips of a parallelogram as illustrated in Fig. 5.10. One point of the parallelogram is fixed, two points are attached to capacitor plates that move if the capacitor is charged. The remaining point is displaced. The displacement at the last point is much greater than that of the capacitor plates. First prototypes demonstrated a displacement of 5 μm.

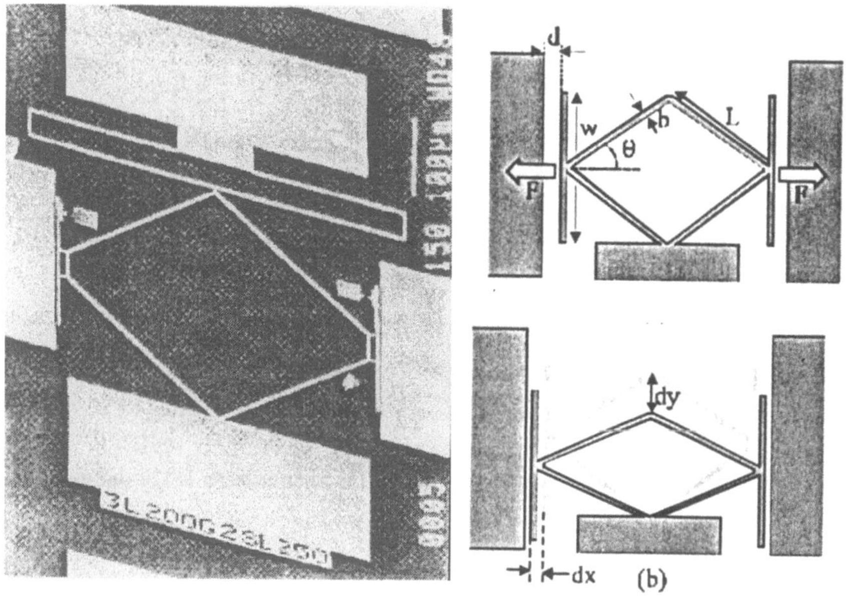

Fig. 5.18. Design of an electrostatic actuator (Takeshima 1991).

Rotating Micromotors

Although the micromotor is much farther away from any application that makes direct sense at this time, the result of the attempts to realize the first rotation micromotors were most spectacular. The groups first active in development of micromotors are at MIT, AT&T Bell labs and Berkeley. Later developments at the University of Wisconsin in Guckel's group resulted in a magnetic micromotor using the LIGA process (Guckel 1991). Also at Georgia Tech (Ahn 1991) a surface micromachined magnetic motor has been presented using polysilicon as the mechanical material. A different technology using electroplating has been described by a Japanese group (Furuhata 1993) In Europe, de Rooij's group at the University of Neuchâtel is currently developing an electrostatic micromotor. For more details we refer to a recent excellent review article (Mehregany 1991) on electrostatic micro motors.

The first polysilicon micromotors were quite small due to residual stress problems in the thin film, from which the films tend to take the shape of a potato chip. Larger motors can be fabricated by increasing the thickness of the structural layer.

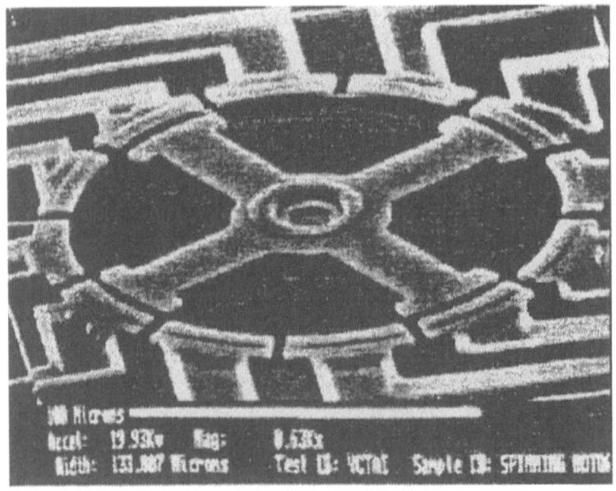

Fig. 5.19. Side drive polysilicon variable capacitance motor. The rotor diameter is 120 μm, air gaps are 2 μm wide (Duhler 1992).

Fig. 5.20. Top drive polysilicon variable capacitance motor. The rotor diameter is 300 μm, air gaps are 2 μm wide (Mehregany 1990).

Basically there are three types of variable capacitance micromotors. Side drive motors and top drive motors are shown in Figs. 5.18 (Tai 1990) and 5.17 (Mehregany 1990). Top-drive motors have the potential of producing a larger torque, but they are considerably less stable than side-drive motors due to large vertical forces. The torque in these motors is produced by the electric field between the activated stator poles and the rotor poles. The gap in top-drive motors is defined by the thickness of a PSG thin film, which is comparably easy. In side-drive motors the gap can be defined by photolithography, or by the lateral thickness of a PSG sacrificial layer. Since for successful operation the gap spacing must be below 2 μm, this type of motor puts high demands on the fabrication equipment and care of fabrication. In Fig.5.21 we show a typical process sequence.

The third type of variable capacitance motors is the harmonic side-drive motor ("wobble motor"). In this motor, the rotor roles ("wobbles") on the stator; the motor operation relies on normal electric forces attracting the rotor, which is electrically grounded. As the excitation of the poles rotate, the rotor will follow. Since the material for micromotors (polysilicon) is conductive, the rotor must not touch the excited stator poles. Therefore the design of micro wobble motors requires a bearing on which the rotor rolls. An example is shown in Fig.5.22.

The voltage needed to drive the motors lies in the range of 10-100 V. This voltage corresponds to an electrical field of more than 10^8 V/m, well in agreement with the expectations of the breakdown extrapolated from the Paschen relationship. The motive torque of the side drive motors is typically 10 pNm (10^{-11} Nm). This is five times smaller than the torque that is needed to move the micro manipulator that is discussed in paragraph 5.3.1.

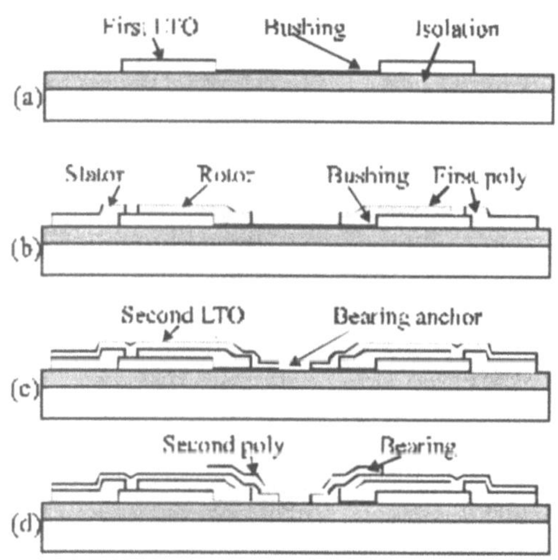

Fig. 5.21. Basic process sequence for a side drive motor: (a) after patterning the first PSG layer ("LTO", Low Temperature Oxide); (b) after patterning rotor and stator; (c) after patterning second sacrificial layer; (d) completed device after sacrificial layer etching (Meregany 1990).

Fig. 5.22. A typical flange baring wobble motor. The rotor diameter is 100 μm, air gap is 2.5 μm wide (Duhler 1992).

Typical revolution speeds are in the range of 10,000-20,000 rpm. The revolution times are so small that the rotors do not move smoothly, rather they tend to flip from one energy minimum position to the next. For wobble motors the maximum speed is ca. 700 rpm.

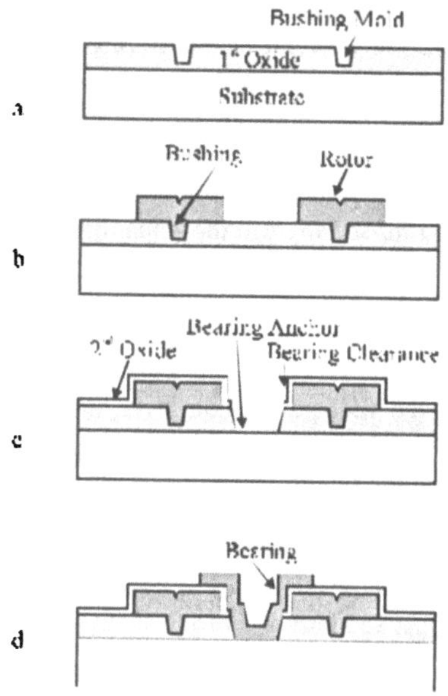

Fig.5.23. Center-pin process: (a) after the bushing mold is patterned; (b) after the rotor is patterned; (c) after the bearing anchor is patterned; (d) device before sacrificial layer etching.

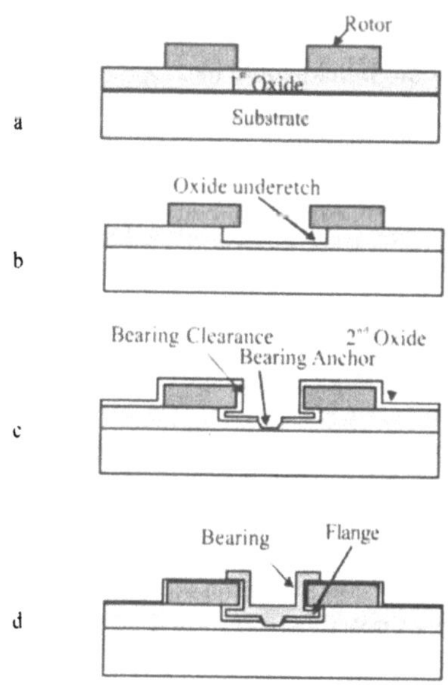

Fig. 5.24. Flange fabrication process: (a) after the rotor is patterned; (b) after isotropic oxide etching; (c) after the bearing anchor is patterned; (d) device before sacrificial layer etching (Mehregany 1991).

The performance and lifetime of the micromotors described so far vary widely. With the first motors just a few slow stepping motions could be achieved, at driving voltages of more than 100 V. One of the problems was a strong electrostatic attractive force between the rotor and the underlying substrate. This problem is now solved by changing the fabrication process, that now includes a conductive plate (polysilicon) on the substrate to shield the rotor from the substrate. The performance and lifetime have now been improved significantly. Motors rotating for many days have been fabricated, some have even rotated for so long that the actual lifetime is not yet known.

In order to reduce the friction between rotor and substrate the contact surface between rotor and substrate must be reduced as much as possible. Two fabrication schemes were proposed to achieve this: the so called "center-pin process" (Fig.5.23), in which the rotor slides on the bushings which are formed during deposition of the first polysilicon layer, and the "flange process" (Fig.5.24): a flange is formed by isotropic underetching of the oxide after deposition of the first polysilicon layer; during operation the rotor slides on the flange.

Wear limits the life time of micromotors and other sliding micromachines in at least two ways. The first obvious way is the wearing down of structures to dimensions where they fail to function. The other way may be more serious: dust and particles produced by wear can contaminate the structures, causing additional (perhaps fatal) friction and short circuits. Little is known about wear of microstructures.

Recently, Mehregany and Duhler demonstrated that polysilicon wobble motors can also be operated in demi water and in silicone oil (Tarow 1992).

A discussion on design rules for side drive motors can be found in (Tang 1990, Tarow 1991). The main issues are:

- the number of rotor poles to stator poles should be 2:3;
- there should be as many poles as possible;
- stator pole width over stator pole pitch should equal 0.6;
- the radius of the bearing should be as small as possible (the radius is limited by photolithography);
- the radius of the rotor should be as large as possible (limited by warpage);
- the gap spacing should be as small as possible, certainly below 2 μm;
- the bearing clearance must be small compared to the gap spacing;
- the distance between the rotor and the shield plate (i.e. the height of -the bushing in center-pin motors) should be as large as possible.

Generally, the mass of the rotor is so small that inertia forces can be completely ignored, since they are small compared to surface forces such as dry friction and sticking. This makes quite a difference with ordinary motors.

In order to reduce the dry friction in motors and other rotating devices such as the gear train shown in Fig.5.22, the possibility of levitation of the rotors was considered using electrostatic and fluid flow forces. The feasibility of these ideas are demonstrated in (Tang 1990, Kumar 1991, Huang 1991). However, combination of motors and the levitation technology has not been demonstrated yet.

In our own labs we succeeded in the fabrication of an electostratically driven axial gab wobble motor which is capable to drive micromechanical structures (Legtenberg 1996). The special aspect of this work is the anchoring of the axis. This is shown schematically in Fig. 5.25.

The process starts with the deposition of a 1 μm thick low stress silicon nitride layer. Low stress Si_xN_y is obtained if one grows this material under excess of silicon; too much silicon leads to a compressive stress. Legtenberg used in an LPCVD furnace 10 sccm $SiCl_2H_2$ and 18 sccm NH_3 at 850 °C and a pressure of 200 mTorr. Then the underelectrode material is deposited (polysilicon, heavily doped with boron). The process parameters are: Deposition: LPCVD at 590 °C, 50 sccm silane flow at 250 mTorr, doping: solid source drive in for 1 hour at 1150 °C.

After stripping the boron silcate glass in BHF a second low stress LPCVD nitride is deposited to isolate the rotor from the stator. Contact windows are opened by RIE using CHF_3/O_2 plasma. A first part of the sacrificial layer is deposited now (PECVD oxide). In a photolithographic step the ball bearing is defined by etching through the layer package (RIE using CHF_3/O_2), the photoresist is stripped in an oxygen plasma. At the location of the bearing the silicon surface is now open, and the ball bearing is etched in an SF_6/N_2 plasma by isotropic RIE. A second sacrificial layer is deposited by LPCVD from TEOS. This layer covers conformally the ball bearing.

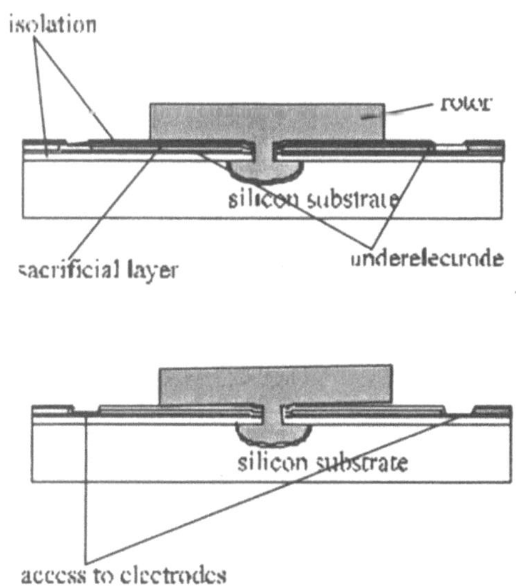

Fig. 5.25. Schematic of an axial gab wobble motor (redrawn from R. Legtenberg 1996). Above prior sacrificial layer etching, below after release and opening of the bond pads.

The next step is deposition of the structural layer and its patterning by photolithography. The structural layer consists of a heavily doped polysilicon layer, deposited and doped similarly to the electrode layer described above, except its thickness, which is 2 μm, and a 6 μm thick amorphous silicon layer sputtered in argon. This layer is annealed at 450 °C for one hour to reduce the residual stress in the amorphous layer. Patterning is done by anisotropic RIE in an SF_6/O_2 plasma.

The process is finished by sacrificial layer etching and opening the contact windows.

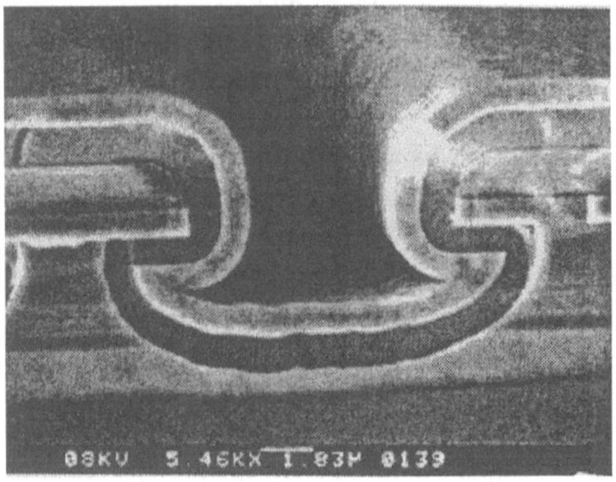

Fig. 5. 26. Cross section of a slider bearing, similar to the ball bearing used for the wobble micro motor (Legtenberg1996)

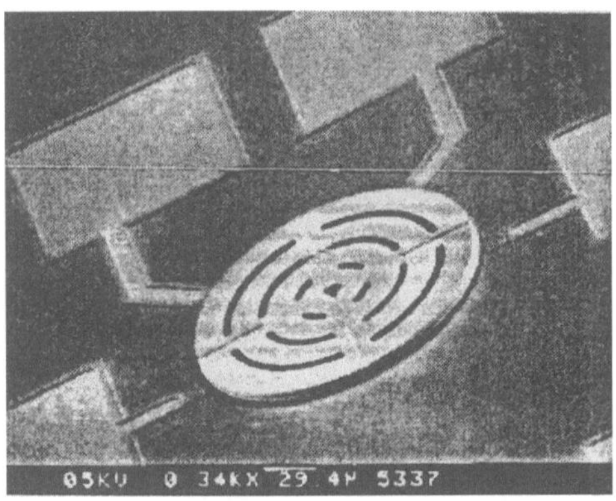

Fig. 5. 27. SEM image of an axial gab wobble motor, showing the rotor and bondpads to four stator electrodes. The rotor diameter is 200μm, and its thickness is 8 μm. (Legtenberg1996)

Results of the process are seen in Fig. 5.26 (ball bearing [3]) and 5.27 the top view of the device.

By introducing tooth to the rotor, other structures can be driven, such as gears or sliders. These structures have been fabricated in a process similar to the one that we described above, but it is more complicated because more masks are required. A few examples are shown in Fig. 5.28 through

Fig. 5. 28. A gear train with ball bearings and an axial gab wobble motor (Legtenberg1996)

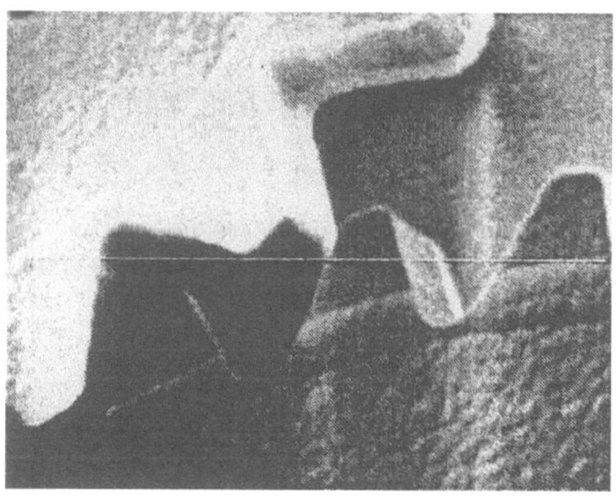

Fig. 5. 29. Close-up of the gear teeth (Legtenberg 1966a)

[3] Actually this is not the ball baring but a slider bearing - one needs very good luck to break the wafer right through a baring with a radius of 10 μm!

Fig. 5.30. An axial gab wobble motor driving a linear slider (Legtenberg 1996a)

Fig. 5. 31. Close-up of the slider from Fig. 5. 30, showing the bearing at the upper left of the image (Legtenberg 1996a)

There are types of electrostatic transducers that are capable of walking. We show two exa mples (Tas 2000, Akiyama 1995).

The principle of Tas' actuators is to move a part of a mechanical construction a relatively large distance and to transfer the deflection to a much smaller one. The deflection ration gives an amplification of the force. In Tas' microwalker a plate is connected to two electrodes, The is pulled down electrostatically and a lateral tension in the plate results. Having one of the electrodes attached (also by electrostatic forces) to the substrate, the other electrode moves. By claming the defelcted electrode and releasing the first one, the construction as a whole moves. This sequence is outlined in Fig. 5.32. Fig. 5.33 shows drwaings of the walker, and 5.34 an optical microscopic image. The transmission is non-linear. Maximum forces are in the range of a few mN.

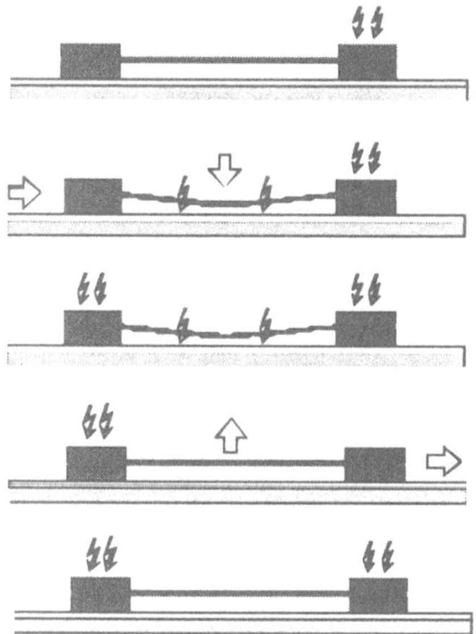

Fig. 5.32. Walking scheme of the microwalker (Tas 2000)

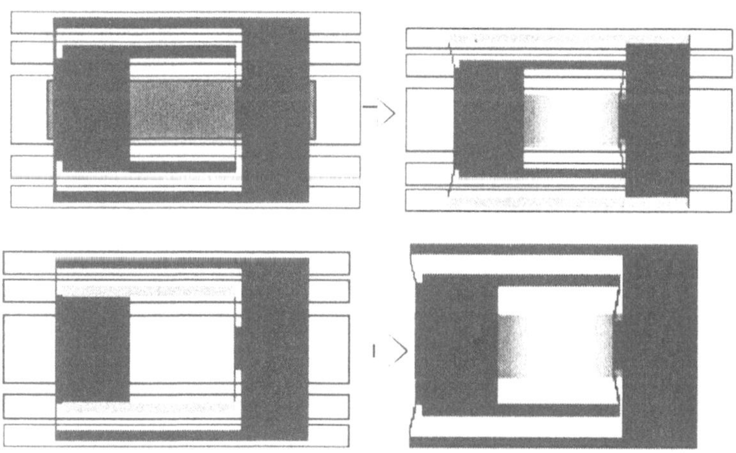

Fig. 5.33. Topview of a microwalker

Fig. 5.34. Optical microscopic image of a microwalker (Tas 2000)

Akiyama (1995) introduces the following idea. Similar to Tas' work he constructed a plate which is attracted by electrostatics forces to the substrate, a step keeps the plate from touching the nsubstrate completely, so the plate bends and pushes the step forward a very small distance (Fig. 5.35.). Again we have a mechanical transmission built in the actuator. An implementation of this principle in a rotating micromotor is shown in Fig. 5.36. Because use is made of friction forces in this actuator it has been called the scratch drive

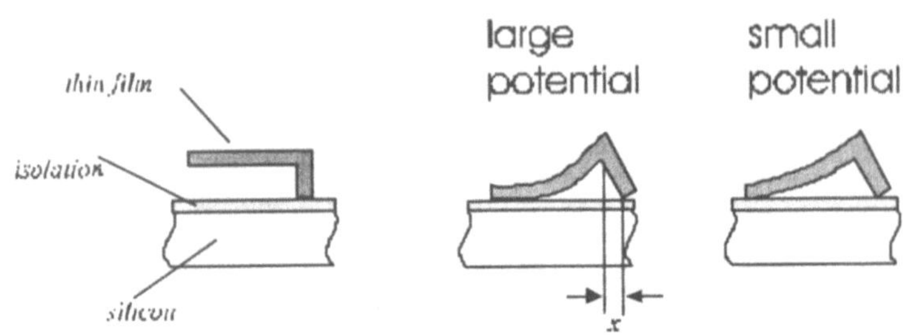

Fig. 5.35. Principle of a micromotor (Akiyama, J.MEMS 2 (1995) 104-110)

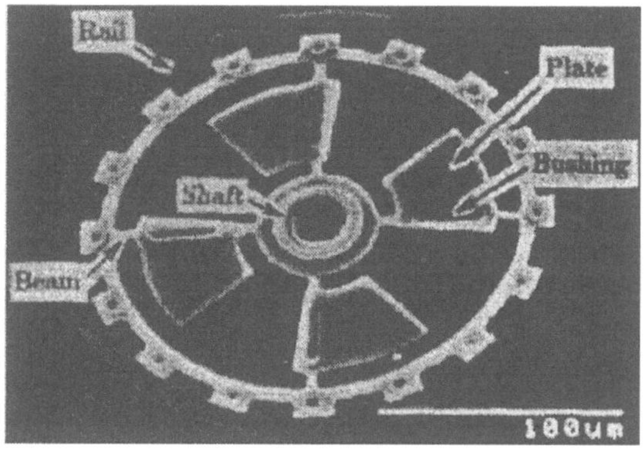

Fig.5.36. SEM image of a rotating micromotor using the scratch drive principle.

Acknowledgements

Contributions of Han Gardeniers, Meint de Boer, Gijs Krijnen, Remco Wiegerink, Erwin Berenschot and Theo Lammerink are gratefully acknowledged.

CHAPTER V

HEALTH MONITORING SYSTEMS

1. Structural health monitoring: concepts and procedures
2. Fiber optics sensors systems
3. Fiber optic elements
4. Strain measurement inside composite materials by fiber optic Bragg gratings
5. Fiber optic sensors for resin flow and composite cure monitoring

J. A. Guemes
Professor
ETSI Aeronauticos
Univ. Politecnica Madrid
28040 Madrid- SPAIN
Tel: 34 91 336 6327
fax: 34 91 336 6334
e-mail: aguemes@dmpa.upm.es

1. Structural Health Monitoring: Concepts and Procedures

A. J. Guemes

ETSI Aeronauticos, Univ. Politecnica Madrid, Madrid- SPAIN

Abstract Structural Health Monitoring is an emerging technology, dealing with the development of techniques and systems for the continuous monitoring, inspection and damage detection of structures, with minimum labor involvement. The ultimate goal of a SHMS is to eliminate current schedule-based inspections and replace it with condition-based maintenance. SHM was envisaged as one of the initial applications of the Smart Structures concepts, as sketched at figure 1 (Ref 1).

1.1 Introduction

A few vehicles incorporating these technologies have already been built, such as the Fast Patrol Boat hull structure, by the Norwegian Defense Research Establishment (Ref 2), or the reusable Space launcher, by NASA (Ref 3). Aircraft industry is actively working on it (Ref 4), but also civil engineering applications are fostering the developments, in both aspects of Load Monitoring (Ref 5) and Damage Assesment (Ref 6). This is a non-exhaustive list of pioneering works, a wider relation may be found at refs 7 to 10.

An essential aspect of SHM, in commonality with every Smart Structure, and which make the difference to conventional NDE, is that sensors and actuators of the system are built-in to the structure, their useful life must be as long as the structural life itself. To get significant results, a large number of sensors is required, working in a near continuous mode. The huge amount of data produced must be automatically processed by the system, and reduced to meaningful information to the user, such as overloading, initiation of damage, accumulated fatigue on the structure.

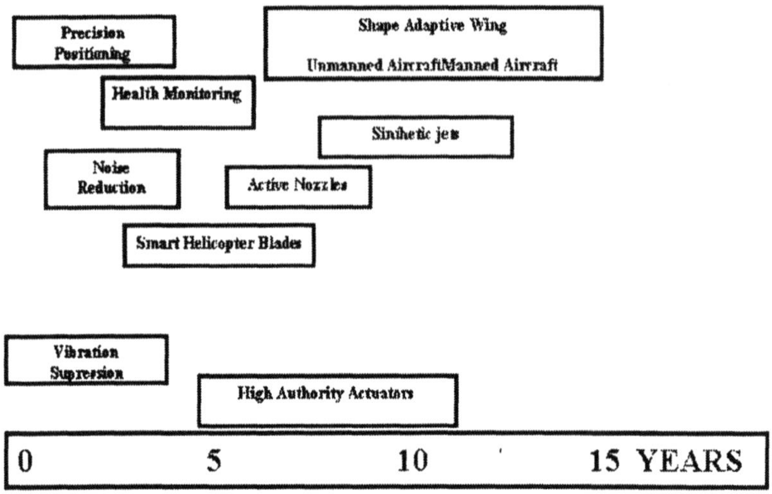

Figure 1. Estimated timeline for insertion of smart structures technologies in real systems

Also, as pointed out by F.K. Chang (Ref 8), traditional NDE tends to use direct measurements to determine the physical condition of the structures, the accuracy of the diagnosis rely on the equipment resolution. The SHM techniques would use the change in the measurements at the same location at two different times, to identify the condition of the structure, then the history data is crucial for the technique. The SHM is dependent only upon the sensors and interpretation software, hence miniaturization is potentially feasible.

Two complementary approaches are followed to attain the global SHM objectives: Structural Usage Monitoring (SUM), and Damage Assesment (DAMAS), their technologies are somewhat different, and are introduced in the next paragraphs.

1.2 Structural Usage Monitoring

Structural Usage monitoring is the continuous survey of strains at different critical points of the structure during its whole life. From these data, occurrence of overloads may be identified and avoided, by warning the operator of any excess. Pioneering works may be traced back to 1989, with the suggesting article: "Smart Structures with nerves of glass", by R. Measures (Ref. 11). The ISIS program, in Canada, merging the efforts of several Universities, industries and Research Centers, have demonstrated the feasibility for long term monitoring of bridges and other civil infrastructures.

Within the aerospace industry, many aircraft components are retired based on flying hours, independently of the flying conditions; this is the so-called safe-life methodology, the flying hours limit calculated from estimated usage spectrum, the S-N material data, and a scattering factor to remain inside the conservative prediction. This approach lead to a premature retirement of many components, meaning higher spare and maintenance labor costs. Modern combat aircrafts incorporates a system to calculate the stresses at critical parts from the flight data parameters, like the g factor, air speed and many others, allowing a more efficient fatigue management of the fleet (Ref 12). The overall procedure is sketched in figure 2.

Figure 2. SHM system for aircraft structures

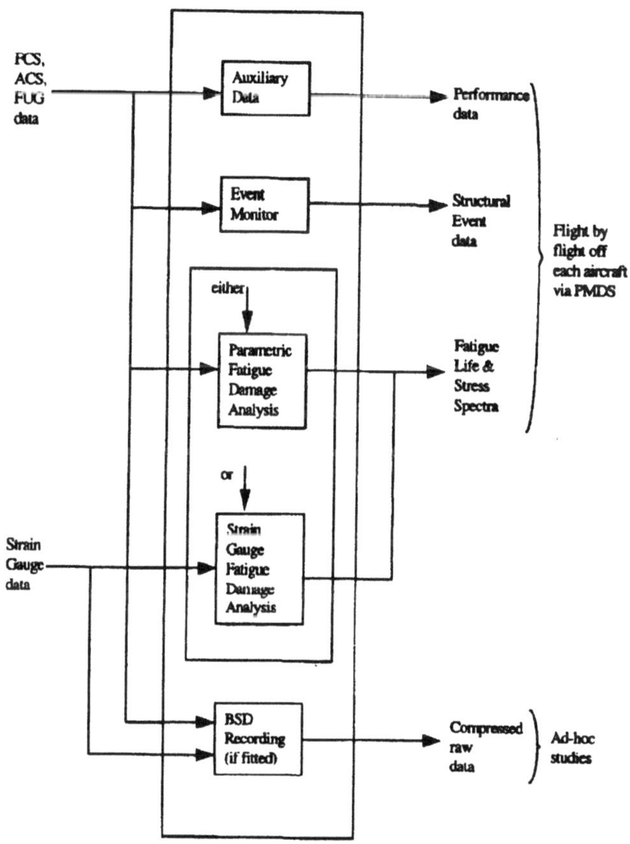

A further step in the same direction will be to embed a fiber optic sensor network into the structure, so the strain field may be accurately mapped. Fiber optic sensors overcome the limitations of currently used strain gages: FOS are insensitive to EM interference and zero drifting, they can be easily multiplexed, so a large number of strain sensors can be installed on the structure with very little hardware overhead. The strain history could be used to replace parts based upon actual fatigue. Maintenance work would be reduced to replace parts only when necessary, periodic inspections would be substituted by on-condition maintenance, diminishing the downtimes of the vehicle and improving the life-cycle costs.

The safety of the vehicle would not be impaired, the continuous monitoring has the potential to identify any excessive usage of the vehicle beyond the predictions, causing a premature aging and a life risk.

The sensor technology is currently well established and demonstrated, as will be discussed in subsequent lessons. The current limiting aspects before full industrial implementation are the development of data processing algorithms able to handle in real time the flow of raw data, and the demonstration of the long term reliability of the global system. A good review of main issues and alternatives is given at Ref 4.

1.3 Damage Assessment

Deals with the detection and identification of incipient damage in structures. Cracks will change the vibration response of the structure, and also the way elastic waves propagate through it. Correlating these changes, which can be measured experimentally, with the damage occurrence, and locating and quantifying the damage, is a really challenging problem, but significant progress have been achieved during last years.

Several methodologies are envisaged as potentially useful, their applicability to a particular structure being related to the size of the critical crack to be detected, and to the excitation source of the structure. For rotating machinery, and for civil engineering applications, the vibration-based methods are the best adapted.

Detectable crack size by modal methods is too large to be acceptable for aircraft and other thin shell structures, and other methods are sought. Some promising procedures are Lamb waves perturbation, acoustic emission, and electromechanical impedance. A brief review will be given next, a detailed description of recent advances may be found at Ref. 9 and 13.

1.4 Vibration-based methods.

Rotating machinery is naturally excited under operation at the rotating speed, the presence and growth of peaks at certain frequencies, such as multiples of shaft speed, can be associated to a certain kind of damage at the bearings. Detailed charts are available to identify the damage type, based on vibration signature analysis (Ref 14). But in general structures, the excitation is never so clean, the damage site is ordinarily unknown, and the method is not so straightforward.

For large civil structures, a certain number of sensors like accelerometers, adequately distributed through the structure, will capture the dynamic response of the structure to natural or artificial excitation. The starting hypothesis of the method is that any damage such as fatigue cracks, delamination etc. causes local changes of stiffness. Corresponding to the changed stiffness distribution, a shift of the dynamic characteristics like eigenfrequencies and mode shapes, or FRF's occur. These changes can be used to diagnose the locations and the extents of damaged areas by means of the model solving an inverse problem.

However, the application to complex structures will produce large equation systems with nany unknowns. Besides, in case of small deviations between the reference model and fresh data observed, it is extremely difficult to statistically decide whether the difference is due to damage or due to measurement and modeling precision, changing excitation, and so on. Also, the modal identification process typically requires user interaction and some engineering skill.

Its solution turns out to be the key point. Before and during the solution procedure, it is necessary to reduce the whole parameter set and to select those parameters that are relevant to the problem. The final set of parameters should concentrate on the damaged regions of the structure.

The application of the damage identification to the structure is carried out in the following steps:

1. Updating of the original FE-model by means of the data of the undamaged structure in order to obtain a good reference model

2. Damage localization: Formulation and solution of the system of equations using regularization methods and a parameter selection strategy to find the multiple locations of damage.

3. Calculation of the extent of damage: Only for the localized areas are considered, fitting the relevant parameters to the experimental data of the damaged structure using nonlinear optimization.

Recent works done in the frame of the COST F3 Action (Ref 15), show that the damage can be localized and quantified. Two benchmarks were extensively analyzed: a two-story steel structure, damaged by a simulated earthquake (Ref 16), and the Z24 bridge on a highway in Switzerland, under different levels of induced damage, by settlement of a pier from 20 to 90 mm. (Ref 17).

1.5 The electro-mechanical (E/M) impedance method.

It is an effective technique for structural health monitoring, damage detection, and failure prevention near high risk areas, like joints, repairing patches, etc.. The method utilises small-size active sensors intimately bonded to an existing structure, or embedded into a new composite construction. Piezoelectric (PZT) wafer transducers have been widely used to this purpose. Experimental demonstrations have shown that the high-frequency impedance spectrum is directly affected by the presence of damage or defects in the monitored structure. The electro-mechanical model of sensor-structure interaction based on complex frequency-dependant dynamic stiffness is sketched in Figure 3.

The method has been applied for checking the integrity of CFRP patches for repairing civil infrastructure, for detecting skin disbonds in helicopter rotor blades, and for damage analysis of laminates. An extensive review of developments and results is given at Ref. 18

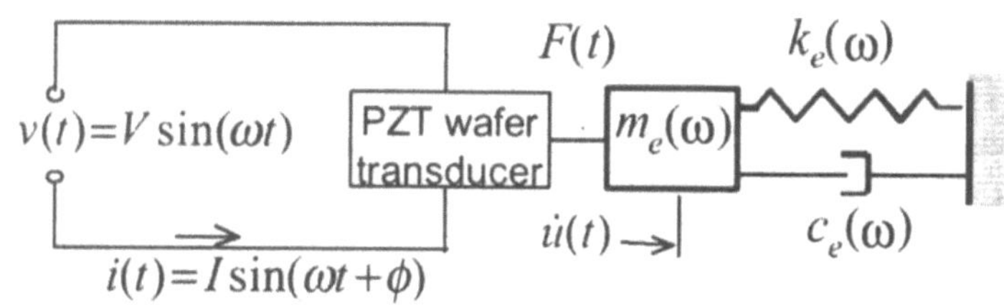

Figure 3: Electro-mechanical interaction between the PZT transducer and the structure through the point-wise structural impedance.

1.6 Wave Propagation models for Structural Damage Identification

Modal based techniques, or vibration analysis, looks for phenomena with a characteristic time (period of the vibrations modes) much longer than the time used by the sound to travel trough the structure. the waves are stationary. The detectability of the procedure is limited to how a local defect may change the global response of the structure.

At much shorter time scale, some authors have investigated on how to produce a burst at a point of the structure and how to detect it at another position. This transient signal may be modified by the presence of an obstacle at or near the path between the actuator and the sensor. This is the essence of the ASDS (Active Sensing

Diagnosis System), figure 4, which is today one of the most active lines of research, potentially useful for damage detection in plate and shell structures, such as oil storage tanks, or aircraft skins. The extreme sensitivity of piezoelectric wafers to dynamic strains, and its capability to work as sensors or actuators, make feasible to integrate a network of PZT's as a layer inside of the laminate for interrogation purposes (Ref. 19).

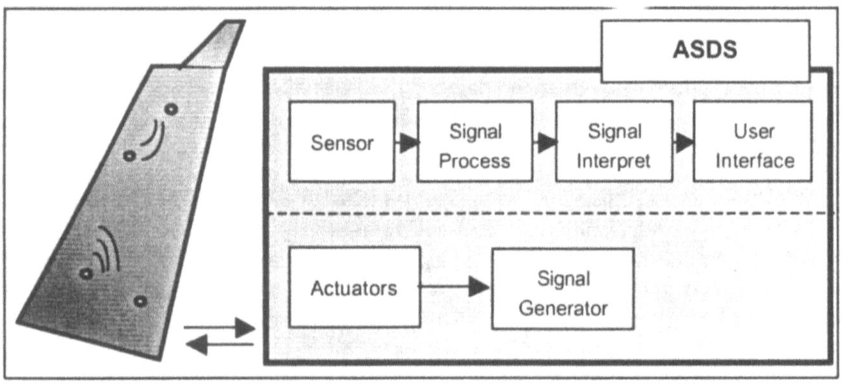

Figure 4: Schematic of active sensing diagnostic system.

These elastic pulses are known as Lamb waves, spreading out from the source with a group velocity dependent on the thickness and stiffness of the material, and on the frequency employed. Different modes are possible, and unless a careful selection of the source parameters is done, the output is so messy that nothing can be identified (Ref 20). Even if a single mode is selected, Lamb waves are always disperse, its energy spread out in both space and time as the signal propagates away from the source. Dispersion is negative from an inspection point of view, because it limit the distance between emitter and receiver, and because the widening of the pulse diminish the resolution of the system. A systematic procedure for the mode selection and transducer design, for a better sensitivity, excitability, detect ability and selectivity, may be found at Ref 20.

Figure 6 shows the intensity-time of the elastic wave at the actuator, and this wave passing through one of the sensor, delayed because of the flight time, widened because of the dispersion, and with and additional scattering, due to reflection at some defect nearby.

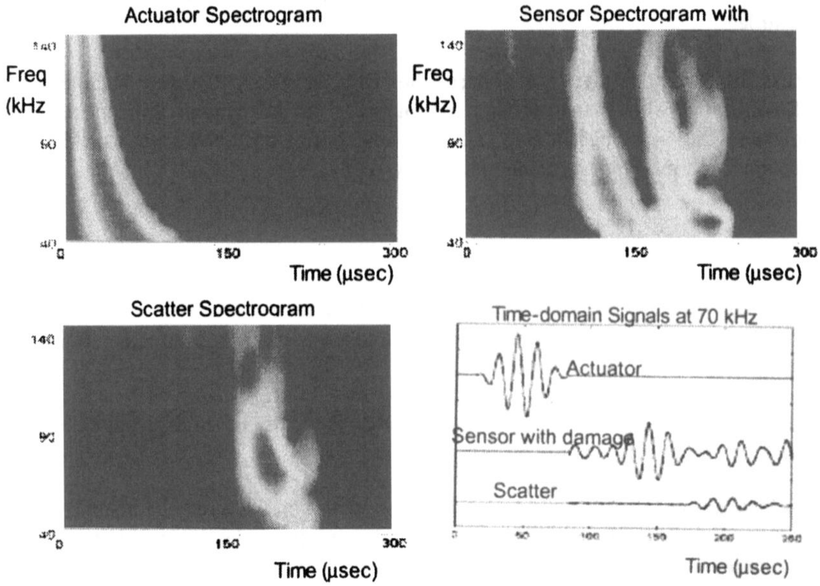

Figure 5: Signal spectrograms and time histories at 70 kHz.

The analysis is better done in the time-frequency domain. By comparing accumulated records, the occurrence of damage may be assessed.

1.7 Acoustic emission

Acoustic emission testing is a technique that "listens" to structures under stress. A new crack or flaw in a stressed structure will emit sound waves due to the rapid release of energy. Sources of acoustic emission include fracture and plastic deformation, impacts, or delaminating and fiber breaks in composite materials. These waves propagate through the structures and can be collected through a network of piezoelectric sensors. The sensor response and front-end filters remove frequencies below about 100 kHz, which includes most audible noise. The result is that acoustic emission can be used to monitor a structure for active damage even when ambient noise levels are extremely high. By comparing the arrival time of a signal at different sensors within the sensor matrix, the actual location of the flaw site can be defined.

AE appeared in 1950's, but up to recently it was not easy to apply this technology outside the laboratory, mainly because data storage and processing techniques were not available. Conventionally, signal processing focused on a time domain approach using the count and count rate, which are the number of times per second when the amplitude exceeds a threshold. However, the number of AE counts is depending not only on fracture mechanisms, but also on events such as the friction between fracture surfaces, causing a great variability on the results. Some correlation was found between the frequency of the signal and the specific failure process, such as delaminating, fiber breakage, which allows a better knowledge of the damage, if broadband sensors were used. It was only with advanced processing techniques, such as the wavelet transforms, or Artificial

Intelligence algorithms, such as Neural Networks, when a quantitative location and description of damage have been feasible.

AE has been successfully demonstrated on board the DC-XA Delta Clipper Technology Demonstrator as a Structural Health Management experiment. The AE system utilized was a commercially available unit that was modified for autonomous control and re-designated **AEFIS**, which stands for Acoustic Emission Flight Instrumentation System.

1.8 Conclusions

Structural Health and Usage Monitoring aims to develop a sensor matrix and a data processing system, integrated into the structure, able to detect incipient damage and to monitor the strain at different critical locations.

Operative systems would improve the safety and maintenance costs of high responsibility structures, for many different engineering applications.

Sensor technology for local strain monitoring is now available and proven. FOS may be bonded onto the metallic surfaces, or embedded into composite or concrete, structures, lasting as long as the structure, without recalibration requirements. It has been demonstrated on bridges, aircraft and ship hull structures. Information processing is a main issue for these applications, data need to be drastically compressed to the useful information to the user. Benefits from implementing this technology are a better understanding of the actual loads experience in service by the structure. When fatigue is a design driver, a direct recording of the strain history allows a more efficient maintenance protocol.

Concerning damage detection and assessment, there are several different procedures according the kind of structure to be inspected. Vibration-based methods concentrate most of the work done up to now, applied to civil engineering structures, but also to other application, like the Truss Structure for the Space Station.

Critical cracks in aircraft structures are much shorter than what would influence the global vibration modes. Sensor arrays, mainly of piezoelectric wafers, can be embedded into the structure, recording transient the elastic waves produced by opening cracks or by cross talk between piezoelectrics. Main issue is again to find the suitable algorithms to extract useful information from the high flow of raw data collected.

References

1. C.R. Crowe, J.M. Sater. *'Smart Aircraft Structures'* AGARD CP-600, Vol 1 (1997)

2. G. Wang, K. Pran *'Ship hull structure monitoring using Fiber Optic Sensors'*, included at Ref 9. (Jun 2000)

3. L. Melvin, et al., *'Integrated Vehicle Health Monitoring for Aerospace Vehicles'*, included at Ref 7 (Sept 1997)

4. S. Rizcalla, B. Benmokrane, G. Tadros, *'Structural Health Monitoring of Bridges with Fiber Optic Sensors'*. at Ref 9

5. R. Pascual, '*Model based Structural Damage Assesement Using Vibration Measurements*',Doctoral Thesis, Univ. Liege (1999)

6. F.K. Chang, ed., '*Structural Health Monitoring. Current Status and Perspectives*' Proceedings of the Inttnal. Workshop on SHM, Stanford, 1997. Published by Tecnomic Publ. Co.

7. F.K. Chang, ed., '*Structural Health Monitoring.* ' Proceedings of the Inttnal. Workshop on SHM, Stanford, 1999. Published by Tecnomic Publ. Co.

8. J.A. Güemes, ed., '*European COST Conference on System Identification & Structural Health Monitoring*', Proceedings of the Conference, Madrid, 2000.

9. M.D. Gilchist, J.M. Dulieu-Barton, K. Worden, ed. , '*Damage Assesment of Structures*', Proc. Intnal Conf. DAMAS 99. Dublin 1999. Trans Tech Publications Inc.

10. R. Measures, '*Smart Structures with nerves of glass*', Progress in Aerospace Sciences, vol. 26, 289-351 (1989)

11. S.R. Hunt, I.G. Hebden, '*Validation of the Eurofighter Typhoon Structural Health and Usage Monitoring System*', at Ref. 9

12. S.W. Doebling, C.R. Farrar, M.B. Prime, D.W. Shevitz, '*Damage Identification and Health Monitoring of Structural and mechanical Systems from changes in their Vibration haracteristics: A literature Review*', Los Alamos National Laboratory Report LA-13070-MS (1996)

13. C.R.Farrar, T.A.Duffey, '*Vibration-based Damage Detection in rotating Machinery*', at ref. 10

14. C.P. Fritzen, K. Bohle, , '*Damage Detection in Structures with multiple cracks using computational models*', at Ref.9

15. E. Görl, M. Link, '*Identification of Damage parameters of a Full Scale Steel Structure damaged by seismic Loading*', Ref 9

16. J. Maeck, B. Peeters, G. De Roeck '*Damage Identification on the Z24 Bridge using Vibration Monitoring Analysis*', at Ref. 9

17. V. Giurgiutiu, C. Rogers, '*Recent Advancements in the Electo-Mechanical Impedance Method for Structural Health Monitoring and NDE*', SPIE Conf. on Smart Structures, Vol 3229. San diego, CA (1998)

18. C.S. Wang, F.K. Chang, '*Built-in Diagnostics for Impact Damage Identification of Composite Structures*', at Ref. 8

19. P.D. Wilcox, R.P. Dalton, M. Lowe, P. Cawley, '*Mode and transducer Selection for Long range Lamb Wave inspection*', at Ref. 10

20. R.D. Finlayson, et al., '*Health Monitoring of Structures with acoustic emission and acousto-ultrasonics*', at Ref. 9

2. Fiber Optics Sensors Systems

A. J. Guemes

ETSI Aeronauticos, Univ. Politecnica Madrid, Madrid- SPAIN

Abstract Fiber optic sensors are reliable devices for long tenn strain monitoring, working similarly to conventional strain gages, overcoming most of its limitations. Different kind of optical sensors are discussed in this lecture, with the optoelectronic systems for interrogating these sensors.

2.1 Introduction

Fiber optic technology started in 1970's, for long distance telecommunications, experiencing an exponential growth during the last decade. Sensing applications are an small spin-off from this technology, taking profit of developments done in optoelectronic components and concepts.

Our discussion will be mainly restricted to strain and temperature measurements, even fiber optic sensor may be found for many other applications. A global view may be found at Ref. 1 and 2. As a wide definition, a sensor is an artifact able to transform a certain physical or chemical magnitude into readable information. The optical fiber itself is a pipe for light, transmitting the information. It may also be sensitive to changes in the external environment surrounding the fiber, like temperature. In fact, the backscattered radiation at an O.F. is proportional to its temperature; by means of an equipment called OTDR it is possible to measure the light retrodispersed along the fiber length and consequently the temperature distribution. The fiber would act simultaneously as a distributed sensor and as the vehicle for information, all along its length. This seems to be the ideal approach, but accuracy and spatial resolution are not good enough for many applications. It is more common the idea of local sensor, either by including into the optical path an external device (extrinsic sensor), or by modifying the fiber during a short length (about 1 cm), to make it sensitive to the strain or any other parameter. These are the so-called instrinsic sensors, which would go undetectable to the naked eye.

For Smart Structures, local intrinsic sensors are the most appreciated, because of simplicity, minimal perturbation to the host material and accuracy of reading. Multiplexing capability, or the possibility to 'write' several sensors in the same optical fiber, is another highly desired feature. When compared to the 4 leads required by each electrical strain gage, a clear advantage is found.

2.2 Classification of Fiber Optic Sensors

The external action to be monitored must change at the sensor some of the magnitudes that define the optical wave: intensity, phase, wavelength, and polarization. Based on these changes, different kind of optical sensors have been developed.

Intensity-based sensors

These are the simplest devices among the FOS, and consequently they were among the first attempts, and still are in use as proximity sensors, and for damage detection. A stable light source, white or monocromatic, an optical fiber, preferably multimode for higher light power transmission and a sensitive photodetector compose the whole system.

Commercially available proximity sensors (Ref 3) are used for non-contact monitoring of the displacement of rotating shafts with accuracy better than 3 microns. They work under the principle of correlating the distance between the cleaved fiber optic end and the moving surface with the amount of light reflected by this surface, and captured again by the optical fiber (Figure 1)

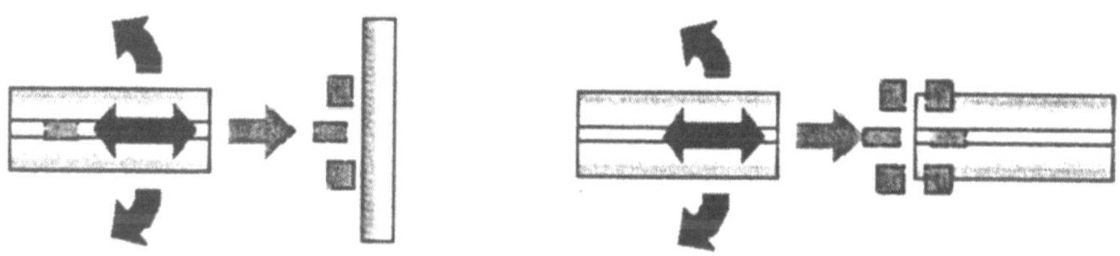

Fig. 4a Fig. 4b

Figure 1 Fiber optic proximity sensors

Microbending sensors were popular intrinsic sensors for pressure monitoring, they were based on intensity measurements (Figure 2). It is known that light is guided inside the optical fiber because its incidence angle is higher the critical reflection angle at the core-cladding interface, that is always true except for very sharp bends. If the O.F. is placed between two rough surfaces, a higher pressure promotes higher bending, and consequently higher optical losses. If adequately calibrated, they would afford a measurement of the transversal load on the fiber, by using a local plate, or even the natural roughness of the composite fabrics. Early works were done with this concept (Ref 4-6), which is currently abandoned; fluctuations of optical power at the light source, connectors, temperature, etc., make the system rather unreliable. Still the microbending losses must be keep in mind when embedding optical fiber into composite fabrics, because they can fade out the optical signal.

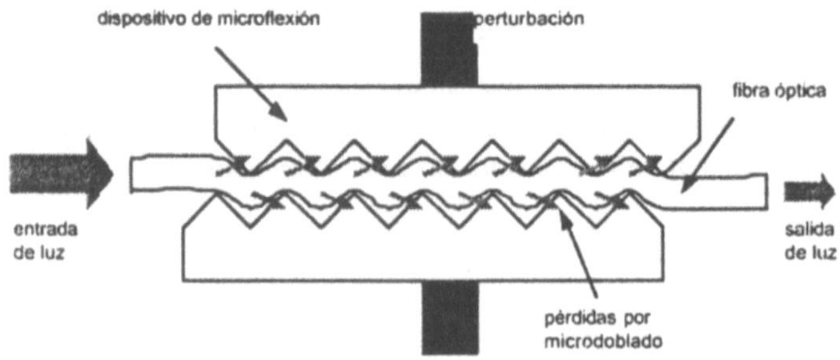

Figure 2 Microbending sensor

If a crack appears in the structure, which had a continuous optical fiber embedded into the material, the crack will eventually cut the optical fiber, which would be detected without incertitude. This was the approach at some papers dealing with damage detection, but the concept seems to lack robustness for practical uses in engineering.

Phase modulated optical fiber sensors, or Interferometers.

Interferometry is the most accurate laboratory technique for precise distance measurements. If a continuous monocromatic light wave is divided in two beams(by a partial mirror when dealing with conventional optics, or by a coupler in the case of optical fibers), and then these waves travels through different paths before being recombined, any slight diference of path-lengths will cause a delay from one wave to the other, and consequently their electromagnetic fields would not summed up, even they may counteract, and the final result is that output intensity may even cancel. It must be bear in mind that there is not display where the optical wave versus time can be shown, as with conventional low frequency electrical signals; only its average power can be detected.

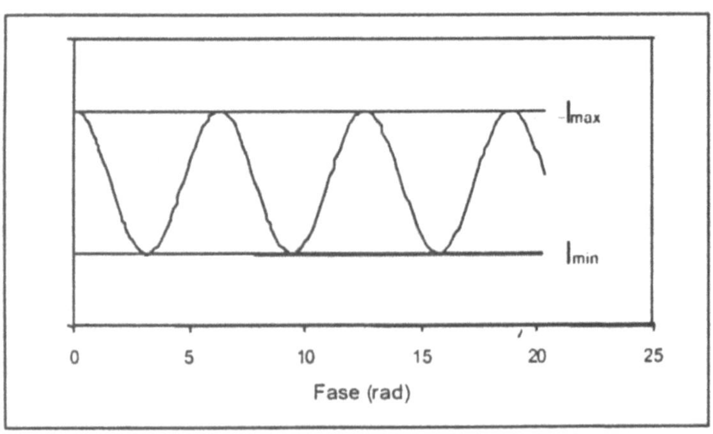

Figure 3. Power output from interferometric merasurements

Interferometers output will move from the power input level to zero level, as sketched in figure 3, each time the length of any of the two paths increment or decrement in half wavelength (380 nanometers, if the red light of a He-Ne laser was used). With an equipment adequately tuned, and if the experiment is carefully carried out, changes in length of the order of 10 nanometers (the size of an oxigen atom!!) can be detected. This extreme sensitivity to any perturbation in the optical path of any of the two arms makes measurements difficult outside the laboratory environment, a local temperature change may cause several maxima and minima drifting. The sensor is the whole length of the optical path of one arm, it is an 'averaging distributed sensor'. The measurements are relative to the length when the equipment is put ON, any interruption will lose the reference; there is no direct indication if increment or decrement is happening, a quadrature technique need to be used to identify it.

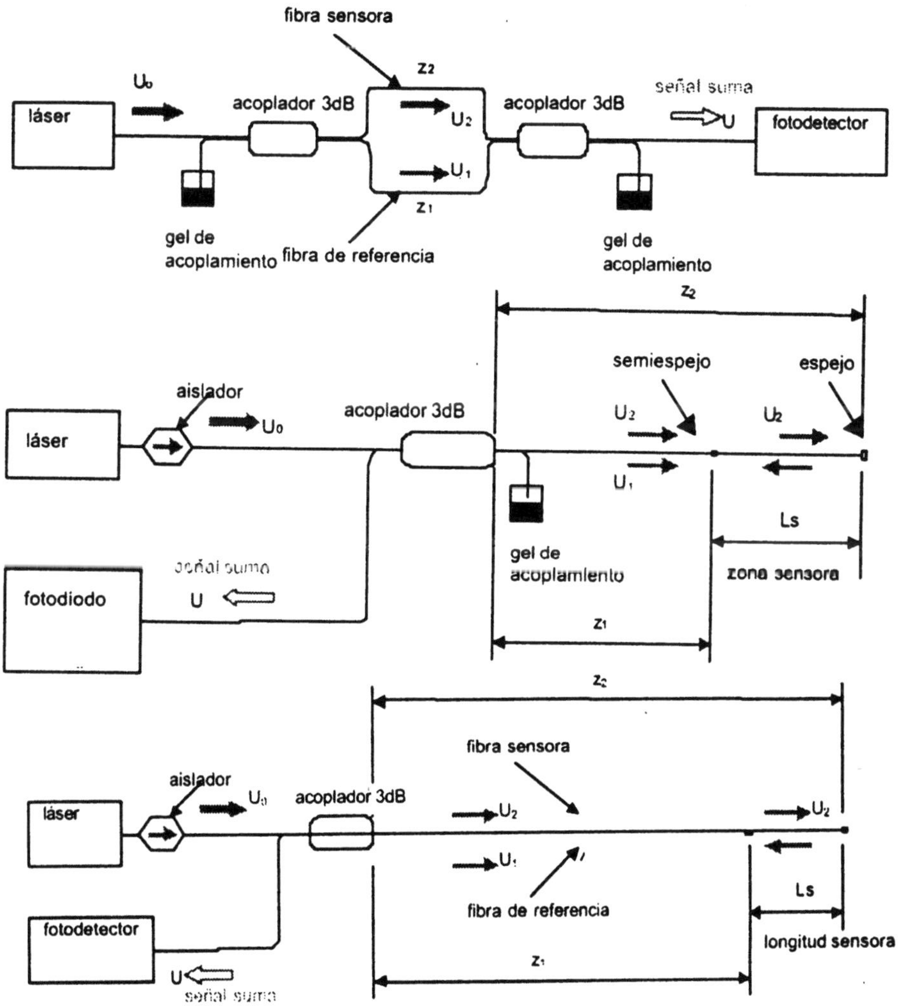

Figure 4. Typical interferometric arrangements

The Figure 4 sketch the most common optical fibre interferometric architectures, known as Mach-Zender. Michelson, Fabry-Perot, from the names of classical optical interferometry. At the Michelson and Fabry-Perot systems, the light is reflected at the end of the optical fibers, traveling back and recombining again. Fabry-Perot is special in the sense that light travels most of the time by a single fiber, interference is caused among the waves reflected at a partial mirror, and the waves that pass this mirror and are reflected back later (Figure 5). A detailed discussion of each of these systems is given at Refs 7 to .

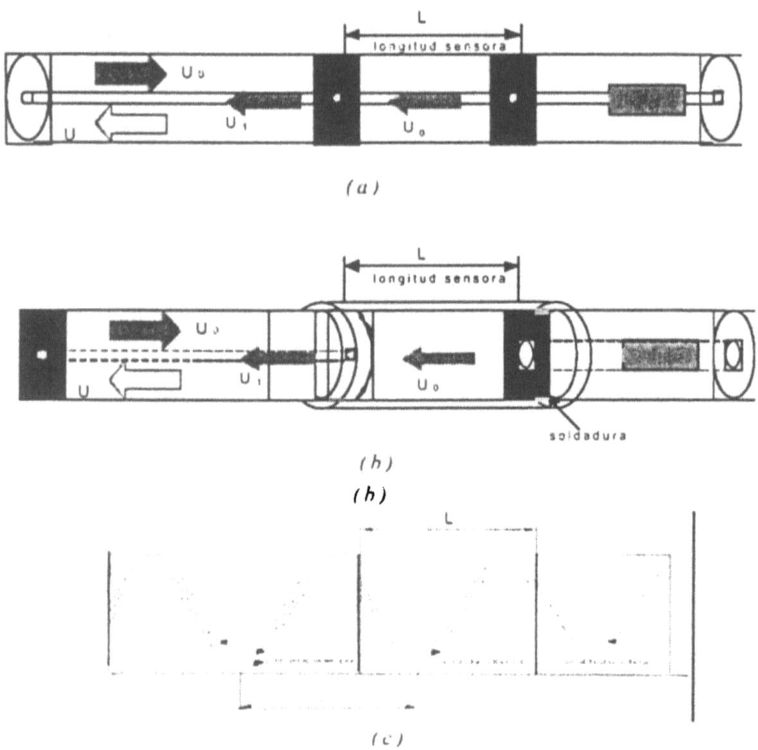

Figure 5 Intrinsic and extrinsic Fabry-Perot sensing heads

For Smart Structures, a single optical fiber was preferred, because only one ingress point into the structure is required. A lot of work was done for developing a rugged sensor head. Soldering optical fibers after partial coating of the fiber ends did intrinsic sensors. With the development of the Bragg grating sensors, this line of work was abandoned. Extrinsic Fabry-Perot heads, known as 'microcavities' are still commercially available; a precision machined wafer is bonded at the end of the optical fiber, leaving an small air gap, making miniature pressure and temperature sensors.

The Mach-Zender arquitecture was used in the european SISCO Project (Ref. 11) for detecting perturbations in the elastic waves propagating through composite materials, aiming to develop a damage detection system. The fiber optic sensing arm was embedded into the laminate, any defect located between the elastic waves source and the optical fiber would change its perturbed length. The complexity of the response has precluded any practical application.

The Michelson arrangement is employed in the 'white light interferometry ', a widely used technique for strain monitoring in civil engineering, which is commercially available by a Swiss company, under the name SOFO (Ref 12). Instead of using a continuous wave monocromatic light, a short pulse is launched through the two arms, the arrival times will be different if their lengths are different, and the photodetector would receive two pulses. If the reference arm can be mechanically stretched to get both pulses coincident, the increase of length in the sensing arm would be known, and from it, its averaging strain is obtained.

Wavelength based sensors, or Bragg gratings

These are a class of local, intrinsic, absolute, multiplexable, interruption immune fiber optic strain sensors, which have concentrate most of the attention since its discovery, at the beginning of the 1990's.

Basic idea is to engrave, at the core of the optical fiber and for a short length (about 1 cm.), a periodic modulation of its refractive index. It will behave as a serie of weak partial reflecting mirrors, which by an accumulative phenomenon called diffraction, will reflect back the optical wavelength that is exactly proportional to their spacing. The diffraction law, first established by Bragg and widely used for crystal structure analysis, simplifies under normal incidence to the simple expression:

$$\lambda_b = 2\overline{n_e}\Lambda_0$$

where 7 is the pitch, or period of the modulation, and n is the average refractive index. The Bragg grating behaves as a very narrow optical filter, as shown at the figure 6. When illuminating the optical fiber with white light, and looking to their spectral composition after the grating, a narrow peak 0.1 nm wide is missing at the transmitted signal, which exactly would be the only frequency that could be detected as reflected back.

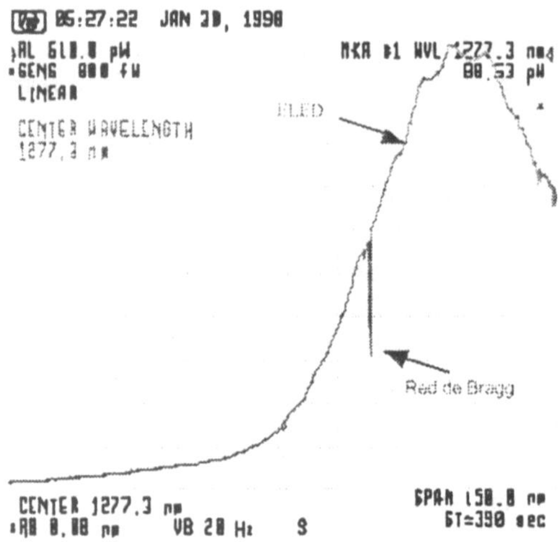

Figure 6 Spectra emitted by an ELED after passing through a Bragg grating

Everything that would change the pitch or the refractive index of the grating will move this wavelength; particularly, longitudinal strain and temperature will do it, and the shifting is shown at figure 7. The response is nearly linear with temperature and strain, shift of peak wavelength is about 1 nanometer per 100 Celsius degrees, or per 1000 microstrains. Any Optical Spectrum Analyzer (OSA) will be able to detect these changes, and transform it into readable information (figure 8).

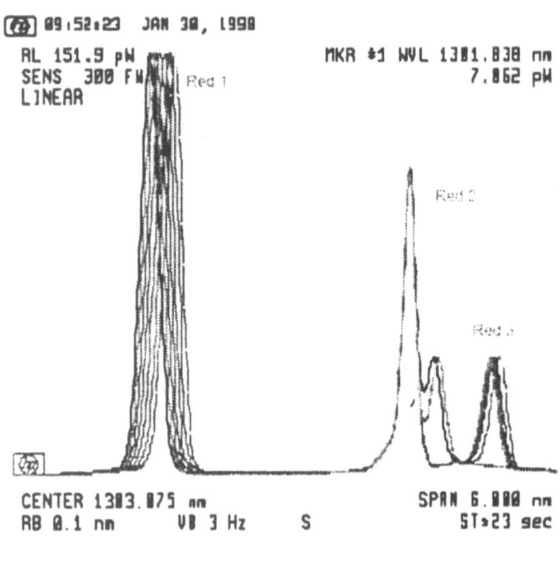

Figure 7 Reflected peaks of the Bragg grating submitted to diferent strains

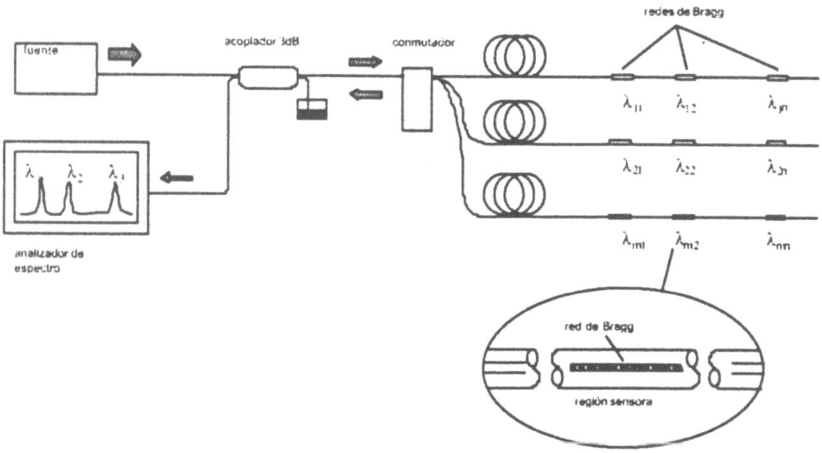

Figure 8. A multiplexed reading system for Bragg gratings.

Furthermore, commercially available white light sources have an spectral width around 40 to 60 nanometers. As the maximum excursion in wavelength is 10 nm for 10000 microstrains, several Bragg gratings, centered at different wavelengths, can be written at the same optical fiber, and interrogated at the same time. Multiplexing is easily implemented. The fact that the information is wavelength encoded makes the sensor very stable to aging, allowing absolute measurements of strain after long terms without recalibrating, a common nightmare with standard strain gages.

2.3 Final Remarks

Fiber optic sensors have followed a fast development during the last two decades, before attaining their status. Criteria to assess the potential suitability of any optical sensor may be [13]:

1. Intrinsic in nature, for minimum perturbation and stability
2. Localized, so it can operate remoteluy with insensitive leads.
3. Able to respond only to strain field, disceming changes of direction
4. All-fiber for operational stability
5. Single optical fiber, single ended for ease of installation and connection
6. Sufficiently sensitive with adequate measurement range.
7. Interrupt immune and capable of absolute measurements.
8. Amenable to multiplexing
9. Easily manufactured, and adaptable to mass production.

Bragg gratings sensors seem the best-balanced solution.

References

1. E. Udd, *Fiber Optic Sensors: An Introduction for Engineers and Scientists*, Wiley Interscience, New York. 1990.

2. E. Udd, *Fiber Optic Smart Structures*, Wiley Interscience, New York, 1994.

3. PHILTEC INC., commercial information (www.philtec.com)

4. C. K. Asawa, S. K. Yao, R. C. Steams, N. L. Mota, J. W. Downs, J. W. *High-sensitivity fibre-optic strain sensors for measuring structural distortion*. Electronics Letters, vol. 18, p. 362-364, 1982.

5. J. S. Schoenwald, P. M. Beckham *Distributed fiber-optic sensor for passive and active stabilization of large structures*. Review of Progress in Quantitative Nondestructive Evaluation, (eds. D. O. Thompson, D. E. Chimenti), pp. 565-572, 1987.

6. E. Udd, J. P. Theriault, A. Markus, Y. Bar-Cohen, *Microbending fiber optic sensors for smart structures*. Fiber optic smart structures and skins II; Proceedings of the Meeting, *SPIE-986*, pp. 478-482, Boston, MA, 1989

7. D. Kersey, D. A. Jackson, M. Corke, *A simple fibre Fabry-Perot Sensor*. Optical Communications, vol. 45, 1983.

8. M. Corke, A. D. Kersey, D. A. Jackson, *All fibre Michelson thermometer. Electronics Letters*, vol. 19, pp. 471-473, 1983.

9. D. A. Jackson, A. Dandridge, S. K. Sheem, *Measurement of small phase shifts using a single mode optical fiber interferometer*. Optical Letters, vol. 4, pp. 179, 1988.

10. S. W. Case, J. J. Lesko, B. R. Fogg, G. P. Carman *Embedded extrinsic Fabry-Perot Fiber Optic Strain Rosette Sensors*. Journal of Intelligent Material Systems and Structures, vol. 5, no. 3 pp. 412-417,1994.

11. W.S.Staszewski, s.g. Pierce, K. Worden, B. Culshaw, *'Cross-Wavelet Analysis for Lamb wave damage detection in Composite Materials using Optical Fibres'*, DAMAS'99, Dublin

12. D. Inaudi, S. Vurpilot, N. Casanova, P. Kronenberg, *Structural monitoring by curvature analysis using interferometric fiber optic sensors*. Smart Materials and Structures, vol. 7, n. 2, pp. 199-208, 1998.

13. R.d. Turner, T. Valis, W. Hogg, R. Measures *Fiber Optics Strain sensors for Smart Structures*. J. of Intelligent Material Systems and Structures, vol. 1 pp. 26-49 , 1990.

3. Fiber Optic Elements

A. J. Guemes

ETSI Aeronauticos, Univ. Politecnica Madrid, Madrid- SPAIN

Abstract In this section, the different elements of a fiber optic sensing systems are discussed. Beside the particular interest of some reader to build up his interrogation system, this basic information is required to understand the trade-offs, and for estimating what is likely to be available in the future.

3.1 Optical Fibers

Optical fibers are cylindrical dielectric guides for light waves made out of high purity, low loss optical materials, usually silica, but optical fiber of plastic and other transparent materials are also commercially available. The refractive index of the core (about 1.46 for silica) is slightly higher than surrounding material or cladding, by the presence of dopants. Optical rays arriving at the interface with angles higher than total reflectance angle, defined by the Snell law, will follow total reflections and will travel along the fiber, locally confined at the core. Only when the fiber is locally bent with sharp radius, light may escape. Externally the of uses to be protected from mechanical scratches with a plastic coating, and frequently several optical fibers are bundle together, and assembled with high strength mechanical fibers, such as Kevlar, to make a robust product known as optical cable, that can withstand rough industrial manipulation.

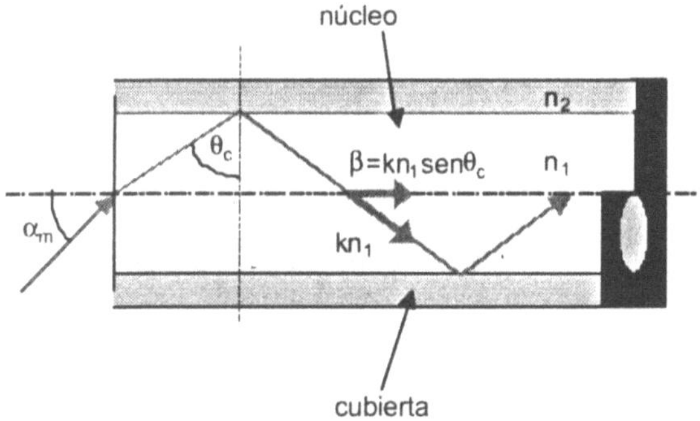

Figure 3.1. Esquematic representation of an optical fiber.

Optical fibers classify in two main groups.

Monomode OF, of small core (about 10 μm, depending on the intended optical wavelength). The electromagnetic waves travelling by the core must satisfy the Maxwell equations; the cylindrical contour conditions only allow for a discrete number of solutions, V, dependent on the fiber diameter, and the wavelength:

$$V = \sqrt{2\Pi^2 a^2 \frac{n_1^2 - n_2^2}{\lambda^2}}$$

When V is less than 2.4, only one mode (2 ortogonal polarizations) propagates through the optical fiber, it is called monomode.

Monomode OF offer some advantages for optical communication, such as:

- Optical attenuation is smaller for monomode OF, due to the lower difference of refractive index between core and cladding, requiring a lower concentration of dopants.
- Multi mode OF will suffer group dispersion due to different modes travelling at different speed.

The multimode OF have the only advantage of a larger core (30 to 100 microns). The larger core makes simpler the alignment with the optical sources and connections. When only used as light pipes, they are preferred, as in many medical applications. For sensing purposes they can only be used for intensity based methods.

Optical power loss is extremely small with current to OF, in the order of 0.03 dB per kilometre, meaning that 100 km will only divide optical power by two; this is an important requirement for telecommunication. Optical power is minimised near the wavelength of 1300 nm or 1550 nm., where minimum Rayleigh dispersion and minimum infrared absorption is found. From a sensing point of view, optical losses at the fiber are irrelevant, but still 1550 is the preferred windows, because optoelectronic components are more easily available.

3.2 Optical Connections

Continuity from the light source to the photodetector is required at any optical circuit, like in any electrical circuit. Working with near-infrared invisible light, checking the optical continuity is not an easy task. Only OTDR (Optical Time Domain Reflectometry) would afford some information, with a very limited distance resolution. Good quality connectors are required, otherwise high optical power losses may occur.

Connection may be grouped in three classes, namely:

1. Mechanical connectors, to interconnect optical fibers coming from the structure, to the interrogation equipment. If optical fibers have to be mechanically connected, its cores need to be precisely aligned, and this requirement is much more stringent for monomode OF.

To avoid uncontrolled backscattered radiation, each flat end of both fibers is polished, and a matching gel is sometimes used to minimise losses. Several standards like FC, ST, SC and SMA are being used by the telecom industry with high reliability.

2.Fusion splices, or permanent joint between two fibers.

It is usually done for joining the optical sensors to an standard optical fiber, allowing to put the sensor at any desired position inside the structure. After stripping and cleaving the two ends of the OF, they are precisely positioned and approached under a microscope, an electrical discharge melts locally the material, causing a permanent joint. Fully automated systems are available for this operation.

3. Couplers, or connections that redistribute the energy between two or more fibers. When the core of two optical fibers are brought very close a part of the energy travelling by one fiber may pass to the other, by the evanescent wave of the cladding. The most standard coupler is the so called 3 dB or 2x2, meaning that the power from any input fiber divides symmetrically, and the opposite for the light travelling back.

3.3 Active Optoelectronic Components

Light sources

Light emitting diodes (LEDS) are very common devices used as omnidirectional broad band light sources, good for displays but not useful for sensing purposes.

Its typical central wavelength is related to the semiconductor junction (700 nm for Ga P, 900 nm for Ga As, 1200-1700 for In Ga As), with a bandwith around 40 to 80 nm.

As a component for interrogation systems, pig-tailed edge emitting diodes are used. Light emission characteristics are as before, but a more efficient coupling to the optical fiber is achieved. Power level in the range of milliwatts is now available, being still is a limiting factor.

Laser diodes works with a double heterojunction, promoting stimulated emission, the semiconductor crystal acts as optical resonator.

Their optical outputs is much higher, and are recommended for highly multiplexed sensor systems, but a tunable capability at sufficient speed rate is required.

Light detectors

Photodetectors convert an input of photon flux to an output electrical current. PIN junction diodes are popular for sensing applications. Again, the operating wavelength is related to the semiconductor material, but specificity is not as high as photodiodes.

Trade-off is done among sensitivity and response time. Every optoelectronic component is quite sensitive to temperature, both in lifetime and optical response. Most of actual components are stabilised by Peltier cells.

3.4 Bragg Grating Sensor Manufacturing

Doped optical fibers are photosensitive, its characteristics change when exposed to intense UV light. To engrave the periodic pattern of refractive index, a sequence of very close light spots needs to be impinged over the core of the fiber, for a short time. Best way to procedure this regular pattern is by a diffracting procedure of monocromatic light, either by holographic interference or by the phase mask procedure.

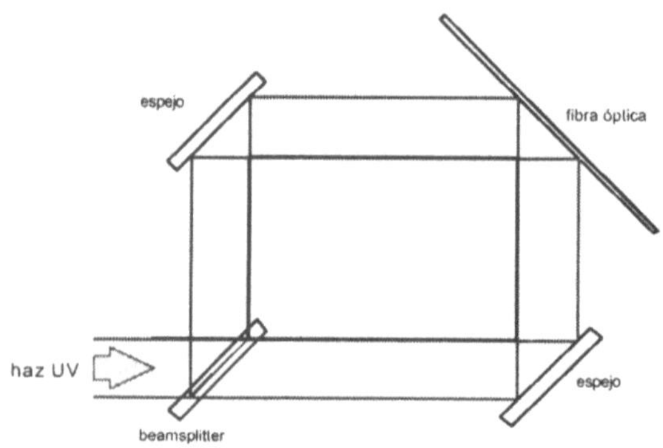

Figure 3.2. Holographic interference diagram.

Holographic interference has the advantage that the pitch of the Bragg grating can be easily changed by changing the angle of the two incident rays, but a highly coherent source is required. Phase mask procedure is easier to use, but different phase masks are required for different pitch. Figure 3.4 shows a picture of the facility available at UPM for manufacturing Bragg gratings by this procedure.

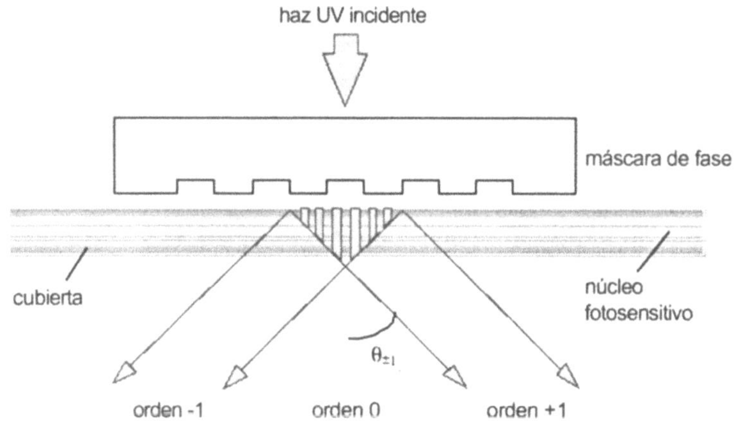

Figure 3.3 Interference pattern produced by a phase mask.

Figure 3.4. Installation at UPM to write Bragg gratings.

References

1. B. A. Saleh, M.C.Teich, *'Fundamentals of Photonics'*. John Wiley &Sons, Inc (1991)
2. Newport's Photonics Catalog (www.newport.com)

4. Strain Measurement Inside Composite Materials by Fiber Optic Bragg Gratings

A. J. Guemes

ETSI Aeronauticos. Univ. Politecnica Madrid, Madrid- SPAIN

Abstract This section summarizes the results to analyse the behaviour of fibre Bragg gratings when submitted to longitudinal non-uniform strain fields, and to decode the strain field applied to the grating by integrating its spectral information.

4.1 Introduction

A main advantage of fibre optics is their capability to be embedded inside the composite materials which allows the measurements of internal strains distribution, besides of being protected from the environment. The optical fibre can easily be included into any fresh laminae during lay up, and cured together, causing a permanent bond, ensuring that strain at the optical fibre are the same as in the adjacent laminae. Internal residual stresses are known to be produced when curing laminates at high temperatures, and this biaxial stress field will modify the response of the grating. This change needs to be understood before any practical use of embedded Bragg gratings.

The simplest behaviour of the grating is a drift of the peak wavelength when submitted to a pure longitudinal strain, which was discussed in previous lessons. This article deals with fibre Bragg gratings and composite materials in configurations prone to complex but realistic conditions: fibre Bragg gratings submitted to non-uniform or/and non-unidirectional strain and stress fields.

These conditions result respectively in a distortion or/and a splitting of the typical spectral signal of the grating, in such a way that the central wavelength drift of the grating doesn't gives enough information about the real mechanical fields around the grating. In these cases, the use of conventional "peak locators" is not useful, and the results given by these instruments would be useless, being necessary to obtain useful information to have the full spectral response of the grating and a clear understanding of the problem. After an exhaustive revision of the bibliography, it is not difficult to find experimental results showing erratic strain data resulting from complex spectral response of the sensors involved in the experiment.

Several research groups have done important efforts to analyse the behaviour of fibre Bragg gratings when submitted to longitudinal non-uniform strain fields, and to decode the strain field applied to the grating by integrating its spectral information. Results obtained by us are summarised in this document, to analyse non-uniform strain fields in different mechanical tests: bonded and bolted joints, fracture mechanics problems, etc.[1,2,3].

However, the situation that complicates the suitability of fibre Bragg gratings is their response to non-unidirectional stress. The first sign of this effect appears in fibre Bragg gratings embedded in a thick graphite/epoxy laminate cured at high temperature. Several gratings used to monitor strain and temperature during the cure process split in twin spectra during the cooling of the laminate. Bending of the fibre shows an exchange of optical power between the two sides of the twin spectra, pointing out a birefringence of the fibre promoted by the presence of strong residual stresses inside the laminate [4].

A transverse stress field applied to the grating promotes a birefringence that modifies the is otropic optical structure of the fibre, generating two principal directions of polarization, with two different refractive indexes. The fibre Bragg grating verifies two different Bragg conditions, and the optical spectrum of the grating splits in two. By using a fibre Bragg grating written in the core of a standard low-birefringence optical fibre, immersed in a general stress field, it would not be possible to separate the components of the stress field using the spectral information of the grating. These phenomena have been experienced by many authors and several methods have been proposed to avoid, or minimise them: use of special fibre, encapsulate the gratings, etc.

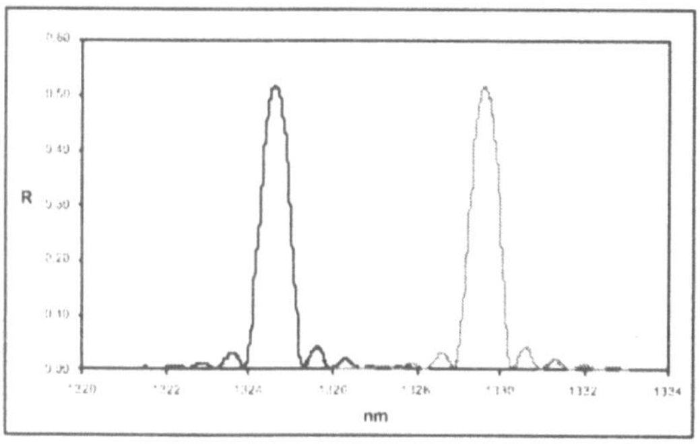

Figure 4.1. Spectra of a fibre Bragg grating bonded in an aluminium beam: unloaded (narrow line) and submitted to a uniform longitudinal strain of 5000 me (thick line). The "ideal" behaviour of a grating.

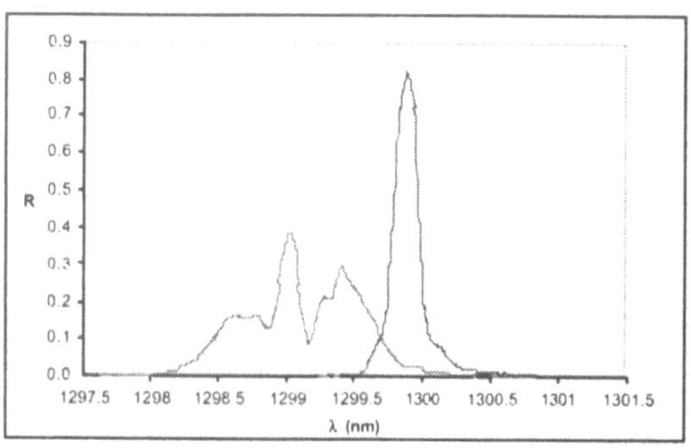

Figure 4.2. Spectra of a fibre Bragg grating embedded in a unidirectional composite laminate near a bonded joint: unloaded (narrow) and submitted to longitudinal load (thick). The distortion is due to strain gradients.

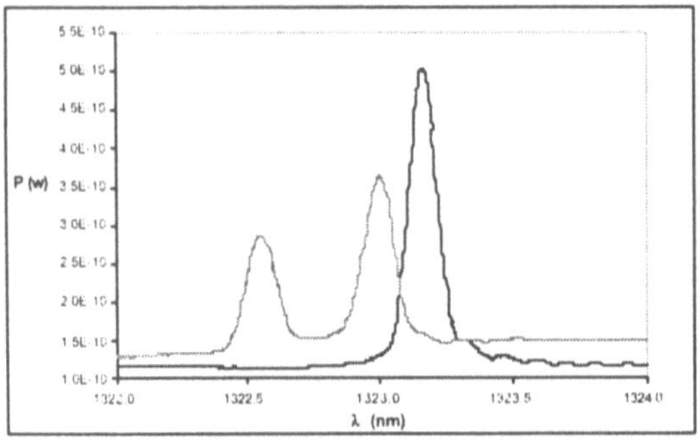

Figure 4.3. Spectra of a fibre Bragg grating before (narrow) and after being embedded in a thick quasi-isotropic laminate cured in autoclave (thick); the split red spectrum shows the effect of strong transverse stresses over the grating.

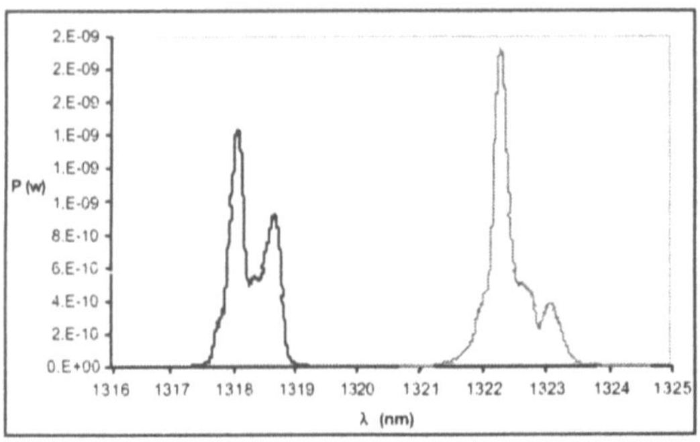

Figure 4.4. Spectra of a fibre Bragg grating embedded in a composite beam made with fabric and cured in autoclave: unloaded (narrow), shows the typical splitting of a biaxial stress state; loaded (thick), shows the effect of a complex non-uniform biaxial stress field.

Some authors have try to solve this problem using a pure optical approach to solve the general tridimensional case[6], by means of fibre Bragg gratings written in a high-birefringence optical fibre; with a precise positioning of the fibre, and solving a simplified photoelastic model, it is possible to resolve the longitudinal and the two transverse components of the stress applied to the sensors. However, the solution is complex, and requires an extremely precise positioning of the fibre, which is almost impossible in the generally rough conditions of a composite laminate, and also requires special fibres and optoelectronic devices for the demodulation.

The plane stress state dominant in laminates allows to simplify the solution of the photoelastic problem. The model considers the longitudinal and transverse stresses applied to fibre Bragg gratings written in low-birefringence fibre and embedded in real manufacturing conditions. It has been validated using simple laminate models and finite element methods, and shows a good agreement between experimental results and stresses calculated by the lamination theory.

4.2 Photoelasticity in a Plane Stress State

The core of a regular single-mode optical fibre is isotropic, and a uniform refractive index independent of the direction of propagation can be considered. In this case, the spectrum of a fibre Bragg grating written in the core of the fibre shows a single peak centered in a wavelength given by the Bragg condition $\lambda_b = 2\overline{n_e}\Lambda_o$, where $\overline{n_e}$ is the average effective refractive index along the grating, and Λ_o is the uniform pitch of its diffractive structure.

In a high birefringent optical fibre, the core of the fibre presents an orthotropic symmetry, with two well-defined polarization axis whose directions are defined by the index ellipsoid of the fibre [5]. A coordinate system can be defined with its main axis oriented in the longitudinal direction of the fibre, named as x_1, and two directions normal to it and parallel to the polarization axis, x_2 and x_3. If a fibre Bragg grating is written in the core of a high birefringent fibre, the intensity spectrum of the grating presents two well-defined peaks correspondent to the refractive indexes in the two polarization axis of the fibre, n_2 and n_3.

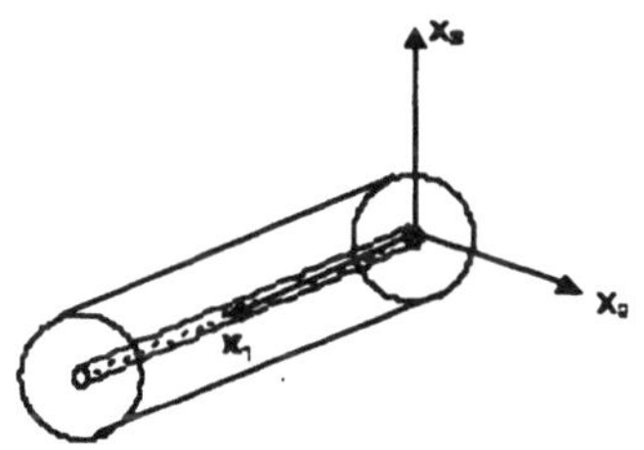

Figure 4.5. Single mode optical fibre and coordinate axis parallel to the principal polarization axis.

A tridimensional strain field applied to a regular single-mode optical fibre with an inscribed Bragg grating, will promote an effect of birefringence. Two different polarization axis with different effective refractive indexes have to be then considered [7], and the perturbation on the initial refractive index of the grating can be written as:

$$\delta\left(\overline{n_{e,2}}\right) = -\frac{\overline{n_e}^3}{2}\left[p_{11}\varepsilon_2 + p_{12}(\varepsilon_1 + \varepsilon_3)\right] \qquad (1)$$

$$\delta\left(\overline{n_{e,3}}\right) = -\frac{\overline{n_e}^3}{2}\left[p_{11}\varepsilon_3 + p_{12}(\varepsilon_1 + \varepsilon_2)\right] \qquad (2)$$

where p_{11}, p_{12}, p_{13}, p_{23} and p_{33} are the elements of the strain-optic tensor of a homogeneous orthotropic material, that define the optical response of a medium submitted to mechanical deformation, and ε_i are the components of the strain field applied to the medium, in contracted notation, $\varepsilon_1 = \varepsilon_{11}$, $\varepsilon_2 = \varepsilon_{22}$, $\varepsilon_3 = \varepsilon_{33}$, $\varepsilon_4 = \gamma_{23}$, $\varepsilon_5 = \gamma_{13}$, $\varepsilon_6 = \gamma_{12}$. The perturbation in the Bragg wavelength due to a strain field applied to the fibre will be given by the expression:

$$\delta\lambda_p = 2\left(\delta\left(\overline{n_e}\right)\Lambda_o + \overline{n_e}\delta\Lambda\right) \quad (3)$$

where $\delta\Lambda = \Lambda_o\varepsilon_1$ and ε_l is the strain in the direction of the optical fibre. Therefore, the shift of the Bragg wavelength of a diffraction grating in the principal polarisation axis of the fibre will be given by the expressions:

$$\frac{\delta\lambda_2}{\lambda_0} = \varepsilon_1 - \frac{\overline{n_e}^2}{2}\left[p_{11}\varepsilon_2 + p_{12}\left(\varepsilon_1 + \varepsilon_3\right)\right] \quad (4)$$

$$\frac{\delta\lambda_3}{\lambda_0} = \varepsilon_1 - \frac{\overline{n_e}^2}{2}\left[p_{11}\varepsilon_3 + p_{12}\left(\varepsilon_1 + \varepsilon_2\right)\right] \quad (5)$$

where λ_o is the Bragg wavelength previous to the perturbation.

4.2 Fibre Bragg grating submitted to a uniform load applied over two opposite generatrix of the fibre.

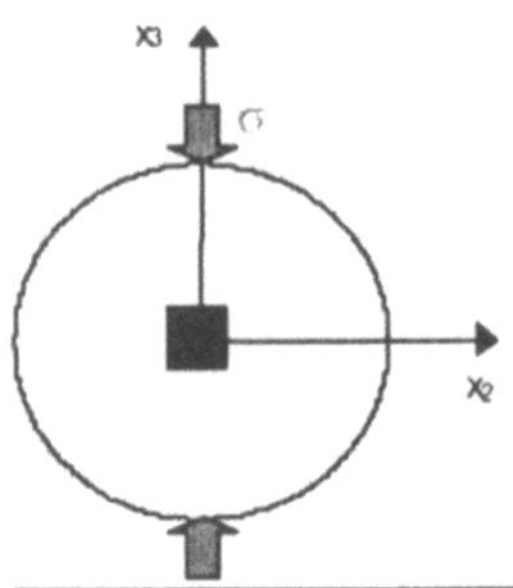

Figure 4.6. Diametral load applied to the optical fibre.

The transverse load applied promotes a splitting of the spectrum of the grating in two twin spectra, as is shown in Figure 4.7.

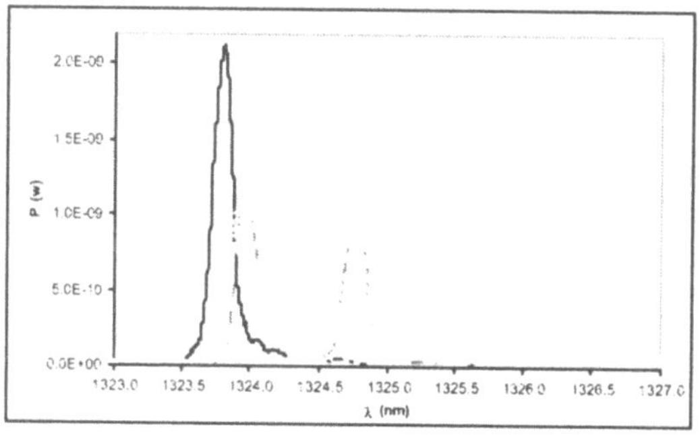

Figure 4.7. Spectra of a unloaded grating, and under two different levels of transverse load

For this specific load configuration, a plain strain state in the plane x_2-x_3 is a reasonable hypothesis; it is possible then to assume that $\varepsilon_1 = 0$ y $\varepsilon_2, \varepsilon_3 \neq 0$. Therefore, and considering that the shear stresses are negligible, it is found that $\sigma_1 = \nu_f(\sigma_2 + \sigma_3)$. The components of the strain in the principal directions of the fibre ε_1, ε_2, ε_3 are then:

$$\varepsilon_1 = 0 \tag{6}$$

$$\varepsilon_2 = \frac{1+\nu_f}{E_f}\left[(1-\nu_f)\sigma_2 - \nu_f\sigma_3\right] \approx \frac{1}{E_f}(\sigma_2 - \nu_f\sigma_3) \tag{7}$$

$$\varepsilon_3 = \frac{1+\nu_f}{E_f}\left[(1-\nu_f)\sigma_3 - \nu_f\sigma_2\right] \approx \frac{1}{E_f}(\sigma_3 - \nu_f\sigma_2) \tag{8}$$

and the relative shifts of the Bragg wavelengths in the axis x_2 y x_3 could be written as follows:

$$\frac{\delta\lambda_2}{\lambda_0} = -\frac{\overline{n_e}^2}{2}\left[p_{11}\varepsilon_2 + p_{12}\varepsilon_3\right] \tag{9}$$

$$\frac{\delta\lambda_3}{\lambda_0} = -\frac{\overline{n_e}^2}{2}\left[p_{11}\varepsilon_3 + p_{12}\varepsilon_2\right] \tag{10}$$

Substituting ε_2, y ε_3 in (9) and (10) results:

$$\begin{pmatrix} \frac{\delta\lambda_2}{\lambda_0} \\ \frac{\delta\lambda_3}{\lambda_0} \end{pmatrix} = \begin{bmatrix} S_1 & S_2 \\ S_2 & S_1 \end{bmatrix} \begin{pmatrix} \sigma_2 \\ \sigma_3 \end{pmatrix} \tag{11}$$

where

$$S_1 = -\frac{1-\nu_f}{E_f}\frac{\overline{n_e}^2}{2}\left[(1-\nu_f)p_{11} - \nu_f p_{12}\right] \tag{12}$$

$$S_2 = -\frac{1-\nu_f}{E_f}\frac{\overline{n_e}^2}{2}\left[(1-\nu_f)p_{12} - \nu_f p_{11}\right] \tag{13}$$

We estimated the array of coefficients S using published data: typical values of 70 GPa and 0.17 for the elastic modulus and the Poisson's coefficient of the fibre, the strain-optic coefficients determined by Bertholds and Dandliker[8], $p_{11} = 0.113$ and $p_{12} = 0.252$, and a average refractive index after the writing of the grating in the optical fibre used, $\overline{n_e} = 1.4496$. The matrix S^{-1} allows calculating the field of transverse sircsses in the core of the optical fibre from the measured wavelength shifts of the two peaks of the spectrum of the grating:

$$\begin{pmatrix} \sigma_2 \\ \sigma_3 \end{pmatrix} = 70 \begin{bmatrix} 1.7447 & -6.5048 \\ -6.5048 & 1.7447 \end{bmatrix} \begin{pmatrix} \frac{\delta\lambda_2}{\lambda_0} \\ \frac{\delta\lambda_3}{\lambda_0} \end{pmatrix} GPa \tag{14}$$

This model shows an excellent correlation with experimental results. Figure 9 compares the values of transverse stresses obtained using a finite elements model (figure 8 shows the value of the transverse stresses in the two principal directions of the fibre, σ_2 and σ_3, for a transverse load applied to the fibre, in the configuration shown in figure 6), and those estimated using the Bragg wavelength of the splitted gratings and applying the proposed photoelastic model, showing an almost perfect coincidence in the value of the component of the transverse stress in the direction of the applied load, σ_3.

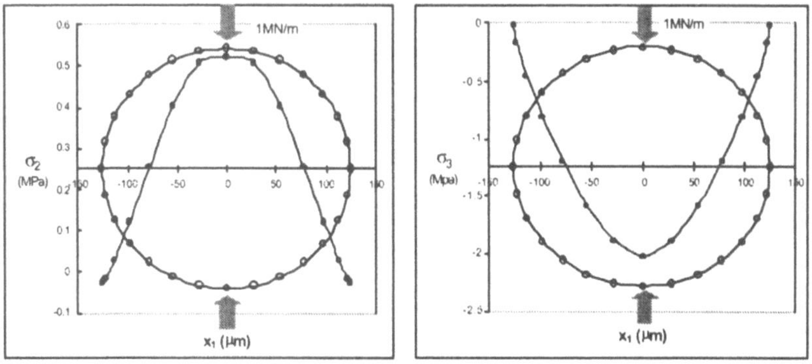

Figure 4.8. Transverse stresses in the two principal directions of the fibre, σ_2 and σ_3, for a uniform transverse load applied on the fibre (finite elements model)

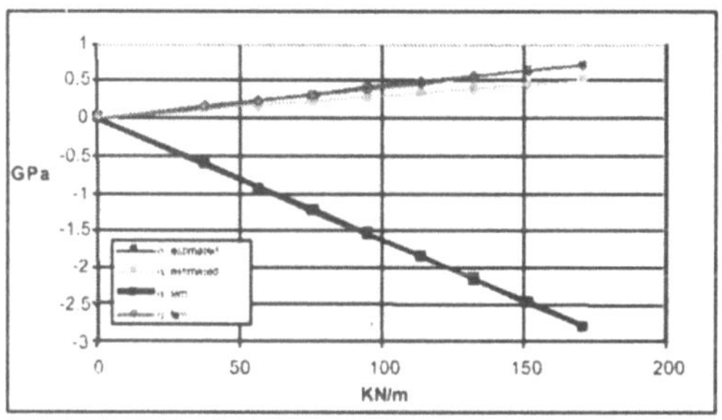

Figure 4.9. Comparison between the transverse stresses obtained using the proposed model and the values obtained using FEM

4.3 Fibre Bragg grating embedded in a composite laminate;

In this case, the optical fibre is submitted to a uniformly distributed tridimensional load. This fact would considerably difficult the solution of the photoelastic problem, unless the special configuration of this materials is considered. In most cases, a composite laminate has a clearly biaxial configuration, with a thickness several orders of magnitude smaller than the dimensions in the plane of the plies, and the loads are always applied in the plane of the laminate (see Figure 4. 10).

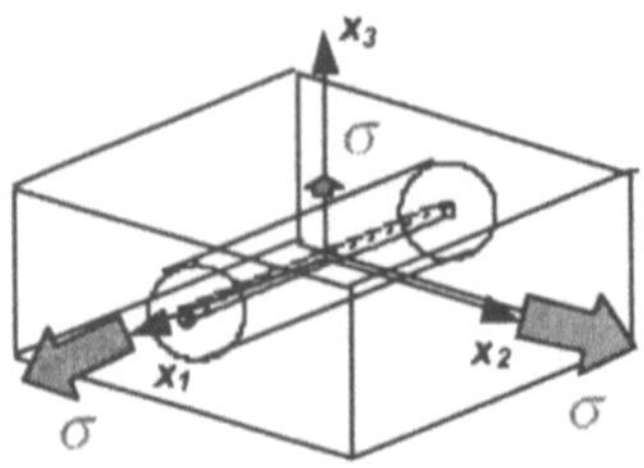

Figure 4.10. Element of laminate with an embedded optical fibre, and common coordinate system.

In this case, and with the objective of simplifying the solution of the problem of calculate the strain an stress field in the proximity of a Bragg grating inscribed in the core of an optical fibre embedded in a laminate, it is reasonable to assume a plane stresses state in x_1-x_2, which implies that $\sigma_1, \sigma_2, \sigma_6 \neq 0$, $\sigma_3 = \sigma_4 = \sigma_5 = 0$ (in our experiments, we arrange for $\sigma_6 = 0$). The components of the strains field in the principal direction of the fibre ε_1, ε_2, ε_3, are then:

$$\varepsilon_1 = \frac{1}{E_f}(\sigma_1 - \nu_f \sigma_2) \tag{15}$$

$$\varepsilon_2 = \frac{1}{E_f}(-\nu_f \sigma_1 + \sigma_2) \tag{16}$$

$$\varepsilon_3 = -\frac{\nu_f}{E_f}(\sigma_1 + \sigma_2) \tag{17}$$

Substituting this equalities in the expressions (4) and (5) it is possible to obtain the value of the relative shifts of the Bragg wavelength of the split peaks correspondent to the two principal axis of polarization x_2 y x_3 as a function of the longitudinal stress σ_2 and the transverse stress in the plane of the laminate σ_2:

$$\begin{pmatrix} \frac{\delta\lambda_2}{\lambda_0} \\ \frac{\delta\lambda_3}{\lambda_0} \end{pmatrix} = \begin{bmatrix} U_1 & U_2 \\ U_1 & U_3 \end{bmatrix} \begin{pmatrix} \sigma_1 \\ \sigma_2 \end{pmatrix} \tag{18}$$

where

$$U_1 = \frac{1}{E_f}\left\{1 - \frac{\overline{n_e}^2}{2}\left[(1-\nu_f)p_{12} - \nu_f p_{11}\right]\right\} \quad (19)$$

$$U_2 = \frac{1}{E_f}\left\{-\nu_f - \frac{\overline{n_e}^2}{2}\left[p_{11} - 2\nu_f p_{12}\right]\right\} \quad (20)$$

$$U_3 = \frac{1}{E_f}\left\{-\nu_f - \frac{\overline{n_e}^2}{2}\left[(1-\nu_f)p_{12} - \nu_f p_{11}\right]\right\} \quad (21)$$

As in the previous case, it is possible to estimate the array of coefficients U using the same coefficients: 70 GPa and 0.17 for the elastic modulus and the Poisson's coefficient, p_{11} = 0.113 and p_{12} = 0.252, and $\overline{n_e} = 1.4496$. Expression (22) allows then calculating the field of transverse stresses in the core of the optical fibre given the measured wavelength shifts of the two peaks of the spectrum of the grating embedded in a composite laminate with a planar configuration.

$$\begin{pmatrix}\sigma_1\\ \sigma_2\end{pmatrix} = 70\begin{bmatrix}2.7020 & -1.4526\\ 5.8514 & -5.8514\end{bmatrix}\begin{pmatrix}\frac{\delta\lambda_2}{\lambda_0}\\ \frac{\delta\lambda_3}{\lambda_0}\end{pmatrix} GPa \quad (22)$$

The most immediate result that can be induced from this expression is that distance between the two peaks of the splitted peak is only affected by the transverse stress applied to the fibre, and is proportional to it (Figure 4.11):

$$\sigma_2 \cong k\left(\frac{\delta\lambda_2}{\lambda_o} - \frac{\delta\lambda_3}{\lambda_o}\right) \quad (23)$$

where $k \cong 70 \times 5.8514\ GPa$ is the coefficient of proportionality, λ_o is the Bragg wavelength of the non-perturbed grating, and $\delta\lambda_2$ and $\delta\lambda_3$ are the wavelength shifts of the splitted peaks.

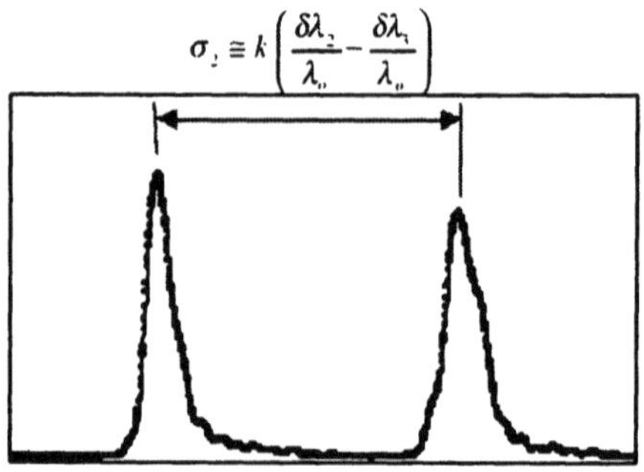

Figure 4.11. The distance between the two peaks of the splitted peak is only affected by the transverse stress applied to the fibre, and is proportional to it.

It is also possible to deduce an expression of the longitudinal strain of the grating as a function of the displacement of the wavelength of one of the peaks and the distance between the two peaks, $\delta\lambda = \delta\lambda_2 - \delta\lambda_3$, that is:

$$\varepsilon_1 = 1.25\frac{\delta\lambda_2}{\lambda_0} + 0.46\frac{\delta\lambda}{\lambda_0} \tag{24}$$

In order to prove the validity of these results, two different experimental tests have been implemented. The first one, very simple, uses the well-known thermal expansion of the aluminium to prove the importance of the effect of transverse stresses over fibre Bragg gratings in apparently very simple applications. The second test, with four gratings embedded in different positions in a graphite/epoxy laminate, will allow evaluating the effect of the residual stresses generated in a laminate during the cure processing

4.4 Experiments

The use of fibre Bragg gratings to monitor strain distributions in bonded joints in aluminium elements using high temperature adhesive film was one of the first applications in which the importance of transverse stress in their spectral response appears clearly. The following test has been configured as an illustrative example of this behaviour, and shows the necessary precautions that have to be taken when using this technology in applications in which temperature or external loads can promote transverse strain fields over the gratings.

Figure 4.12 shows a bare fibre Bragg grating 10 mm long written in photosensitive fibre and submitted to thermal stabilisation at 300C during 2 hours has been bonded in a beam of AA5083 aluminium using an epoxy adhesive film. The adhesive has been cured using a standard cycle shown in the graphic of Figure 4.13 a.

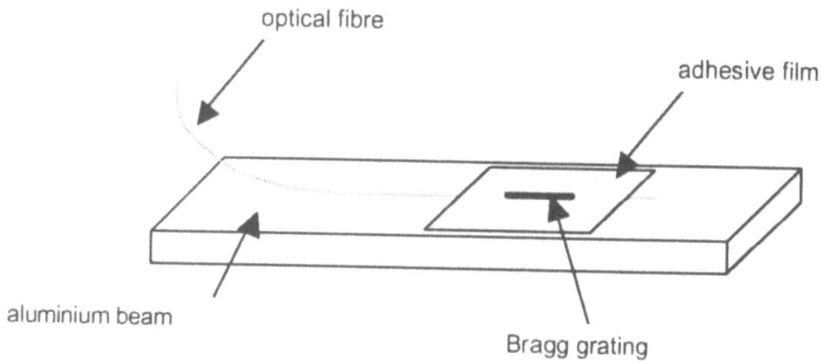

Figure 4.12. Schematic of aluminium beam with a fibre Bragg grating bonded using adhesive film

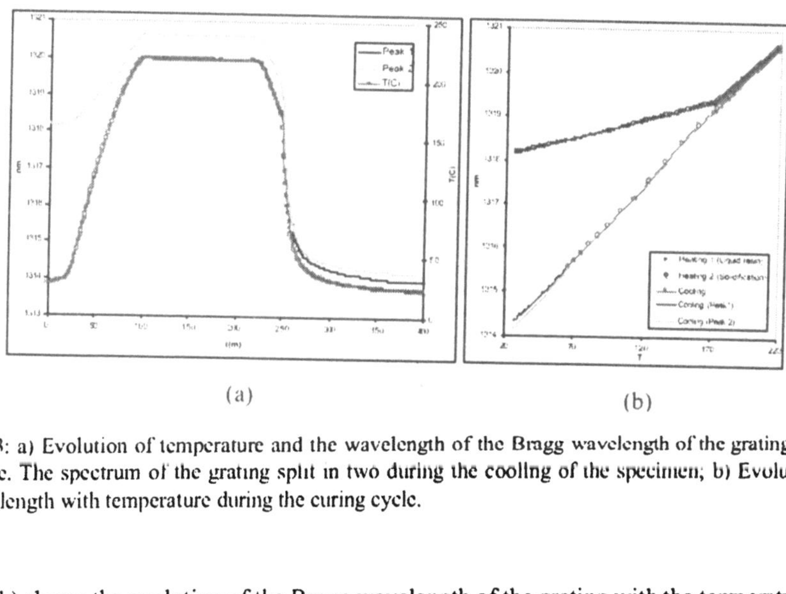

(a) (b)

Figure. 4.13: a) Evolution of temperature and the wavelength of the Bragg wavelength of the grating during the curing cycle. The spectrum of the grating split in two during the cooling of the specimen; b) Evolution of the Bragg wavelength with temperature during the curing cycle.

Figure 13 b) shows the evolution of the Bragg wavelength of the grating with the temperature during the curing cycle. Initially, during the first part of the heating stage, the wavelength of the grating increases according the curve wavelength/temperature of a free grating (the curve of thermal caracterization of the grating adjusted at UPM for the series of gratings is:

$$T = T_0 + 1.671 \cdot 10^5 \left(\frac{\Delta\lambda_b}{\lambda_{b0}}\right) - 2.404 \cdot 10^7 \left(\frac{\Delta\lambda_b}{\lambda_{b0}}\right)^2 + 4.608 \cdot 10^9 \left(\frac{\Delta\lambda_b}{\lambda_{b0}}\right)^3 - 3.643 \cdot 10^{11} \left(\frac{\Delta\lambda_b}{\lambda_{b0}}\right)^4 \quad (25)$$

valid from 0 to 600C, with a deviation of ±1.6C. This curve includes both effects: free expansion of the optical fibre and changes in the refractive index).

At 170C, there is a discontinuity in the slope promoted by the fast gelification of the resin, which drags the grating following the expansion rate of the specimen. During the cooling, the wavelength decreases with this last slope, but the grating split in two at 70C (the discrimination of two different peaks depends on the resolution of the spectral analyzer used) due to the thermal transverse strain of the beam.

To prove the validity of the results, a simple calculus can be done: the Bragg wavelength of the grating is given by the well-known Bragg condition:

$$\lambda_b = 2\overline{n}_e \Lambda_o \tag{26}$$

The wavelength shift of a grating bonded to an aluminium specimen with a linear expansion coefficient α_{Al} is given by:

$$\delta\lambda_b = \delta\lambda_{bT} + \lambda_{bo}(\alpha_{Al} - \alpha_f)\delta T \tag{27}$$

where $\delta\lambda_{bT}$ is the wavelength shift of the free Bragg grating with the temperature, given by expression (25), and α_f is the linear expansion coefficient of the optical fibre. The values of linear expansion coefficient for the AA5083 aluminium and the optical fibre are known ($\alpha_{Al} = 23.5\times10^{-6}K^{-1}$ and $\alpha_f = 5.5\times10^{-7}K^{-1}$), and the value estimated for the final Bragg wavelength of the grating after the cooling is 1314.20 nm, whereas the empirical result obtained extending the tendency line of the cooling step in figure 13 b) is 1314.19 nm for a cooling from 180C to 20C.

In this case, it is difficult to find an immediate quantitative relationship between the transverse shrinkage of the aluminium specimen and the shift of the spectrum of the grating due to the complex host/adhesive/sensor interaction.

A graphite-epoxy laminate specimen made of 24 plies of unidirectional T300/F155 prepreg tape with the following configuration: $[[90]_8[0]_4]_s$, 150 × 125 × 8 mm, cured in a hot plates press at 180C in an quasi-isothermal cycle (the specimen was introduced in the press with the plates at the curing temperature. Four bare fibre Bragg gratings were embedded in the laminate parallel to the reinforcing fibres, two in the middle plane (gratings 1 and 2) and two in the outer laminae (gratings 3 and 4). Spectra of the four gratings in different moments of the process and numerical data are shown in figure 14 and table 1.

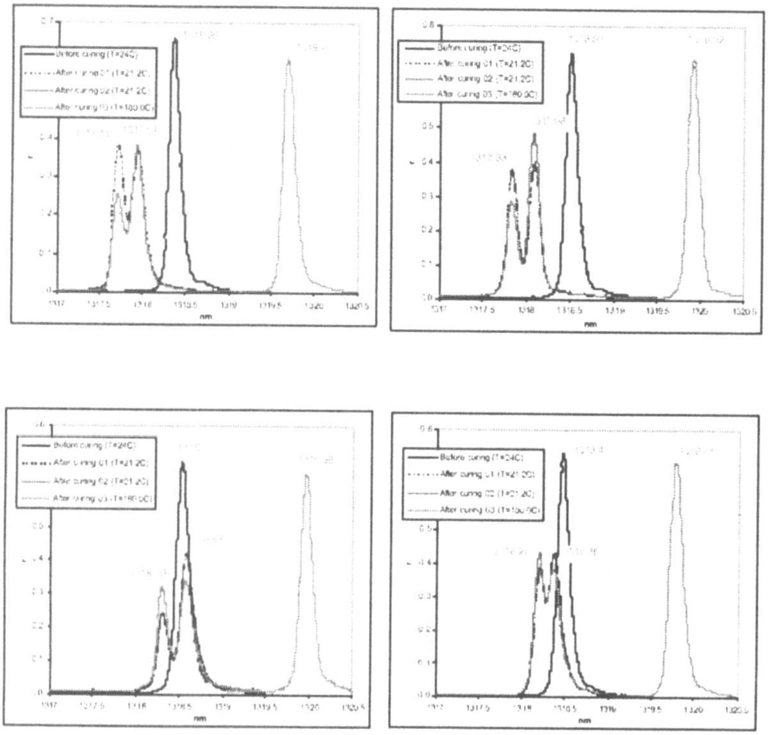

Figure 4.14. Spectra of the four gratings in different moments of the process: grating at 24C before curing process, grating at 180C after curing process, and grating at 21.2C after curing process.

Table 4.1. Numerical data of the experiment

Grating n°	λ_0 before curing (24C)	λ_0 after curing (180C)	λ_3 (peak 1) after curing (21.2C)	λ_2 (peak 2) after curing (21.2C)	$\delta\lambda = \delta\lambda_2 - \delta\lambda_3$
1	1318.36	1319.70	1317.71	1317.94	0.23
2	1318.50	1319.92	1317.83	1318.08	0.25
3	1318.53	1319.96	1318.30	1318.57	0.27
4	1318.45	1319.78	1318.20	1318.36	0.16

To use these results to calculate the residual strains and stresses generated during the curing process, it is necessary to compensate the effect of the temperature in the optical properties of the optical sensors. The wavelength shift of a grating due to the effect of the temperature over its refractive index is:

$$\delta\lambda_{bT} = \lambda_{bo}\varsigma_f \delta T \tag{28}$$

where ζ is the thermo-optic coefficient of the fibre. This expression can be also written as

$$\delta\lambda_{bT} = \delta\lambda_{bTf} - \lambda_{bo}\alpha_f \delta T \tag{29}$$

where $\delta\lambda_{bTf}$ is the wavelength shift of the grating given by the curve wavelength/temperature (25) for the considered temperature increment. As a first result, this value, 1.27 nm for a temperature increment from 24C to 180C is roughly the same that the shift of the four gratings during the heating, as could be expected after a quasi-isothermal curing process.

With these data, and using the expressions deduced from the proposed photoelastic model, the following values of strains and stresses are calculated:

Table 4.2. Residual strains and stresses calculated using the photoelastic model proposed

Grating n°	σ_{1f} (Mpa)	σ_{2f} (Mpa)	ε_{1f} (? ?)	ε_{2f} (? ?)
1	-15	71	-385	1057
2	-19	78	-453	1155
3	13	84	-20	1167
4	24	50	-87	705

where an "f" has been added as a subscript to remark than this results are the strains and stresses in the core of the grating. Stresses and longitudinal strain in the laminate and the fibre have to be roughly the same, but the difference of one order of magnitude in the stiffness' of the couple sensor/host implies important differences between the transverse strains given by the lamination theory and those calculated in the grating. Data obtained applying the lamination theory are:

Table 4.3. Residual strains and stresses calculated using the lamination theory

Laminae	σ_1 (Mpa)	σ_2 (Mpa)	ε_1 ($\mu\varepsilon$)	ε_1 ($\mu\varepsilon$)
Middle plane	-58	30	-530	4410
Outer laminae	-15	29	-190	4070

4.5 Conclusions

When Bragg gratings are embedded in thick non-unidirectional laminates cured at high temperature, biaxial stress state arises around the Bragg grating sensor. After the cooling of the part, spectra of embedded gratings shows strong splitting due to residual transverse stresses applied to the sensor.

This paper proposes a model that exploits the dominant biaxial configuration of composite structural elements, considering a plane stress state to simplify the solution of the photoelastic problem and resolve the longitudinal and transverse stress applied to fibre Bragg gratings written in low-birefringence fibre and embedded in real manufacturing conditions.

The model has been validated in two steps: using f.e.m. methods to prove the validity of the optic and photoelastic coefficients used, and the classical laminate theory to compare the results obtained using fibre Bragg gratings embedded in a laminate with those given by the theory. In spite of that experimental errors induced by the coefficients used, and the resolution of the instruments, the photoelastic model proposed offers data qualitative and quantitatively coherent with expected results.

References

1. M. Leblanc, S. Y. Huang, and R. M. Measures, Fiber optic Bragg intra-grating strain gradient sensing. *Smart Structures and Materials 1995: Proceedings of the Meeting, SPIE-2444*, pp. 136-147, 1995.
2. M. Leblanc, S. Y. Huang, M. M. Ohn, A. Guemes, and A. Othonos, Distributed strain measurement based on a fiber Bragg grating and its reflection spectrum analysis. *Optical Letters*, vol. 21, n. 17, pp. 1405-1407, 1996.
3. J. A. Güemes, S. Díaz-Carrillo, and J. M. Menendez, Measurement of strain distribution in bonded joints by fiber Bragg Gratings. *Smart Structures and Materials 1998: Proceedings of the Meeting, SPIE-3330*, p. 264-271, San Diego, CA, 1998.
4. J. M. Menéndez, and J. A. Güemes, Strain measurements inside thick CFRP laminates at the vicinity of bolted joints. *Smart Structures and Materials 1999; Proceedings of the Meeting, SPIE-3670*, p. 184-194, Newport, CA. 1999
5. B. E. A. Saleh, and M .C. Teich, *Fundamental of Photonics*, Wiley Interscience, 1991.
6. C. M. Lawrence, D. V. Nelson, and E. Udd, Measurement of transverse strains with fiber Bragg gratings, *Smart Structures and Materials 1998: Proceedings of the Meeting, SPIE-2444*, pp. 136-147, 1995.
7. G. B. Hocker, Fiber-optic sensing of pressure and temperature, *Applied Optics*, vol. 18, n. 9, 1979.
8. A. Bertholds, and R. Dandliker, Determination of the strain-optic coefficients in single mode optical fibers, *Journal of Lightwave Technology*, vol. 6 n. 1, p. 17-20, 1988.

5. Fiber Optic Sensors for Resin Flow and Composite Cure Monitoring

A. J. Guemes

ETSI Aeronauticos, Univ. Politecnica Madrid, Madrid- SPAIN

Abstract This section focuses on the developments done at the author Institution on Fiber Optic Sensors (FOS) to monitor resin flow and cure; other partners were simultaneously working on dielectrics and ultrasonic sensors, enabling a compari-son of results and characteristics of the different kinds of sensors.

5.1 Introduction

Improvements in quality of composite parts, accompanied of production costs and time-cycle reductions, may be attained by combining reliable real time sensors to efficient control algorithms, based on accurate models of the processes and/or ma-chine learning methodologies.

This is the global approach for the PERFECT Project (Process Efficient Regulation for Economical Composite Technologies), a Brite-Euram project that started in December 97, involving a number of aeronautical European industries and research centres. The project focuses on autoclave and RTM processes, as the most cur-rently used technologies for manufacturing thermoset composites structural components.

Thermocouples are nowadays the only used sensors during the curing cycles; more information on the evolution of key properties, like degree of cure or viscosity, is required before any practical attempt of 'smart processing' may be implemented, to take into account raw materials variability, different tooling and autoclave loading, etc.

5.2 Resin Flow Sensors for RTM Process

RTM process is a highly automated procedure that promises significant costs re-ductions of composite parts. However, the high fiber volume required for aerospace structures implies very dense preforms, which difficult the resin flow; if not adequately programmed, resin get gelled before filling up completely the mould. Numerical models of the resin flow are available, with different levels of accuracy, but presently, to a large extent, developing a mould for RTM is still a costly empirical process solved by trial and error.

The proposed system is a tool to get some insight of the process sequence, by monitoring the resin flow front movement. Our system is based in an array of stan-dard optical fibres. The only proposed alternative approach is based on an array of electrical wires, checking the change of electrical resistance at the cross points due to the resin arrival [1].

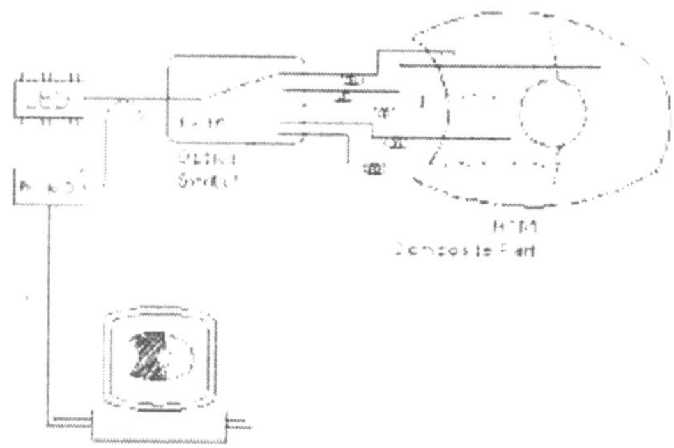

Figure 5.1. Schematic of the experimental set-up to monitor resin flow

Figure 5.1 sketches the system under tests for monitoring the resin flow front during resin infiltration. An array of 16 optical fibers is distributed in the dry preform, ending at different predefined positions. Light is launched into the fibers through an optical switch (100 msec. commuting time). About 5% of this light is backre-flected at the flat end-face of the optical fiber, according to Fresnel law, when air is at the interface. This reflection drops to almost zero when resin arrives, due to the better matching of refractive indexes. A computer controls the switch position among the different optical fibers, comparing the time of resin arrivals to the model predictions.

If adequately oriented, optical fibers may be easily embedded into the laminate without any detrimental effect on its mechanical strength. Ingress/egress of the op-tical fiber from the composite structure is still a main issue if the sensors must remain operative after curing, and the part need to be trimmed in all its edges. This is not the case for process monitoring, and optical fibers can ingress the lay-up through the venting ports, in RTM, or through the sealing paste in the autoclave process.

The proposed system is robust and adaptable to any part geometry; fibers will re-main embedded into the structure without any detrimental effect. They may later be used for cure monitoring or health monitoring purposes, if the optical fiber were prepared also for that mission. Tests were satisfactory; only the occasional presence of bubbles in the resin front causes troubles in the measurements.

5.1 Fiber Optic Sensors for Cure Monitoring

The cure process of thermoset composites may be accurately tracked in a laboratory environment by a variety of instrumental techniques, but still there are not robust procedures to be used in a factory environment. This issue has been throughfully discussed at the literature [2], [3], [4]. The development of reliable optical sensors has the attractiveness that light may go in and out of the part through optical fibres, allowing a remote control with little interferences.

Several fibre optic based techniques have been proposed for resin cure monitoring: evanescent wave interactions, transmission spectrum analysis, Fresnel reflection, fluorescence, FT Raman spectrum measurement, etc. Changes in the IR spectrum of the resin are a good indicator of the cure evolution, and remote FTIR spectros-copy seemed the most attractive approach, because the information is spectral, and not subject to changes due to intensity loss. Standard optical fibers are not trans-parent to mid-IR, but still sensitive changes are found at the near IR band (4000 to 9000 cm-1), for epoxy and BMI resins. Commercial probes based on diffuse re-flection are available, but they are too bulky to be embedded. We have tried to develop a transmission-based single-fiber optic sensor, without success up to now because of poor repetitivity.

An alternative approach makes use of the changes in the refractive index of the resin that happen during the curing stage, due to the resin shrinkage. For a typical epoxy resin, refractive index increases from 1.45 to 1.52 for a fully cross-linked resin; temperature produces similar changes that need to be compensated. Several techniques may be used to detect this change [5]. Fresnel reflection is probably the simplest technique. A light source, a photodetector, an optical coupler and a cleaved optical fibre, are enough to monitor the evolution of the refractive index of the resin.

A sketch of a basic device is shown in Figure 5.2: Light coming out from an optical source is launched into an optical fibre through a coupler. The tip of the optical fiber, cleaved very precisely, is immersed in the medium under analysis, in our case, resin. The interface core-resin acts as a sensor: most of the light that arrives to this surface pass through it, but a part of it is reflected back, in accordance with Fresnel law. The light reflected back pass again through the coupler and is detected by a photodiode. The relationship between detected power and refractive index is relatively simple.

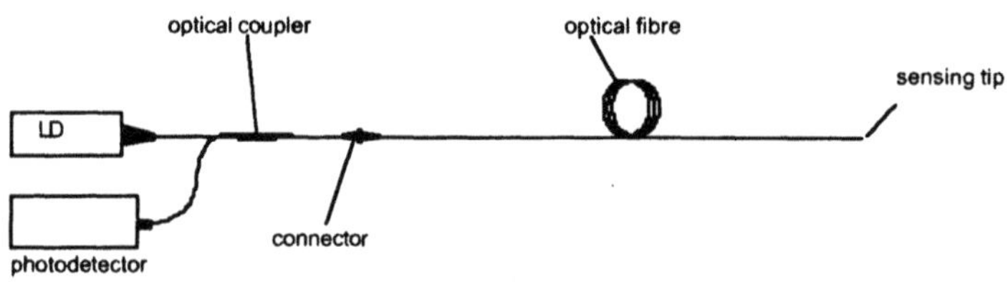

Figure 5.2. Schematic of a Fresnel based curing monitoring system

It has been demonstrated an association between the refractive index with the de-gree of cure and the temperature of the resin [6], so it would be possible to use this technique to monitor the evolution of the cure processing. The use of standard 125 mm diameter singlemode optical fiber allows reaching remote zones of the compos-ite part under analysis with extremely slight mechanical interference.

Unfortunately, this technique shows the same problem as any other technique based in light intensity measurements. Power drifts of the light source, unstable connections, output power variability in optical switches, mechanical perturbations in the optical path, due to macrobending and microbending of the optical fibre em-bedded into a composite laminate during the cure process limit its applicability to the highly controlled laboratory conditions. Furthermore, it would be necessary to have an accurate reading of the temperature at the place where the tip of the fibre is embedded, to compensate its effect on the refractive index of the resin.

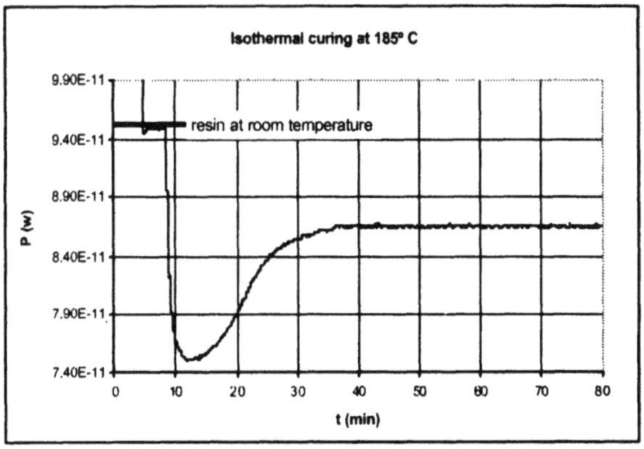

Figure 5.3. Optical signal of a Fresnel based system during an isothermal curing

5.3 Self-referenced refractive index and temperature monitoring probe

To overcome the previously mentioned issue, we have developed and patented a sensing probe that works succesfully to discriminate and compensate the effect of any perturbations on the signal coming from the sensing head. Intracore fibre Bragg gratings, optical devices which have proven their versatility and feasibility in many applications in the field of fibre optic based smart structures, having become al-most an standard as embeddable strain and temperature sensors, are used in this case as very precise references to remove undesirable perturbations on the signal reflected back by the core-resin interface.

Hardware set-up and operating principles The system proposed and developed by the UPM is similar to the basic scheme represented in Figure 5.2, but includes two fibre Bragg gratings and the photodetector is substituted by an optical spectrum analyzer (OSA), which allows not only to detect the total power reflected back by the different devices of the optical path, but to separate the light reflected at differ-ent wavelengths. The Bragg gratings operate in this scheme as narrowband mirrors, which backreflect a constant percentage of the power launched through them. Therefore, the gratings are used to detect which is the amount of light power that reaches certain point of the optical path.

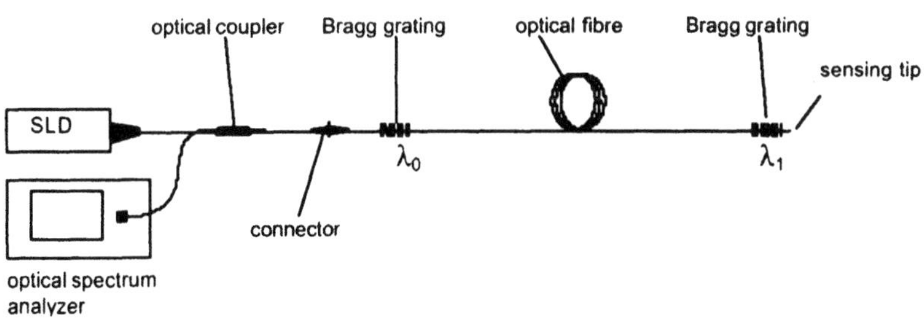

Figure 5.4. Schematic of a fibre Bragg grating referenced system

The way in which the system operates is the following: Bragg grating 1 operates as the main reference of the system. While the power of the optical source remains stable, the optical spectrum analyzer detects the sum of the backreflections along the optical path, the reflection in the fibre-resin interface of the sensing head, and the spectrum of the grating 1. If any ex-ternal perturbation is applied to the optical fibre, it causes a power drift that affects in the same way the power reflected by the grating and to the signal re-flected in the fibre-resin interface. Therefore, it is possible to discriminate the variations of the signal due to the refractive index changes from those due to external perturbations.

Bragg grating 0 operates as an auxiliary reference, and allows compensating power drifts due to instabilities of the source, or variable backreflections in the optical path due to optical switches, or unstable optical connections.

Figure 5.5 shows a schematic of the spectrum received by the optical spectrum ana-lyzer, and the contribution of each part of the optical system to the signal detected.

An important additional advantage of the system proposed is that the same fibre Bragg grating placed near the cleaved tip to operate as internal reference, acts as a local temperature sensor, which allows calculating very accurately the effect of the temperature on the refractive index.

Figure 5.6 shows an example of the optical detected by a conventional reflectometer, distorted by power drifts of the light source and external perturbations in the opti-cal fibre, and the signal compensated by a double-grating compensating system: n1, n2, n3 and n4 correspond to the refractive index of four different fluids. The system is able to correct signal deviations which would correspond to refractive index variations higher than 0.5.

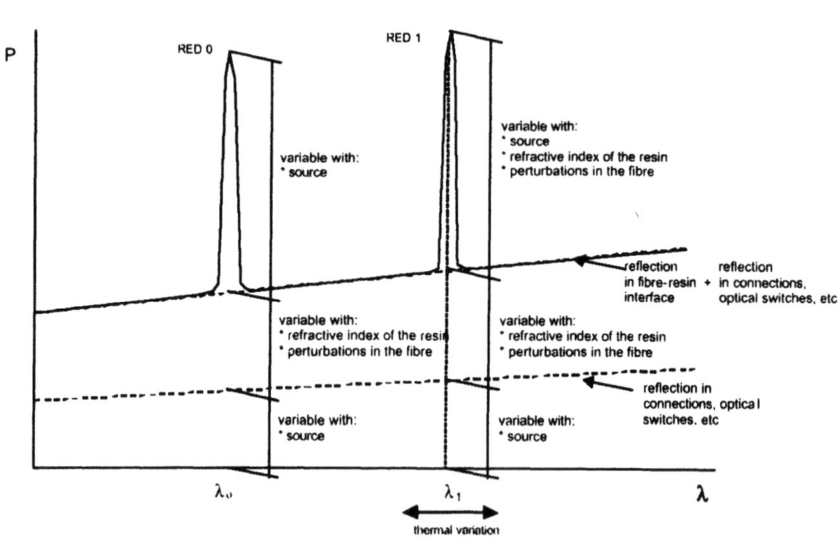

Figure 5.5. Schematic of the different contributions to the spectrum received by the OSA.

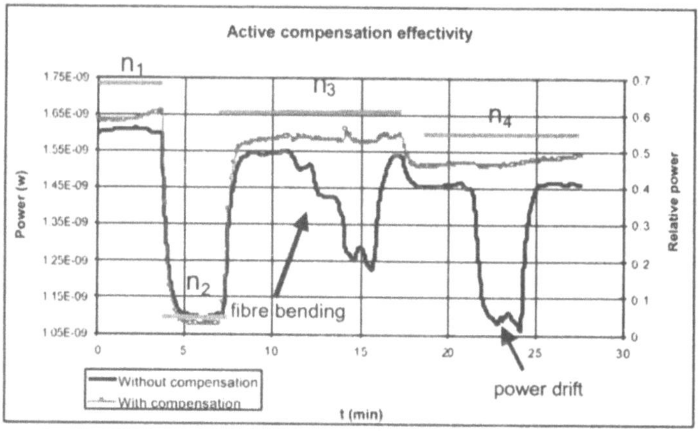

Figure 5.6. Signal detected by a conventional reflectometer and signal compensated by a double-grating compensating system.

This technique is compatible with fiber Bragg grating strain monitoring systems and RTM resin flow detection optical fiber based systems, which allows the use of a single multitask equipment to monitor in-situ the manufacture of composite ma-terials. The reference Bragg gratings, embedded into the structure, can be finally used as conventional strain sensors.

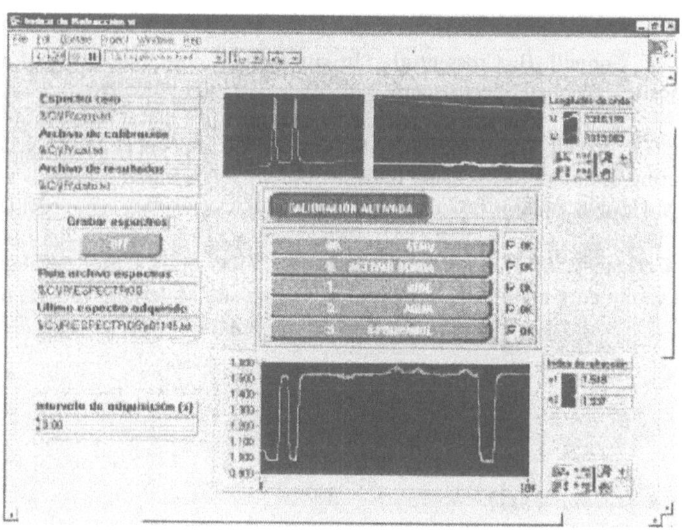

Figure 5.7. Window of the Labview based interface used for refractive index detection

5.4 Conclusions

Considerable research and development effort is being devoted to improve the quality and the reliability of the processes for manufacturing advanced composites, and to transform fixed and standardized cycles into intelligently controlled and optimized operations.

Optical fiber sensors has revealed as useful tools in several stages of the quality control of the manufacturing of composite parts: resin flow, refractive index and temperature monitoring, strain measurement. The systems and processes presented in this document allow an easy integration of these techniques in standard manufacturing processes, while the slight influence of embedded optical fibers in the mechanical properties of the composite material allows that fibers remain embedded during the active life of the part.

References

1. C.W. Lee and B.P. Rice, "Resin transfer Process Monitoring and Control", SAMPE Journal, vol. 34, n. 6, 1998.

2. D.D. Shepard and K.R. Smith, "A new ultrasonic measurement system for the cure monitoring of thermosetting resins and composites", Journal of Thermal Analysis, vol. 49, n. 1, p. 95-100, 1997.

3. G. Maistros, D. Bofilios, G. Cracknell and A. Milburn, "Intelligent curing of composites: Integration of dielectric cure monitoring with a knowledge-based system for efficient component recovery", Proceedings of the 19th International SAMPE Europe Conference of the Society for the Advancement of Materials and Process Engineering, p. 733-743, Paris, 1998.

4. S. Cossins, M. Connell, B. Cross et al., "In situ near-IR cure monitoring of a model epoxy matrix composite", Applied Spectroscopy, vol. 50, n. 7, p. 900-905, 1996.

5. P.A.Crosby, G.R.Powell, G.F.Fernando, C.M. France, R.C. Spooncer and D N Waters, "In situ cure monitoring of epoxy resins using optical fibre sensors", Smart Materials and Structures, vol. 5, n. 4, p. 415-428, 1996.

6. Y. M. Liu, C. Ganesh, J.P.H. Steele and J.E. Jones, "Fiber optic sensor development for real-time in-situ epoxy cure monitoring", Journal of Composite Materials, vol. 31, n. 1, p. 87-102, 1997

MIX
Papier aus verantwortungsvollen Quellen
Paper from responsible sources
FSC® C105338

If you have any concerns about our products,
you can contact us on
ProductSafety@springernature.com

In case Publisher is established outside the EU,
the EU authorized representative is:
Springer Nature Customer Service Center GmbH
Europaplatz 3, 69115 Heidelberg, Germany

Printed by Libri Plureos GmbH
in Hamburg, Germany